# The Network Technical Guide

# The Network Technical Guide

| | |
|---|---|
| Contributors: | Douglas T. Anderson |
| | Pat Dawson |
| | Sandra Honomichl |
| | Michael Tribble |
| Research Assistance: | Sandra Honomichl |
| | David Vehrs |
| Cover Illustration: | Perry Bergman Graphics |
| Cover Design: | Anni Wildung |
| Interior Illustrations: | Michael Tribble |

Copyright © 1997 by Micro House International Inc. All rights reserved, including the right of reproduction in whole or in part, or in any form.

**Disclaimer**
Micro House International provides this publication "as-is" without warranty of any kind. Information in this publication is subject to change without notice and does not represent a commitment on the part of Micro House International.

**Trademarks**
The Micro House logo is a trademark of Micro House International Inc.
*All other trademarks have been used for informational purposes only and are property of their respective holders. Most hard drive and controller card names are trademarks of their respective holders.*

 **Micro House International, Inc.**

P.O. Box 17515
Boulder, CO 80301-9679

First Edition February 1997

ISBN 1-880252-31-7

Printed in the United States

# Table of Contents

**Trademarks** .................................................................................................. x

**Introduction** ................................................................................................ xi
    About this Book ..................................................................................... xii
    How To Use This Text ........................................................................... xiii
    Conventions........................................................................................... xiii
    Information Level Symbols .................................................................. xiv

**Chapter One:  Network Basics**.................................................................. 3
    What is a Network? ................................................................................. 3
        LANs and WANs ............................................................................ 3
    The Evolution of Networking Technology ............................................ 5
    Basic Network Elements ......................................................................... 6
        Workstation .................................................................................... 6
        Network Interface Adapters ......................................................... 7
        Cabling ............................................................................................ 7
        Servers ............................................................................................. 8
        Inter-Network Connectivity Devices .......................................... 9
            Hubs ......................................................................................... 9
            Network Segmenting/Connecting Devices ....................... 10
        Peripherals .................................................................................... 10
    Network Features .................................................................................. 10
        Shared Peripherals ....................................................................... 10
        Networked Applications ............................................................. 10
        Remote Access ............................................................................. 11
        E-Mail ........................................................................................... 11
        Internet/Intranet .......................................................................... 11
    Chapter Review ..................................................................................... 12

**Chapter Two:  Communications Standards** ........................................ 15
    Overview ................................................................................................ 15
    Standards Organizations ...................................................................... 16
    The OSI Model ...................................................................................... 18
        Application Layer ........................................................................ 19
        Presentation Layer ....................................................................... 19
        Session Layer ................................................................................ 20
        Transport Layer ............................................................................ 20
        Network Layer ............................................................................. 20
        Data Link Layer ........................................................................... 21
        Physical Layer .............................................................................. 21

Data Exchange Across OSI Layers ............................................................. 21
OSI Model as an Industry Standard .......................................................... 25
Media Access Control and Signaling Standards ................................................ 28
Data Link Functions ............................................................................... 28
Data Framing ................................................................................ 30
Accessing a Shared Medium ............................................................ 31
Flow Control ................................................................................ 32
Detection of Frame Errors .............................................................. 33
Physical Layer Functions ........................................................................ 34
Signal Transmission ..................................................................... 34
IEEE LAN Standards ............................................................................ 34
Chapter Review ........................................................................................... 36

# Chapter Three: Local Area Networks ................................................. 39

General LAN Categories ............................................................................ 39
Topology ............................................................................................... 39
Linear Bus Topology .................................................................... 39
Star Topology .............................................................................. 41
Ring Topology ............................................................................. 42
Special Case Topologies ............................................................... 43
Bandwidth Distribution ......................................................................... 44
Shared Bandwidth ........................................................................ 44
Dedicated Bandwidth ................................................................... 45
A Word About LAN Categories ............................................................ 45
LAN Standards .......................................................................................... 46
Ethernet ................................................................................................ 46
Token Ring ........................................................................................... 50
Token Ring Multistation Access Units ......................................... 57
ARCNET ............................................................................................... 61
FDDI ..................................................................................................... 65
SAS and DAS FDDI .................................................................... 69
Emerging LAN Standards ..................................................................... 70
ATM ..................................................................................................... 70
100VG-AnyLAN (100BaseVG) ................................................... 71
Chapter Review ........................................................................................... 73

# Chapter Four: LAN to LAN Connectivity ........................................... 77

Inter-LAN Connectivity Overview ............................................................. 77
Segmenting LANs ................................................................................. 77
Connecting Geographically Separate LANs ......................................... 78
WANs ........................................................................................................... 78
Telecommunications Overview .................................................................. 79
Switching .............................................................................................. 79

|  |  |
|---|---|
| Non-Switched (Leased) Lines | 81 |
| Circuit Switching | 81 |
| Packet Switching | 82 |
| Cell Switching | 84 |
| Telecommunications Media | 85 |
|    Guided Media (DSx Channel Hierarchy) | 85 |
|    Unguided Media | 86 |
| Telecommunications Standards and Services | 87 |
|    STM and ATM | 87 |
|    Leased Lines | 90 |
|    Circuit-Switched Networks | 91 |
|       PSTN | 91 |
|       ISDN | 91 |
|       Switched-56 | 92 |
|    Packet-Switched WANs | 93 |
|       X.25 | 93 |
|       Frame Relay | 94 |
|    Cell-Switched WANs | 94 |
|       ATM | 94 |
|       SMDS | 95 |
|    Internet | 96 |
| Inter-LAN Connectivity Devices | 97 |
|    Network Connecting/Segmenting Devices | 97 |
|       Bridges and Switches (Low-Level Segmenting) | 97 |
|       Gateways and Routers (High-Level Segmenting) | 99 |
|    Remote Access Adapters | 100 |
|       Modems | 100 |
|       ISDN Terminal Adapters | 101 |
|       T1/E1 Adapters (CSU/DSU) | 101 |
| Chapter Review | 102 |

## Chapter Five: Cable Guide .................................................................. 105

|  |  |
|---|---|
| Overview | 105 |
| Cable Types | 105 |
|    Twisted-Pair | 105 |
|    Coaxial | 107 |
|       Baseband Coaxial | 107 |
|       Broadband Coaxial | 108 |
|    Fiber-Optic | 109 |
|    Wireless | 111 |
| Cabling Considerations | 112 |
|    Upgrading | 112 |
|    Cost | 113 |

Flexibility .................................................................................................... 113
Performance.............................................................................................. 114
Safety........................................................................................................ 114
Security..................................................................................................... 115
Shielding and Interference........................................................................ 115
Conclusions .............................................................................................. 116
Chapter Review ................................................................................................. 117

## Chapter Six: Network Operating Systems................................................ 121
Overview ........................................................................................................... 121
Background .............................................................................................. 121
Common NOS Functions ......................................................................... 122
Control Network Access ................................................................ 122
Network File Handling................................................................... 123
Shared Applications ....................................................................... 123
Sharing Peripherals......................................................................... 123
Network Operating Systems and the OSI Reference Model ................... 124
Local and Network OS Interaction .......................................................... 125
Popular NOSs .................................................................................................... 125
Novell NetWare........................................................................................ 126
Microsoft Windows NT............................................................................ 126
Banyan VINES ......................................................................................... 127
UNIX ........................................................................................................ 128
Artisoft LANtastic .................................................................................... 129
Network Operating System Trends.................................................................... 129
OS/NOS Integration ................................................................................. 129
Intranet...................................................................................................... 130
Chapter Review ................................................................................................. 132

## Chapter Seven: Hardware Settings .......................................................... 133
Overview ........................................................................................................... 135
Jumper and Switch Position Indicators............................................................. 135
Board Element Naming Conventions ............................................................... 136
Additional Hardware Notes ............................................................................... 137
System Resources Dedicated to the Serial Port ....................................... 137
I/O Address............................................................................................... 138
Interrupt Request (IRQ)............................................................................ 138
Boot ROMs............................................................................................... 138
Ethernet AUI Fuse ............................................................................................. 139
Diagnostic LEDs................................................................................................ 139
Jumper Settings.................................................................................................. 140
3COM Corporation................................................................................... 141
Accton Technology Corporation .............................................................. 148

| | |
|---|---|
| Addtron Technology Company, Ltd. | 153 |
| Advanced Interlink Corporation | 161 |
| Alta Research Corporation | 163 |
| American Research Corporation | 165 |
| Ansel Communications, Inc. | 167 |
| AT-LAN-Tec, Inc. | 169 |
| Cabletron Systems, Inc. | 171 |
| CNET Technology, Inc. | 175 |
| Cogent Data Technologies, Inc. | 181 |
| Compex, Inc. | 184 |
| D-Link | 185 |
| Danpex Corporation | 189 |
| Datapoint Corporation | 196 |
| Digital Communications Association, Inc. | 198 |
| Eagle Technology | 200 |
| Edimax Computer Company | 204 |
| GVC Technologies, Inc. | 206 |
| Hewlett-Packard Company | 208 |
| ICL | 210 |
| IMC Network Corporation | 212 |
| Intel Corporation | 214 |
| Katron Technologies, Inc. | 216 |
| Kingston Technology Corporation | 218 |
| Klever Computers, Inc. | 220 |
| Longshine Microsystem, Inc. | 223 |
| Madge Networks, Ltd. | 225 |
| Maxtech Corporation | 233 |
| Microdyne Corporation | 236 |
| Multi-Tech Systems, Inc. | 238 |
| NDC Communications, Inc. | 240 |
| Networth, Inc. | 241 |
| Olicom International | 244 |
| Proteon, Inc. | 249 |
| Pure Data, Ltd. | 252 |
| Racal-Interlan, Inc. | 263 |
| Racore Computer Products, Inc. | 276 |
| Siemens Nixdorf Informationssysteme Ag | 283 |
| Silcom Manufacturing Technology, Inc. | 287 |
| Standard Microsystems Corporation | 290 |
| SVEC Computer Corporation | 296 |
| Syskonnect, Inc. | 298 |
| Thomas-Conrad Corporation | 303 |
| Tiara Computer Systems, Inc. | 317 |

viii  Table of Contents

Top Microsystems, Inc. .................................................................................326
Transition Engineering, Inc. .........................................................................328
TTC Computer Products...............................................................................332
Tulip Computers ...........................................................................................334
Ungermann-Bass Networks, Inc...................................................................337
Unicom Electric, Inc. ....................................................................................339
Xinetron, Inc..................................................................................................343
Zenith Data Systems.....................................................................................345
Zero One Networking...................................................................................354

## Appendix A: Node ID Quick Reference .................................................. 359
Setting an ARCNET or Token Ring Node ID ..............................................359
ARCNET Node ID Quick Reference Table .................................................360

## Appendix B: Network Specifications...................................................... 369
Overview .......................................................................................................369
Ethernet Specifications .................................................................................369
General .....................................................................................................369
Specifications for Ethernet Networks Over Thick Coaxial Cable (10Base5).........370
Specifications for Ethernet Networks Over Thin Coaxial Cable (10Base2)...........373
Simple 10Base2 Guidelines................................................................374
Complex 10Base2 Guidelines ............................................................375
Specifications for Ethernet Networks Over Unshielded Twisted Pair Cable
(10BaseT) ................................................................................................376
Specifications for 100 Mbps Ethernet Networks Over Twisted Pair and Fiber-Optic
Cable (100BaseT)....................................................................................379
Token Ring Specifications............................................................................380
General .....................................................................................................380
Standard Token Ring Network Requirements ......................................381
Token Ring Twisted Pair Connector Pin Assignments .................................386
Token-Ring Cable Types 5, 6, 8, and 9 .................................................390
FDDI Specifications .....................................................................................392
Specifications for SAS FDDI .................................................................392
Specifications for DAS FDDI ................................................................392
ARCNET Specifications...............................................................................393
General .....................................................................................................393
Specifications for ARCNET Networks Over Coaxial and Twisted Pair Cable ......393
ARCNET Active Hubs........................................................................395
ARCNET Passive Hubs......................................................................396
Specifications for ARCNET Networks Over Twisted Pair Cable .........397
ARCNET Star Topology Network .........................................................398
ARCNET Twisted Pair Hubs and Repeaters)...................................399
ARCNET Daisy-Chain Topology Network...........................................400

**Appendix C: Directory of Manufacturers** .................................................. 403
    Disclaimer ........................................................................................................ 403
    Contact Information ........................................................................................ 403

**Appendix D: Bonus CD-ROM Programs** ................................................ 455
    The Micro House Technical Library of
    Network Interface Cards ................................................................................ 458
    FCC ID Finder .................................................................................................. 459
    Micro House Multimedia Product Demonstration ...................................... 460

**Glossary of Terms** ......................................................................................... 461

**Index** .................................................................................................................. 501

## Trademarks

| | |
|---:|:---|
| LANtastic | Artisoft, Inc. |
| VINES | Banyan Systems, Inc. |
| OS/2, Systems Network Architecture (SNA) | IBM Corporation |
| DOS, Windows, Windows 95, Windows NT | Microsoft Corporation |
| NetWare | Novell |

*All other trademarks respective to their owners.*
*Most network interface device model names are trademarks of their respective manufacturers.*

## Limits of Liability and Disclaimer of Warranty

The authors have used their best efforts in preparing this book. These efforts include the compilation and research of information contained within. None of the source information in this guide has been electronically or mechanically reproduced in any way. All specifications and information have been compiled through years of research and testing. The authors and publisher are not responsible for errors or omissions contained in this guide.

# Introduction

Thank you for selecting the **Network Technical Guide**, from the industry leader in computer hardware reference tools.

The purpose of this text is to be a practical introduction to the technology and terminology used in modern network data communications. Whether you are a novice wanting to familiarize yourself with networks, or a technician in need of a handy reference, we hope you will find this text a valuable tool.

This text also contains complete hardware set-up information for hundreds of the most popular network interface adapters produced. This includes Ethernet, Token Ring, and ARCNET. There are also appendices that cover node ID settings and network specifications.

As a bonus, we have included a CD-ROM that contains demonstrations of some of our popular products, including a fully operational version of the acclaimed Micro House Technical Library (MTL). We know you will find the MTL extremely useful and hope the demonstrations spark your interest in our other software products.

The staff at Micro House sincerely hopes that you find the **Network Technical Guide** a valuable tool for your network reference needs!

## About This Book

This book is arranged into six chapters and four appendices:

- Chapter One is a cursory introduction to general data communications.
- Chapter Two covers industry standards
- Chapter Three is a more detailed look at popular local area networking technologies.
- Chapter Four explains wide area data communications.
- Chapter Five details popular network cabling types.
- Chapter Six contains information concerning networking operating systems.
- Chapter Seven offers set-up information on selected network hardware from popular manufacturers.
- Appendix A is a quick reference for setting up node identification numbers.
- Appendix B covers specifications for popular LANs.
- Appendix C is directory of contact information for organizations related to the networking industry.
- Appendix D contains the documentation corresponding to items found on the bonus CD-ROM included with this text.

There is also a comprehensive glossary for fast reference of terms you may not be familiar with.

# How to Use This Text

Our goal for this text is for it to be a useful reference, no matter what level of technical expertise you may possess. We realize no one is interested in reading a reference book cover-to-cover, especially if they are looking for a specific piece of information. With this in mind, we have organized the ***Network Technical Guide*** according to the level of detail the information requires.

Because it is sometimes annoying to sift through explanations of every acronym or technical term, and just as annoying when the author assumes the reader is fluent in computerese, we have used bold type-face to reflect words which are covered in the glossary.

Another handy reference feature is the symbols located next to section heads and important information. If you are looking only for essential information, simply skip the sections that have the Nice-To-Know symbol. Pay close attention to the passages that are marked by the card with a lightning bolt through it and the broken floppy disk – they could save you from a tragic episode.

# Conventions

- The term **computer** refers to the main-board and all of the components native to it (Central Processing Unit, memory, Basic Input/Output System).
- **Network**, **information systems**, **data communications system**, **distributed logic environment** are interchangeable terms.
- The term **peripheral** refers to any input or output device attached to the computer.
- 1**MB**, unless otherwise noted, denotes $10^6$ bytes in this text.
- **Adapter** and **interface card** are interchangeable terms.

# Information Level Symbols

## Warning

Denotes failure to take proper precautions could result in damage to hardware.

## Data Loss

Denotes improper use of procedure could result in permanent loss of data.

## Important

Denotes key technical information the reader should become familiar with.

## Technical Information

Denotes non-essential technical information.

## Nice-To-Know

Denotes anecdotal/historical, non-essential information.

### Back-Up

This symbol is a prompt to back-up the information on your hard drive in preparation for a procedure that could or will destroy data on the drive.

### CD-ROM

The information contained in this passage refers to one or more of the supplemental programs found on the CD-ROM included with this text.

### Micro House Product

The text under this logo refers to one of our excellent products. Please contact our sales staff for details. Address and telephone information can be obtained in **Appendix C**.

**CHAPTER ONE**

# Network Basics

## Chapter One Contents:

Part 1: What is a Network? ............................................. 3

Part 2: The Evolution of Networking Technology ........ 5

Part 3: Basic Network Elements .................................... 6

Part 4: Network Features ............................................... 10

# What is a Network?

A network is a general term used to describe a group of data processing devices that are in communication with each other, as well as the system that enables them to do so. The network encompasses the physical transmission medium as well as the logic that enables a standard method of communication.

## LANs and WANs

*A LAN is a limited group of individual computers that share resources and work together through a dedicated hardware and software connection.*

A **Local Area Network (LAN)** is a limited group of individual computers that share resources and communicate with each other through a dedicated hardware and software connection. LANs have revolutionized the office environment, evolving at a rapid pace in parallel with other computer industry developments. There are many different implementations of LANs – they are as diverse as the businesses that use them. In spite of these variations, the general concept is always the same – to allow individual computers to more efficiently work together as a team. This is done by providing common access to applications, files, and peripherals.

*Figure 1-1: LAN*

The typical use of a LAN is to integrate individual computers together in a relatively limited (local) area, such as an office floor or building.

High-speed data communications across greater distances come under the scope of **Wide Area Networks**, or **WAN**s. A WAN consists of two or more LANs that are linked together. WAN is a very broad definition, encompassing anything from the Internet down to two corporate offices using a proprietary connection.

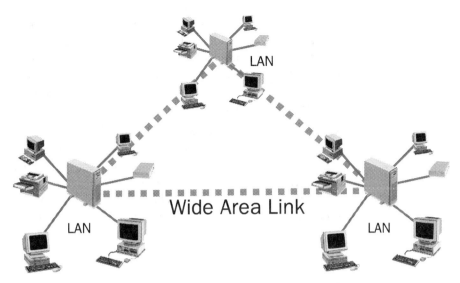

*Figure 1-2: WAN*

Most of the references in this book deal with LAN topics, although there are gray areas that apply to both LANs and WANs.

 # The Evolution of Networking Technology

The nature of data communications is always changing. Early networks involved **centralized processing** environments. Terminals were "dumb" devices that contained no native logic (ability to execute software commands). They simply presented the interface used to access the resources of mainframe or mini computers that did all the processing work.

Desktop computers brought about a revolution in data communications that resulted in a **distributed processing** environment. Initial LANs consisted of local workstations that relied on one or more relatively powerful computers, called **servers**, for network communications and other services such as file storage and handling. Most of the processing work for applications is performed by the local workstation, but network access is controlled by the server. This dependent relationship is known as the **Client/Server Model**. A client is the computer requesting a service, such as access to a file, while the server is the computer providing the service.

Desktop computers and network technology have advanced to where **Peer-to-Peer Model** networks are now common. Peer-to-peer networking means that no intermediary server is required to enable network communications. The client to server relationship between computers is momentary, meaning that any computer on the network can be the interrogator or service provider.

Current network standards are based on a hierarchy of protocols known as the **ISO-OSI Reference Model**. This international guideline specifies seven layers ranging from the physical cabling up to the high-level application interfaces to the network. **Chapter Two** contains extensive details about the OSI model, including examples of where current network elements fit into this scheme.

## Basic Network Elements

In the distant past of data communications, networks involved proprietary, single-vendor installations. One company, such as IBM, provided installation and service, as well as the software. The modern computer industry involves open architectures from high-level software applications down to low-level network interfaces.

Today's multi-vendor network environment means that there are countless variations on the LAN theme, but most LANs will contain some variation of the basic components listed in the following sections.

## Workstation

The most visible component of the LAN is the **workstation**, where the end-user interacts with the network. The workstation in a LAN is considered an "intelligent" terminal because it has its own processing power and can run applications autonomously. Please note that this is vastly different in concept from dumb terminals attached to a mainframe computer, which are basically I/O devices to the mainframe's central processor.

LAN workstations can be customized according to primary use. Some of the ways in which workstations can vary include processor speed, amount of memory, installed applications, and data storage devices. **Diskless workstations** have no native resources for reasons of cost and/or security. These machines are configured with bare-bones attributes and rely almost entirely on network resources. More commonly, workstations will have some local storage capacity and ability to operate in isolation from the network.

Workstations use a **local** or **host operating system** such as DOS, Windows, or OS/2. The local operating system provides the workstation with an independent means of running applications and using local peripherals (hardware not accessed through the LAN).

## Network Interface Adapters

The component which enables an individual network element to interface with the rest of the LAN is known as the **Network Interface Card** (**NIC**) or **adapter**. A NIC typically occupies an expansion slot of the workstation or **file server** and provides the physical attachment to the LAN transmission media (cabling). The NIC performs low-level network accessing and signaling chores. Each element of the LAN in which a NIC is installed is known as a **node**. Every node has a unique address for data routing.

*See Chapter Two for information on **Media Access Protocols**.*

NICs are specific to the type of signaling and **Media Access Control** (**MAC**) protocols used by the LAN. These topics are covered in **Chapters Two** and **Three**.

## Cabling

The physical connection between network nodes is the **cabling**. The LAN cable is an important consideration when setting up the network because it can be the single costliest element and has the greatest effect on the maximum **data transfer rate**, or **throughput**.

Two general terms that are often applied to data transmission media are **baseband** and **broadband**. Baseband is a method of data transmission in which a single frequency occupies the entire available **bandwidth** of the medium. Broadband means that the available bandwidth in the medium is divided into several frequency ranges in order to accommodate separate channels simultaneously.

The term used to describe the sharing of a transmission medium among multiple channels is known as **multiplexing**. The type of multiplexing used depends on the media.

Broadband media is by definition divided into different channels by multiplexing available bandwidth into several frequencies. Each signal is broadcast at a different frequency over the cable simultaneously. This is called **Frequency Division Multiplexing** (**FDM**). Cable television is a common example of FDM.

Although a baseband medium can only hold a single signal frequency at a time, it can be multiplexed. This is accomplished by dividing channels into segments, then interleaving (distributing at even intervals) the segments of different channels into the medium. This is known as **Time Division Multiplexing (TDM)**.

The most popular cable types in use today include **twisted pair**, **coaxial**, and **fiber-optic**. **Wireless** LAN connections, which use infrared or microwave technology, are also available. The major cable types are covered in detail in **Chapter Five**.

> See Chapter One for information on **topology** and Chapter Five for information on LAN cable types.

The physical relationship in which the cables connect the various nodes is known as the **topology** of a LAN. Cables types, topology, and Media Access Control protocols are tied closely together in the configuration of a LAN. Please see **Chapters Two** through **Four** for details on how these elements mesh together.

## Servers

A **server** is a computer that provides file-access and other shared services to the network. Servers are equipped with relatively large amounts of 1) processing power; 2) RAM; and 3) mass storage because all of these resources are shared across the entire network.

> See Chapter Six for information on **Network Operating Systems**.

A server uses a **Network Operating System (NOS)**. The NOS is the "brains" of the network, enabling and controlling workstation access to shared resources. Novell's NetWare, IBM's OS/2, and Microsoft's Windows NT are typical NOSs installed in LANs today. **Chapter Six** covers NOS topics in detail.

There may be several different servers on a LAN, or one server may accommodate several different functions. The most common is the **file server**, which provides network access to the files in its mass storage media. An **applications server** runs programs which are used across the network, such as a database. A **print server** provides access to one or more network printers. Servers may also provide security features that restrict access to network resources according to the level permitted to a specific user.

# Inter-Network Connectivity Devices

Inter-network connectivity devices is a broad category of devices that connect network nodes or segments. They may range from simple pass-through devices to help manage wiring, or computers that provide wide-area data routing services.

## Hubs

> The primary purpose of a **hub**, or **concentrator**, is to provide a central point of connectivity for nodes in a LAN,

The primary purpose of a **hub**, or **concentrator**, is to provide a central point of connectivity for nodes in a LAN, in the process adding an organizational level. A hub which simply provides physical attachments for multiple cable segments is considered a **passive hub**. Hubs that boost the signal in order to accommodate greater cable lengths are called **active hubs**.

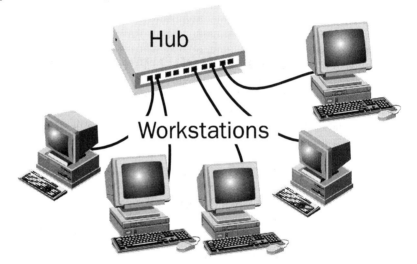

*Figure 1-3: Hub/Concentrator*

Hubs may have a nomenclature specific to the media access protocol used. For instance, Token Ring hubs are called **Multi-station Access Units** (**MAU**). There are also task-specific hubs that are used to provide network segmenting or translation services. These devices are named according to the communications level that they work on.

### Network Segmenting/Connecting Devices

> See Chapter Four for more information on network connecting devices.

The general purpose of network segmenting/connecting devices is to provide a link to separate networks or segments of the same network. These include **bridges, gateways, routers,** and **switches**. Network segmenting devices are typically very complex, requiring extensive internal logic. The basic differences between these devices lie with the communications level at which they work. Please see **Chapter Two** for detailed information on communications levels (the **OSI Reference Model**), and **Chapter Four** for details on bridges, gateways, routers, and switches.

> See Chapter Two for detailed information on the OSI Reference Model.

### Peripherals

The ability to allow network users to share peripherals is one of the primary functions of a LAN. These peripherals include mass storage, printers, and modems.

## Network Features

Network connectivity enhances workplace efficiency by speeding up communication and allowing convenient sharing of resources. This section covers some of the applications that take advantage of network data communication.

### Shared Peripherals

One of the ways that network connectivity increases efficiency is in the use of shared peripherals. Through the use of task-specific servers, multiple users can access printers, storage devices, fax machines, and other peripherals in an orderly manner.

### Networked Applications

Simply accessing shared resources from an application running on a local workstation isn't taking full advantage of the potential of network communication. Programs can be designed to integrate the work of a group of people efficiently – reducing redundant effort and providing a common point of reference. Common networked applications typically involve a shared database.

## Remote Access

Remote access from an outside workstation to office resources is becoming increasingly popular. With special software, a network workstation can be taken over by a remote computer via modem. Some high-bandwidth wide-area connections even allow direct access.

Like WAN connectivity, remote access is a source of potential security problems. Network security is an entire industry in itself and may involve complex hardware and software suites.

## E-Mail

In its simplest form, **e-mail** (**electronic mail**) is straight ASCII text that is transmitted over a local and/or wide-area network. More sophisticated e-mail applications can even support graphics and file attachments. In many offices, e-mail has virtually replaced the use of paper memos.

## Internet/Intranet

> See Chapters Two and Four for more information on TCP/IP and the Internet.

Local Area Networks can easily be configured to transmit **TCP/IP**, a packet-routing protocol that is the language of the Internet. The TCP/IP is an integrated group of protocols designed to be used as a universal communications standard that is platform-independent. Please see **Chapter Two** for more information on TCP/IP.

In the past couple of years, the Internet has had a tremendous impact on the manner in which businesses communicate. It has become the most significant element in accelerating the development of computer hardware, software, and high-bandwidth media.

Although Internet protocols were intended for wide-area communications, they have been adapted for local-area use. Corporate Intranets are rapidly reshaping the very definition of LANs, often blurring the distinction from WANs. Intranets are used as anything from simple bulletin boards to elaborate task management systems. Each Intranet is unique to the corporation that it is implemented by. An interesting aspect of Intranets is that there is no single driving force responsible for their rapid development. As a result, the leading network operating system and workgroup software vendors are scrambling to incorporate Intranet features into their products.

# Chapter Review

- A network is a general term used to describe a group of data processing devices that are in communication with each other, as well as the system that enables them to do so.

- A Local Area Network is currently described as a network that is confined to a limited geographic area, generally within an office building or floor.

- A Wide Area Network is the connection between two or more local area networks.

- Network concepts have evolved from the centralized processing environment of mainframe to dumb terminal to today's peer-to-peer communications.

- Networks give users access to a variety of shared services, reducing inefficient distribution of resources and facilitating teamwork.

- Modern data networks usually involve multi-vendor hardware.

# CHAPTER TWO

# Communications Standards

## Chapter Two Contents:

Part 1: Overview .......................................................... 15

Part 2: Standards Organizations ................................... 15

Part 3: The OSI Model ................................................. 18

Part 4: Media Access Control and Signaling Standards 28

## Overview

The nodes of a network have to be able to communicate in the same language. To that end, various organizations have created hardware and software standards for vendors of network products to follow. These standards detail how information must be passed over the physical medium, as well as how information must be encoded and decoded so it can be understood. This chapter focuses on rules or protocols that specify how information must be exchanged so network nodes can communicate.

Communication protocols are a hidden but critical part of a network. A simple example will illustrate. Imagine that you want to send a message to a user on another computer. You activate your mail software application, type in the message content and a destination address, and press the Send button. If everything is working properly, the addressee is notified of the reception by his mail application, opens up the file and reads the message as you sent it. The process seems simple, yet under the surface a great deal of activity is required for your message to be delivered.

A chain of events must be invoked to package your message before it is transmitted over the physical medium. Your message needs to be formatted for output display, addressed for network routing, broken into appropriate packet sizes, and sequenced before being converted to electrical signals and transmitted across the physical medium. The order and content of these events may vary depending on the network operating system and mail application in use. However, specific rules must be adhered to by each function in the chain, otherwise the information will not reach its intended destination or be inaccurately reproduced.

Both the sending and receiving nodes must use the same rules and functions to package and unpackage a message. The concept is not unlike speech, in which each person must use the same vocabulary, grammar, parsing, and enunciation rules to build sentences so that the other can understand. Within networks, the order and content of communication functions is defined by a specific architecture. The sets of rules that dictate how these functions format and exchange information are communications protocols.

The first part of this chapter is an introduction to international standards. The various standards organizations and their major responsibilities will be presented to give you a better understanding of what some of the acronyms mean.

This chapter also presents a brief overview of communications protocols that are established standards within the industry. These protocols will be introduced with a look at the **Open System Interconnection (OSI) Model**, sometimes called the OSI Reference Model. The OSI model is a network architecture supported by the **International Standards Organization (ISO)** that defines a hierarchical structure of layered communications protocols. The OSI layers will be referred to many times in this chapter, as well as in other parts of this text for purposes of comparison to portions of popular network architectures.

Next, the functions performed by protocols at the lower layers of the OSI model are discussed. These protocols, which are usually hard-coded on network interface cards (NICs), define signaling characteristics and control access to the physical network media.

## Standards Organizations

The primary purpose of standards organizations is to recommend rules and conventions for manufacturers to follow in order to ensure compatibility among their products. In an extremely competitive and dynamic industry, these organizations stabilize product development by establishing common architectures, protocols, electrical and carrier interfaces, and signaling rates. The alternative is to wait for the open market to eventually establish a de facto standard. The legions of stranded Betamax video tape machine owners can attest to benefits of the more methodical approach.

The **Open System Interconnection (OSI) Model** is a network architecture developed as a standard by the **International Standards Organization (ISO)** based in Geneva, Switzerland. Descriptions of this model and its functional layers are available to any vendor wanting to develop a product that interfaces with others using the same model. To date, few vendors have strictly followed this model for developing products. However it has become very useful as a basis against which the industry compares and communicates product functions and protocols. The role of this model as a reference standard may be changing, however. A joint effort of ISO and **International Telecommunications Union (ITU)**, formerly the **Consultive Committee for International Telegraph and Telecommunications** (CCITT), recently established communications protocols for higher layers within the model and developed tests to measure vendor compatibility.

The ITU currently has more influence over European standards. Most ITU standards are concerned with how signals are transmitted across the physical medium.

The **American National Standards Institute (ANSI)** is the United States representative to ISO. Standards established by ANSI cover a variety of aspects within the computer industry, from programming languages to communications. Their network contributions include control procedures, signaling rates, and network architectures. **Fiber Distributed Data Interface (FDDI)** is one prominent example of an ANSI standard. FDDI defines a token-passing data communications scheme implemented with fiber-optic cable.

Other prominent organizations that serve to establish network standards include the **Institute of Electrical and Electronic Engineers (IEEE)** and the **Electronics Industries Association/Telecommunications Industries Association (EIA/TIA)**. A subcommittee of IEEE, known as the IEEE 802 Committee, is responsible for the development of architectural and protocol standards for several networks, which include **Ethernet** and **Token Ring**. The EIA/TIA focuses on electrical standards that include wiring specifications and serial connectors such as RS-232.

In addition to the above membership organizations, several government organizations are active in setting standards for the networking industry. One of these is the **Federal Communications Commission (FCC)** which regulates technical matters such as radio frequency allocation, as well as pricing, profits, and company mergers. This commission monitors carriers both within and across states to protect consumers and encourage open system competition.

# The OSI Model

The goal of the OSI network architecture is to provide a set of guidelines that enables network equipment to be designed and implemented so it can reliably communicate with equipment from other vendors. The OSI layers were derived by identifying functions required to send and receive messages at each node, organizing these functions into groups and then layering them hierarchically. The result is a model in which each layer provides specific functions and interfaces with layers above or below it according to specific rules.

The OSI model defines seven layers of functions, as illustrated in **Figure 2-1**. When a message is sent from one node to another within the model, it is assumed to start at the highest layer, the Application Layer, and descend through each successive layer until it reaches the lowest level, where it is transmitted across the physical medium. Upon arrival at the destination node, the message percolates up the same layers in reverse order, from the Physical Layer to the Application Layer. OSI protocols define functions that must be performed by each layer, as well as how these layers must communicate with one another to exchange a message.

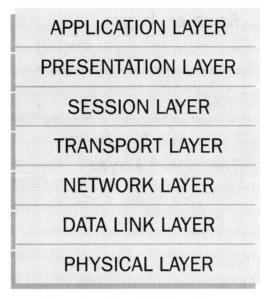

*Figure 2-1: The OSI Model Layers*

The specific functions at each layer of the OSI model are covered in the following sections.

## Application Layer

This is the level at which the user interacts with the network. The user enters all network requests for network services, such as electronic mail, database access, and print jobs at the Application Layer. Commands from operating systems, such as **DOS**, **OS/2** and **UNIX**, are also entered into the system at this level. Normally, this is the only layer the typical user is concerned with or even aware of.

## Presentation Layer

This layer formats the data it receives from the Application Layer for proper output at the destination node. Different computer systems present data differently. If a source computer is using **ASCII** character codes and the receiving computer uses **EBCDIC** character codes, the translation from ASCII to EBCDIC would occur at this level. The layer also handles graphics, encryption, and file formatting and it contains drivers for printers, plotters, scanners, and other devices. Microsoft's Windows and IBM's OS/2 are two examples of environments that perform functions that satisfy the Presentation Layer standard.

## Session Layer

If two applications (or different parts of the same program) need to communicate or share data across the network, the Session Layer of the OSI Model coordinates this interaction. This layer synchronizes dialogs and data exchange. Some of the important functions provided at this level include providing security, file locking, log on, and administration facilities.

## Transport Layer

The Transport Layer is responsible for the quality of source-to-destination node transmissions. At the source node, this layer segments and sequences large blocks of information, multiplexes this information and selects a level of error control. At the destination node, the layer recombines and sequences transmitted segments, providing the appropriate error checking and recovery if a transmission is lost or received incorrectly. **Transmission Control Protocol (TCP)**, NetWare's **Sequenced Packet Exchange (SPX)**, **Internetwork Protocol Exchange (IPX)**, and **NetBIOS** all perform Transport Layer tasks.

## Network Layer

The Network Layer is used in extended networks, such as WANs, to control the physical path along which information is routed. These routes typically involve more than one network and may feature more than one path to the destination node. Network conditions and information priority are among the factors that decide the best route for the transmission. **Internet Protocol (IP)** and **IPX** are examples of protocols that function at this level of the OSI model. Simple homogeneous LANs do not need to implement the Network Layer.

## Data Link Layer

This level is responsible for transferring data from one node to the next along the transmission path. Protocols at this layer format information received from the Network Layer into **packets** and **frames** and control how these data units are transmitted to the next node. Control methods define how the cable is accessed and flow is controlled to prevent data bottlenecks or "traffic-jams" from occurring. "**Accessing a Shared Medium**" on **page 31** offers more detailed information on this topic. **Error detection** and **error correction** functions are also performed at this level. They ensure that the transmitted frames or packets are received correctly, performing such functions as counting the frames to see if there is the correct amount. These functions are not to be confused with error detection and correction functions performed by the Transport Layer, which actually sequences the frames for reconstituting transmitted information.

## Physical Layer

The lowest OSI level performs two functions: encoding and transmission. The Data Link layer defines *what* is to be sent on to the next node on the network and the Physical Layer defines *how* it is transmitted. This level is the hardware layer. It defines how electrical signals are transferred across the physical medium that connects the two nodes, encompass cables, wires, connectors, and electric signaling speeds. The mechanism at the Physical Layer of the sending node blindly receives data bits from the Data Link layer, encodes them into electrical signals, and transmits these signals into the medium. The Physical Layer at the destination node reverses the process, receiving electrical signals, decoding them into data bits, and passing these bits on to the Data Link Layer. Protocols at this level include such items as cable pin-out definitions and hardware interfaces. All devices on a network are ultimately connected through the Physical Layer.

## Data Exchange Across OSI Layers

Two features of the OSI design are critical to open system connectivity. They allow vendors to develop products at different OSI layers and insert these products between those developed by other vendors with the assurance that all products will interact properly.

1. The first feature is standard communication protocols – layers within the OSI network architecture communicate through OSI established rules/protocols.

2. The second feature is layer independence – each layer operates independently of others, performing unique functions not performed by other layers. Information concerning that layer is only accessible to that layer or peer layers at other nodes on the network. This is important because it allows a vendor product at one layer to communicate with itself at other nodes, regardless of the vendor products installed above/below.

Because these features are so closely tied to how data is exchanged across the seven layers of the OSI model, a description of this process is presented here. Consider a message (packet of data) that enters the model at the Application Layer. The message is packaged as data and passed onto each successive layer. As each layer is invoked, that layer "negotiates" for the data package and "wraps" it in additional information before passing it on. As can be seen in **Figure 2-2**, this additional information is added as a header until the Data Link Layer is invoked. At this layer, both a header and footer are added and the package is framed for transmission across the physical medium. This frame is transmitted across the medium as electrical signals at the Physical Layer, where it includes much more information than it did at the Application layer.

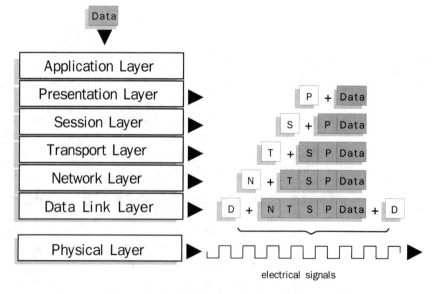

*Figure 2-2: OSI Data Packet Construction*

When the transmitted package arrives at the receiving node, it ascends the seven layers and is unpacked in reverse order (see **Figure 2-3**).

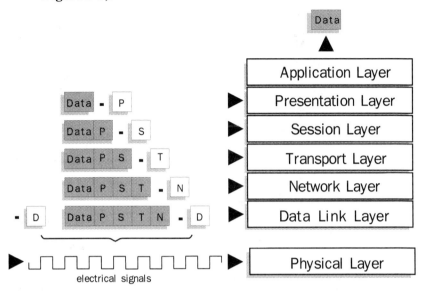

*Figure 2-3: Data Reconstitution*

Each layer unwraps the package it receives from the layer below it, removing *only* the information added by its peer layer in the sending node. For example, the Data Link Layer at the receiving node removes the header and footer added by the Data Link Layer at the sending node, and the receiving Network Layer removes the header added by its peer layer. Each layer sees only the header that concerns its layer, passing the remaining package on as the payload.

Passing the data package in this manner not only allows layers to operate independently of one another, but it also allows **peer-to-peer communications**. Note that this is a different, though similar, concept from Peer-to-Peer Model networking discussed in Chapter One. In peer-to-peer communications, each layer at one node can communicate with its peer layer at different nodes through the information added to the data package. The added header is the communication link.

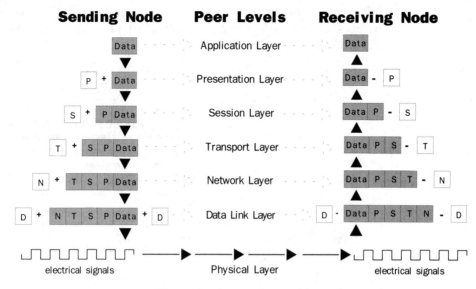

*Figure 2-4: Peer-to-Peer Communications*

To illustrate peer-to-peer communications, consider the interchange between Transport Layers at two different nodes. At the sending node, this layer inserts sequencing information into the headers added to data segments. The Transport Layer at the receiving node uses this information to recombine transmitted packages into a sequential whole. All this communication is not possible, however, unless layers on all network nodes agree to use the same rules to format and unformat their header. This ensures that all nodes know the exact location, size and format of delimiters, time stamps or destination addresses that are contained within the frame header or footer. The need for agreed upon rules brings us to the subject of protocols.

Protocols within the OSI model encompass the body of rules that determine how information is transferred across the interfaces described above. They include sets of rules that define how:

1. the data package is exchanged vertically within the model. This includes interfaces between the seven OSI layers,

2. each layer formats and unformats its header (or footer), and

3. how the data packet is converted to electrical signals and transferred across multiple devices that line the physical path from node to node.

This combination yields integrated **layers of protocols** or a **protocol stack**, as illustrated in **Figure 2-5**.

Higher layers within the OSI protocol stack address how information is to be transferred across layers of software applications. The two lower protocol layers define interfaces on the NIC card, with Physical Layer protocols also encompassing hardware interfaces. All these protocols are critical to promoting open system interconnectivity.

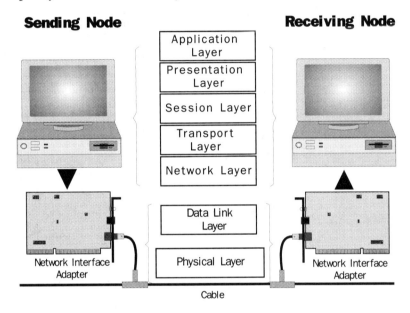

*Figure 2-5: Typical LAN Protocol Stack*

## OSI Model as an Industry Standard

Few networking product vendors rigorously follow the seven layer OSI model. All use a layered architecture that stacks functions in a similar order. However, the specific functions utilized and the grouping of layers vary. Most of these variations are necessitated by the needs of different network architectures. In the end, however, all vendors compare their functions and protocols to those defined in the OSI model.

Here, we present several standard protocol stacks and compare their functional layers to the seven layers of the OSI model. A comparison of protocol layers in **OSI**, **SNA**, and **TCP/IP** architectures is shown in **Figure 2-6**. Layers of similar functions are indicated by horizontal grayed bands.

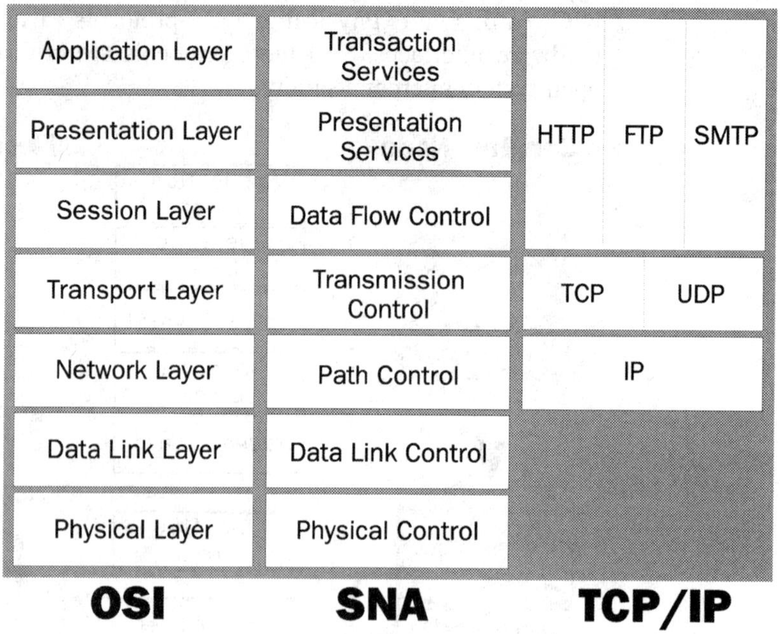

| OSI | SNA | TCP/IP | | |
|---|---|---|---|---|
| Application Layer | Transaction Services | | | |
| Presentation Layer | Presentation Services | HTTP | FTP | SMTP |
| Session Layer | Data Flow Control | | | |
| Transport Layer | Transmission Control | TCP | | UDP |
| Network Layer | Path Control | IP | | |
| Data Link Layer | Data Link Control | | | |
| Physical Layer | Physical Control | | | |

*Figure 2-6: Comparison of Protocol Layers*

**SNA (Systems Network Architecture)**, was introduced by the IBM Corporation in 1973 for use in a mainframe environment. SNA is a de facto standard (established by prominence in the market place rather than standards organizations) that has had considerable influence on later networking technologies. The SNA protocol layers heavily influenced the design of the OSI model. Although OSI layer design is very similar to that of SNA, there are notable differences. This is because SNA was designed to operate in a client-server environment, rather than the peer-to-peer environment more prevalent during the development of the OSI model. Among the differences are that the OSI model has provisions for interfaces that allow dissimilar networks to interconnect, as well as management functions that monitor and control transmission paths to guarantee successful delivery. Unlike OSI, SNA protocol interfaces between layers were not established as standards because the network was only designed to function with products developed by IBM.

> IBM's **SNA** is a precursor to much of the technology used in modern networking.

Our second example, **TCP/IP**, is the de facto communications standard for the Internet, using technology originally developed for the U.S. **Department of Defense** (**DOD**). The standard includes four protocol layers: Link, Network, Transport and Application. TCP/IP is actually a collection of protocols, some of which operate at the same layer. Such a collection is often referred to as a **protocol suite**.

At the network level, the TCP/IP uses **Internet Protocol** (**IP**) for routing services. At the Transport level, the suite includes two protocols: **Transmission Control Protocol** (**TCP**) and **User Datagram Protocol** (**UDP**). TCP coordinates transmission between sending and receiving nodes, whereas UDP transmits data without acknowledgment. Application level protocols are task specific. Some of these include file transfer (**FTP** or **File Transfer Protocol**), remote login, electronic mail (**SMTP** or **Simple Mail Transfer Protocol**), and **HyperText Transport Protocol** (**HTTP**) which is used for the graphical **World Wide Web** (**WWW**). Protocols that operate within the same layer are indicated by vertical dotted lines in **Figure 2-6**.

Novell's **NetWare**, a popular NOS, can also be used to illustrate some of the characteristics of layered architectures. Network operating systems manage network resources, accessing the physical cabling through a protocol stack. Although some operating systems, such as **Windows 95** and **Windows for Workgroups**, are designed to operate in peer-to-peer networks, NetWare operates in a client/server environment with one computer designated as server. Operating systems of this type usually redesign upper level protocols in their stack to include utilities such as messaging services, security management, performance monitoring and information broadcasts. NOS protocol stacks from specific vendors are not typically designed to interact with other NOS vendor's products.

As can be seen from these examples, the OSI model provides a useful way of categorizing and comparing industry protocols and layer functions.

# Media Access Control and Signaling Standards

The remainder of this chapter is focused on industry standards that perform media access and signaling functions. These are generally hardware functions that occur at the lower two layers of the OSI model (Data Link and Physical Layers). These protocols control transmission between adjoining nodes on a network, not to be confused with transmissions between source and destination nodes which are controlled at higher levels within the protocol stack. These lower-level protocols are built into network adapters and their attachments. The protocols interface to higher layers through software drivers that are installed with the adapter.

## Data Link Functions

There are a variety of Data Link/Physical Layer standards on today's market. Each is designed to meet the needs of a specific network architecture. Two simple examples will illustrate the diversity of industry protocols at the Data Link Layer and their dependencies on network characteristics.

As a first example, consider **SDLC (Synchronous Data Link Control)**, the data link protocol for SNA. This protocol was originally designed with the goal of reliably delivering large blocks of data within a client-server environment. To achieve this goal, protocols at sending and receiving nodes negotiate to establish a dedicated two-way connection. Once the dedicated connection is established, protocol stacks at the two nodes communicate throughout on-going data transmissions to ensure that blocks of data are delivered correctly. Flow control includes acknowledgments of transmission or request for retransmission if received data is faulty. Protocols at higher levels in this scheme also participate by providing session-to-session communications and logical error correction, such as recombining blocks into a coherent sequence.

**Token Ring**, one of the more popular LAN access protocols, is quite different. Developed by IBM and ratified by the IEEE 802 committee, Token Ring protocols access the physical medium in a peer-to-peer environment. The relationship between nodes (**topology**) forms a closed ring in which data is transmitted in one direction. Data packets are transmitted from node to node around the ring until reaching the destination node. The needs of the Token Ring architecture dictate totally different data link functions than those described for SDLC.

Despite their diversity, these and other Data Link protocol standards perform common functions that include:

- data framing
- medium access
- flow control (to avoid overloading the receiving node)
- error detection

Some protocol standards perform all four functions; others perform only a subset. Each addresses the needs of a specific network architecture. In general, if the architecture is designed so two connecting nodes negotiate or "handshake" to establish a communications session, the specific protocol usually performs all four of the above functions. Such connections are typical of most WAN transmissions.

If frames are transmitted with no dedicated two-way communications between nodes, network protocols perform only a subset of the above functions. Flow control, which requires communication between nodes, is not possible. Retransmission of faulty frames cannot be coordinated. This type of communication is typical in LANs, where information is simply broadcast into a common medium, and the only the addressee responds.

The following sections provide an overview of the basic functions performed by Data Link protocol standards.

## Data Framing

Framing protocols divide the data packet into predetermined sizes and wrap these packets with header and footer information prior to transmission. The wrapped packets are called **transmission frames** or **frames**. Data link protocols specify *what* information needs to be added to assure accurate frame delivery and *how* this information is to be formatted.

> Specific frame formats for popular LAN standards are presented in **Chapter Three**.

Frames are specific to the type of network they are used in. Flags, address fields and control characters that are added as headers or footers differ, yielding unique frame formats for every network. Specific frame formats for popular LAN standards like Ethernet, Token Ring, **ARCNET**, and **FDDI** are presented in **Chapter Three**. Note that some fields are commonly included in many network frames: delimiters that mark the beginning and ending of frame, a frame sequence number, addresses of sending and receiving nodes and error checking information.

Each network also divides the data packet differently for transmission. Some network protocols transmit data packets of variable length, handling large blocks of data with ease. Other protocols break large packets into a smaller, fixed size for transmission. Fixed-length packets are known as **cells**.

Frames fall into two categories, based on how they are transmitted – synchronous and asynchronous. In synchronous transmission, framing protocols separate data into smaller units and send each unit in synchronization with clock pulses. In asynchronous transmission, no clock is required to coordinate data transmission. Instead, start and stop bits are inserted before and after the data packet to mark its beginning and end.

Most networks today use asynchronous data transmission. Although synchronous data transmission requires less processing overhead, it is more hardware intensive because it requires a means of producing, detecting, and coordinating the timing signals required between two nodes. Asynchronous data transmission is just the opposite. It is less hardware intensive, but requires more processing overhead due to the extra start and stop bits inserted between individual characters. In the past, hardware performance was limited, so the separation of characters and extra bits caused delays. Advances in hardware technology have significantly improved performance to the point that asynchronous transmission is no longer a bottleneck.

## Accessing a Shared Medium

Protocols at the Data Link level also define how a node in the network is to gain admission to the physical cable, sometimes called Media Access Control. The scheme used to accomplish this task lies at the heart of the different types of LANs.

In the **polling** method, a controlling station polls each node to see whether it needs to send data. If data delivery is pending, it polls nodes to see whether they can receive data. Variations on this method allow nodes to send multiple messages per polling and allow node services to be prioritized. Although polling methods are more common in WANs than LANs, they can also be used to control line access on LANs.

> See **Chapter Four** for more details on switching.

In **circuit switching** methods, nodes send messages to the controlling station to request a connection (circuit) when they need to transmit data. The controlling station then establishes a dedicated two-way communications link between the nodes so they can communicate as data is transferred back and forth. The connection is not freed for pending traffic until a request for termination is received. Switching concepts are described in detail in **Chapter Four**.

Smaller LAN environments typically do not require large bandwidth, and transmissions across the network tend to be occur in bursts, meaning that sessions between nodes are fairly short. In this case, minimal logic is needed to provide media access to the nodes in the network. All transmissions may be broadcast throughout the network, to be responded to only by the intended addressees. In this case, a scheme must be implemented in order to resolve the possibility that more than one node is attempting to transmit simultaneously (data collision).

These broadcast or shared-bandwidth access methods fall into two categories: **contention** and **collision-avoidance** (sometimes called **non-contention** or **deterministic**). In contention-type protocols, the possibility of data collisions is allowed to exist. When a collision occurs, the network provides for a means to detect the event and initiate re-transmission. Contention-based MACs are an adequate choice in small peer-to-peer networks because they are inexpensive and easy to implement. There is no requirement for a controlling station to coordinate network traffic. The IEEE 802.3 family of standards, popularly known as Ethernet, formalizes contention-type access methods. Please see **Chapter Three** for more information on Ethernet.

In collision-avoidance methods, all nodes participate to control line access so that the possibility of data collisions are precluded. Token-passing protocols (IEEE 802.4 and IEEE 802.5), for example use a "permission token" that orbits around the network until it reaches a node that needs to transmit. Since only one node can possibly possess the token, collisions are always avoided. Please see Chapter Three for more information on Token Ring.

> Please see **Chapter Three** for detailed information about how Token Ring works.

### Flow Control

The purpose of flow control protocols is to coordinate data packet transmission to avoid overloading the receiving node. Flow control protocols allow the receiving node to tell the sending node to wait until it is ready to receive data again. Such coordination is particularly important when a receiving node is handling hundreds of messages over a short period of time or when the receiving device operates at lower data rates than sending devices.

Flow control is accomplished in several ways. Each method ties up the transmission line during a data transfer operation. The simplest method only supports data flows in one direction at a time. The receiving node error checks each transmitted frame and sends an acknowledgment back to the sending node before the next frame is sent. If a frame is not received properly, it is retransmitted before the line is freed for another transmission. This method, **Stop and Wait Control**, is used by simpler TCP/IP applications.

More complex flow control schemes may use a method like **Sliding Window Control**, which supports full-duplex (simultaneous two-way) data flow and permits data transmission operations to overlap. The sending node transmits data frames with sequence numbers across the line, retaining a list of what was sent. After the receiving node error checks each transmitted frame, it sends back acknowledgment with a sequence number. This allows protocols at the sending node to identify the appropriate transmission on the list and remove it. Newer variations permit messages to be transmitted continuously. Sliding Window Control is used by SNA's SDLC, as well as some TCP/IP applications.

### Detection of Frame Errors

The only error detection performed at the Data Link level is the bit checking of transmitted frames. Three popular error techniques are **Vertical Redundancy Checking (VRC)**, **Longitudinal Redundancy Checking (LRC)**, and **Cyclic Redundancy Checking (CRC)**.

VRC and LRC methods involve parity checking. In VRC, an extra bit is inserted into the data stream so that the total number of binary 1's (or 0's) in a byte is always odd or always even. Every byte in this scheme has eight bits of data and one parity bit. If using odd parity and the number of 1 bits comprising the byte of data is not odd, the $9^{th}$ or parity bit is set to 1 to create the odd parity. Using this technique, a byte of data can be checked for accurate transmission by simply counting the bits for an odd parity indication. If the count is ever even, an error is indicated. In LRC, parity bits are added to each block of characters, rather than to each single character as done in Vertical Redundancy Checking (VRC).

In CRC error checking, error detection information is calculated by dividing the data stream by an agreed upon value. The remainder of the calculation is tacked onto the transmission frame so the receiving node can perform the same operation and compare its calculation to that transmitted from the sending node. A match indicates an error-free transmission.

## Physical Layer Functions

### Signal Transmission

Protocols at the Physical Layer include both encoding methods and definitions of how electrical signals are transferred across hardware interfaces. These protocols accept data bits from the Data Link Layer, encode them as electrical signals and send them over the physical cable.

There are a variety of encoding methods. Each represents a data bit by controlling the electrical current over a fixed interval of time. In **NRZ** (**Non-Return to Zero**) encoding, bits are represented by holding the electrical current constant over the time interval. A low-level current might represent bit "0" and a high-level signal bit "1," or vice versa. **Manchester Encoding**, another common method, employs a 'change' in electrical signal to represent a bit. The transition from a positive signal over the first half of the time interval to a negative signal over the second half might represent the bit "1." Conversely, a transition from a negative to positive signal might represent a "0."

Electrical signals are transmitted over the cable using baseband or broadband transmission techniques, discussed in more detail in **Chapters One**, **Three**, and **Five**.

## IEEE LAN Standards

Media access control and signaling protocols are LAN standards developed by subcommittees of IEEE. They are part of a network architecture that includes only the lowest layers of the OSI model: the Data Link and Physical Layers. These layers are built into the network adapter, and interface with higher layer protocols through software drivers. Higher layer protocols are typically part of the network operating system, such as **Windows 95**, **NetWare**, or **OS/2 LAN Server**.

A comparison of the OSI and IEEE standards is shown in **Figure 2-7**. Notice that the IEEE standards include two layers at the OSI Data Link Layer: **Logical Link Control (LLC)** and **Media Access Control (MAC)**. The LLC layer provides an interface between the family of MAC level protocols that operate below it and network operating systems that operate above it. MAC layer protocols define how all nodes on a LAN communicate with each other. Typically, these protocols only perform a subset of the Data Link functions described in the previous section. Signaling Protocols at the Physical Layer define how data bits are encoded into electrical signals and transmitted.

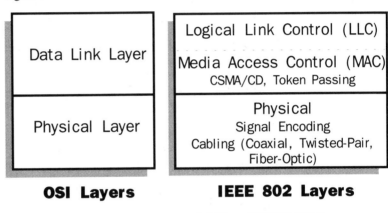

*Figure 2-7: Comparison of OSI and IEEE 802 Layers*

To address the variety of ways in which a physical medium can be accessed and data transmitted, the IEEE developed a number of MAC and Signaling standards. Each answers the needs of a specific architecture. Token Bus (**IEEE 802.4**), Token Ring (**IEEE 802.5**), and the Ethernet family of standards (**IEEE 802.3 Carrier Sense Multiple Access/Collision Detection [CSMA/CD]**) are only a few of the standards developed. The details of these standards are discussed in **Chapter Three**.

# Chapter Review

- Communications standards define how information is formatted and transmitted at various levels across a network.

- Standards organizations recommend rules and conventions to ensure compatibility of network communications products.

- The Open System Interconnect (OSI) model recommends guidelines that enable vendors to develop compatible products. Although vendors rarely adhere to this standard, the model has become a means against which the industry compares and communicates various aspects of its products.

- The OSI model defines seven hierarchical layers that define a specific level of data communications.

- Each OSI layer operates independently to promote open system connectivity. Each layer interfaces with adjacent layers according to specific rules.

- Peer-to-peer communications is the ability of corresponding layers of network nodes to exchange information at their level.

- Popular LAN access standards ratified by the IEEE are implemented at the OSI Data Link/Physical Layer levels. These are hardware-based protocols (residing on network adapters).

- LAN access protocols (media access control and signaling) perform four basic functions: framing, accessing the physical media, flow control and error detection.

# CHAPTER THREE

# Local Area Networks

**Chapter Three Contents:**

Part 1: General LAN Categories .................................. 39

Part 2: LAN Standards .................................................. 46

# General LAN Categories

Because every LAN is implemented under different circumstances, they are all unique. However, all LANs can be typed according to where they fall under three categories: **Topology**, **Bandwidth Use**, and **Media Access Control** method.

## Topology

> The **topology** is the relationship among nodes in a network

**Topology** refers to the relationship among nodes in a network. No matter how the cable is twisted, turned, sent up and down hallways, through ceilings, and into boxes, its basic topology will always fall under one of the three basic categories: **linear bus**, **star** or **ring**.

### Linear Bus Topology

The **linear bus** topology utilizes one continuous, unbroken length of cable. The workstations and server(s) tap the cable using "**T**" **connectors**. The file server may be located anywhere on the bus, it does not have to be at the beginning or end.

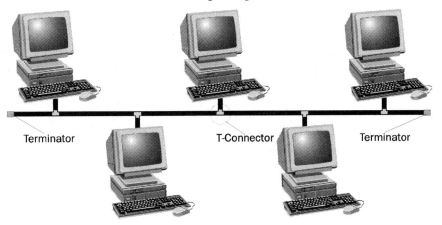

*Figure 3-1: Linear Bus Topology*

A **terminator** must be installed at each end of the cable (see **Figure 3-2**). Terminators are resistors that serve to define the beginning and end of the linear bus.

Each side of the T-connector must be connected to the cable or a terminator. One drawback to this type of topology is that no part of the network will be able to communicate if any part of the cable becomes broken or if either of the terminators are removed. If this happens, the entire network is "down." The main advantage of the linear bus is the minimal amount of cable needed to install a network.

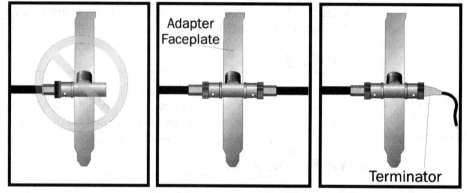

*Both sides of the T-connector must be connected, either to a cable or a terminator.*

**Figure 3-2: T-Connector Configurations**

A variation of the linear bus topology uses a technique called **daisy-chaining**. This topology is used by network interface cards that use twisted pair wiring and two modular jacks, one for cable-in and one for cable-out (similar to the cable arrangement on modems and answering machines). The workstations are chained together to form a linear topology. In this type of topology, a terminator is installed on the unused jack of the node on each end of the linear bus (see **Figure 3-3**).

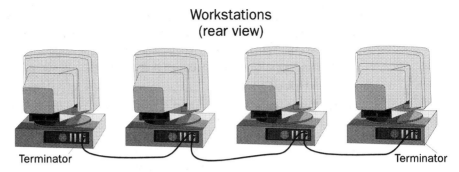

*Figure 3-3: Daisy-Chained Linear Bus Topology*

## Star Topology

The **star** topology uses multiple segments of cable – one for each workstation on the network. The file server must be located at the center of the star. Multiple stars may be linked together to form star clusters.

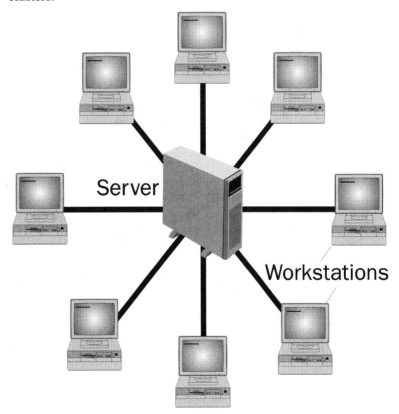

*Figure 3-4: Star Topology*

The major advantage to using this type of topology is reliability. If a cable segment is broken, only the node connected to that segment will lose communications with the network. The main disadvantage to the star topology is the enormous amount of cable required to install a large network.

## Ring Topology

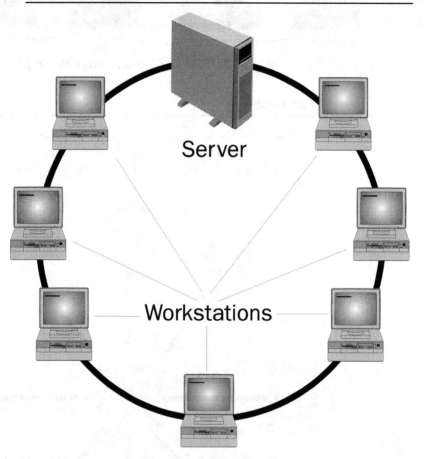

*Figure 3-5: Ring Topology*

Ring topology uses one continuous length of cable to form a loop. The workstations and file server tap this line. The file server may be located anywhere on the ring.

In its basic form, the entire network is down if any part of the cable becomes broken. Most ring LANs use special junction boxes (hubs) known as "**Multistation Access Units**" (**MAU**s) to prevent this occurrence. MAUs close a node connection automatically if the cable connecting a node becomes broken. By closing the connection, the malfunctioning node is isolated from the network, allowing the rest of the network to remain functional.

The special connectors and MAUs this type of network employs often make it more costly in comparison to networks that use other topologies. The main advantage to the ring topology is the minimal amount of cable needed to install a network (see the sections on star and linear bus topologies above).

### Special Case Topologies

It is important to note that even though the cable may go through special boxes, usually called hubs, concentrators, or MAUs, the topologies always remain the same. **Figure 3-6** depicts an example of a star topology utilizing a hub. You'll notice that the star topology remains the same even though the cable goes via the hub.

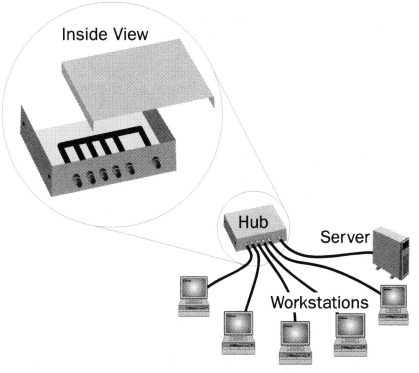

*Figure 3-6: Star Topology Using a Hub*

## Bandwidth Distribution

Another way to differentiate networks is through the way in which they distribute the bandwidth of the transmission medium (cable) among nodes. There are basically two ways this can be accomplished: share the entire bandwidth of the network among all connected nodes, or provide a dedicated path between communicating nodes.

### Shared Bandwidth

In the recent past of computing, the most cost-effective manner of distributing bandwidth was to share it among connected nodes through time division multiplexing (baseband) or frequency division multiplexing (broadband). Sharing bandwidth requires traffic rules to allow the nodes to access the medium in an orderly manner. A Media Access Control (MAC) method is the manner used by a node to gain the use of the physical medium for transmitting information. As stated in chapter two, there are two methods of media access: **contention** and **collision-avoidance**.

Contention-type networks do not resolve the possibility that two or more nodes may attempt to transmit at the same time. However, they do have the facility to determine if that possibility (a data collision) has occurred and attempt a re-transmit. The philosophy is that the chance of a collision occurring is small enough that most transmissions will get through in a reasonable amount of time. The most popular LAN access standard, Ethernet, uses this system.

With a collision-avoidance access method, there is no chance that two nodes will contend for media access at the same time. The most popular collision-avoidance system is token passing, in which a node is allowed to transmit only if it has possession of a an electronic token that is circulated through the network. Collision-avoidance LAN access standards include ARCNET, Token Ring, and FDDI.

### Dedicated Bandwidth

LANs have traditionally shared bandwidth through schemes such as TDM for baseband media and FDM for broadband media. In either case, bandwidth is a finite commodity, therefore the more nodes added to a LAN, the less bandwidth is available per node. The result is slower throughput, as experienced at an individual node. Contention-based MACs compound the problem. As more nodes are added, more collisions occur. This results in degraded network performance as more resources are occupied by re-transmissions.

A quick fix is to divide up LANs into segments that limit inter-network traffic (see **Inter-LAN Connectivity Devices**," **Chapter Four**) giving each node a bigger share of available bandwidth. The logical progression is to give each node dedicated network bandwidth.

Switching is the next step in this evolutionary process. The idea behind switching is to provide a node with its own dedicated network connection, essentially giving it all available bandwidth for the duration of the switched circuit. Newer network standards, such as ATM, are based on switching technology. For more information about switching, please see **Chapter Four**.

## A Word About LAN Categories

Topology, bandwidth use, and MACs are depicted in the preceding sections as separate categories for the sake of simplicity. It is not intended to suggest that these are mutually exclusive categories of networks. Instead, they are provided to help you understand the fundamentals of how networks function. Most LANs contain variations of all these categories.

# LAN Standards

When speaking of "LAN types" or "LAN standards," most people are referring to standards that govern the media access control and signaling systems being employed at the bottom layers (Physical and Data Link) of a network. The media access method is the set of rules used by a node in the network to gain admission to the transmission medium (cabling) and establish communication with another network node. See **Chapter Two** for general information on the OSI layers and media access control (MAC) and signaling basics.

> "LAN standard" usually refers to set of rules used by a node in the network to gain admission to the transmission medium and establish communication with another network node

The following sections cover in detail the most popular multi-vendor LAN access standards. Information on single-vendor (closed architecture) networks, such as IBM's SNA, is better presented by the proprietor/vendor of the technology.

## Ethernet

Introduced in the late 1970s by the Xerox Corporation, **Ethernet** was one of the first LAN architectures. Its economical price, high transmission speed, and wide support for PC-to-mainframe connections made it an early leader. The **IEEE 802.3** Committee formalized the original Ethernet architecture into the **IEEE 802.3 10Base5** specification. Since then, there have been several additions to the 802.3 Committee specifications. They include **10Base2** (or **ThinNet**), **1Base5** (the formal version of AT&T's Starlan), **10BaseT** (twisted pair), and **10BaseF** (fiber-optic). The 802.3 standard continues to be one of the most popular and flexible LAN architectures to this day. The wide variety of cabling options and low price make it an attractive LAN solution for many companies.

> Ethernet specifications may be found in **Appendix C**.

Ethernet is categorized by its contention-type MAC, called **CSMA/CD (Carrier Sense Multiple Access with Collision Detection)**. See **Chapter Two** for more information on Media Access Control methods. Ethernet transmits data in **packets** across the network. A data packet being sent across an Ethernet LAN is broadcast across the entire network, meaning every node will detect it. Each node checks to determine where the packet is addressed to, and is only accepted by the correct addressee. This is the Carrier Sense Multiple Access part of CSMA/CD.

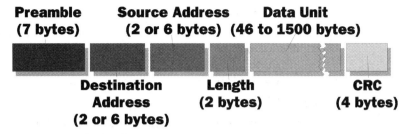

*Figure 3-7: Ethernet Packet*

| Table 3-1: Ethernet Packet Format ||
|---|---|
| Preamble | Used for synchronization between nodes. |
| Destination Address | Can be a single workstation, or one or more groups of workstations. |
| Source Address | Originator of the packet. |
| Type | What format is being used for the Data field. |
| Data | The actual information that is being sent. |
| CRC | Holds the error-checking analysis of the other fields in the packet. |
| **Note**: Each Ethernet packet is 72 to 1,526 bytes in length. ||

The Collision Detection part works as follows:

1. If an Ethernet NIC needs to originate a transmission, it first listens for any other transmissions on the network. If none are detected, the NIC begins its transmission.

2. While transmitting, it listens to the signal level on the network. If the level rises above a preset limit, the NIC determines that a data **collision** has occurred, meaning two nodes are attempting to transmit across the network simultaneously.

3. The NIC then begins to transmit a **collision signal**. When the other NICs on the network receive this signal they stop transmitting and start to listen only.

4. All NICs on the network, including the one that one that originated the collision signal, then wait a random period of time before attempting to transmit. The random interval reduces the possibility that the same nodes will interrupt each other again.

CSMA/CD works reasonably well on networks with a limited number of nodes because the chance of collision is small, and re-transmissions do not occur that often. However, the more nodes attached to a CSMA/CD LAN, the greater the number of collisions. This leads to the network slowing down due to more time being occupied by re-transmissions.

Among the reasons for the popularity of Ethernet are its cost-effectiveness and flexibility. CSMA/CD requires less logic than collision-avoidance LANs, therefore is less expensive to implement. There are variations of Ethernet that support throughput from 10 Mbps all the way to 100 Mbps, using everything from twisted pair to fiber-optic cabling. Because these variations are compatible, Ethernet provides an attractive, cost-effective migration path that allows organizations to extend the use of their legacy hardware.

The different variations of IEEE 802.3 are discussed in the following pages and summarized below:

*Table 3-1: IEEE 802.3 Ethernet Standards*

| IEEE 802.3 Standard | Defines | Throughput |
|---|---|---|
| 10Base2 (ThinNet or Cheapernet) | RG-58A/U or RG58C/U Thin Coaxial Cable | 10 Mbps |
| 10Base5 (ThickNet) | Thick Coax Trunk with AUI Drop Down Wire | 10 Mbps |
| 1Base5 (Starlan) | Unshielded Twisted Pair Wiring (Linear Bus or Star) | 10 Mbps |
| 10BaseT | Unshielded Twisted Pair Wiring | 10 Mbps |
| FOIRL | Fiber Optic Trunk with AUI Drop Down Wire | 10 Mbps |
| 10BaseF/FL/FP | Fiber Optic | 10 Mbps |
| 100BaseT (Fast Ethernet) | Unshielded Twisted Pair Wiring (Star) | 100 Mbps |

The **AUI** (**Attachment Unit Interface**) is used as a universal connector to any Ethernet cabling media through the use of an appropriate external transceiver. For networks that use a main trunk, such as 10Base5, the transceiver taps into the trunk and then connects the NIC by way of a 15-wire drop cable (**Figure 3-8**).

The Network Technical Guide  49

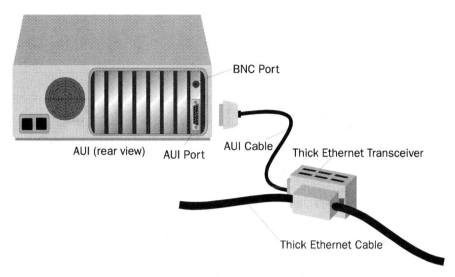

*Figure 3-8: 10Base5 AUI and Transceiver*

The **FOIRL** (**Fiber Optic Inter-Repeater Link**) specification is for NICs that do not have the optic signaling capacities, but do have an AUI port. In this case, the fiber optic transceiver simply converts the AUI signals into the correct pulses of light (**Figure 3-9**).

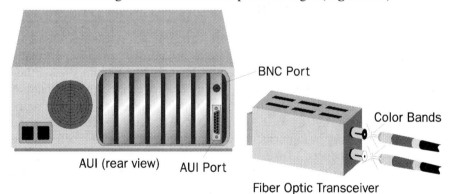

*Figure 3-9: 10Base5*

For 10Base2 or 10BaseT networks, the AUI is used in much the same way as FOIRL networks. The AUI port provides a connection for that type of media if there is not one on the NIC itself.

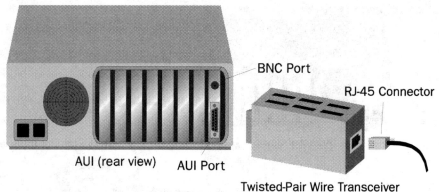

*Figure 3-10: 10BaseT AUI and Transceiver*

The maximum distance between the AUI port and the external transceiver is 164 feet (50m), there is no minimum distance (in some cases the transceiver may even plug directly into the AUI port as shown in Figure 3-10). Another name for the AUI port is the acronym **DIX**, which is a combination of **D**igital, **I**ntel, and **X**erox. These companies were the first to develop the systems that became the Ethernet and AUI port standards.

## Token Ring

> Token Ring specifications may be found in **Appendix C.**

The **Token Ring** network structure was introduced by IBM in 1985 and subsequently adopted as an industry standard, officially named ANSI/IEEE 802.5. Token Ring is less popular than Ethernet, but has had a steady share of the market due to features that have made it attractive for use in high-end applications. Even though it was developed by IBM, and is the first choice for installations using IBM equipment, a wide variety of vendors offer Token Ring products. Token Ring networks are capable of transferring data at maximum rates of either 4Mbps or 16Mbps, depending on the type of cable and other equipment on the network. The differences in IBM cable types are detailed in **Table 3-2**:

*Table 3-2: IBM Token Ring Cabling Variations*

| Type | Uses |
|---|---|
| 1 | Has two shielded twisted pairs of solid core 22 AWG wire. It is used for long, high grade data transmissions within the buildings' walls. |
| 2 | Contains a total of six twisted pairs, two are shielded while the others are outside the shielding. The two 22 AWG solid core shielded twisted pairs are used for the network, while the four 22 AWG solid core unshielded pairs are typically used for telephone systems. |
| 3 | Has four unshielded twisted pairs of 24 AWG solid core wire used for networks or telephone systems. This is used for long low-grade data transmissions within walls, but not as far as Type 1 (due to the shielding properties of the Type 1 Cable). Cannot use this type of cable for 16Mbps Token Ring networks. |
| 5 | Uses two 100μm or 140μm optical fibers in one jacket. |
| 6 | Has two 26 AWG stranded core shielded twisted pairs used for networks only. It is most commonly used to connect network devices to wall jacks and is also used in patch panels and punch-down blocks. |
| 8 | Uses a single 26 AWG stranded core shielded twisted pair, typically flat for under-the-carpet installations. |
| 9 | Has two shielded twisted pairs of solid core 26 AWG wire. It is used for long lengths of cable within the buildings walls (Very similar to Type 1.) |

Token Ring networks are more complex than ARCNET or Ethernet. Features, maximum cable lengths and number of nodes will vary from vendor to vendor, and also depend on specific installations.

Because of these variations, it is recommended that when purchasing Token Ring products, an attempt should be made to have all the devices that will be attached to the network provided by a single vendor. Due to the large number of variations that may exist for Token Ring installations, only a brief overview will be provided here.

The Token Ring network gets its name from the method used to control when a node can transmit. A **Permission Token** is originated by the **Active Monitor**. This token is passed in order from node to node. If a node has data to transmit, it must wait until it receives the token before it may do so.

See **Table 3-3** for details on the terms in this figure.

*Figure 3-11: Token Ring Token*

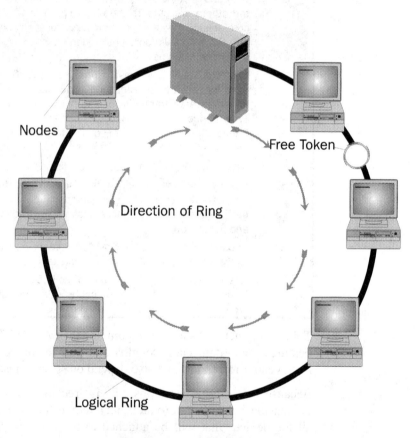

*Figure 3-12: Idle Token Ring*

Once the token has been received from its upstream neighbor, the transmitting node changes the token into a **busy token** and transmits the data in a **frame** to the next station on the ring. In addition to the data, the frame contains the sender's address, and the destination address.

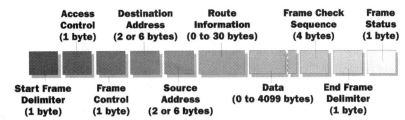

*Figure 3-13: Token Ring Frame*

| Table 3-3: Token Ring Frame Format | |
|---|---|
| Start Frame Delimiter | Marks the start of the frame. |
| Access Control | Used to label the priority of the frame. |
| Frame Control | Defines frame type. A frame can be a Media Access Control (MAC) type, which are read by all stations, or Logical Link Control (LLC) type, which are read only by the destination station. |
| Destination Address | Originator of the packet. |
| Source Address | Intended receiver of the packet. |
| Route Information | Used in bridging applications to describe the route used for information transmitted between different Token Rings. |
| Information | Data being sent. |
| Frame Check Sequence | CRC-32 error checking. |
| End Frame Delimiter | Marks the end of the frame. |
| Frame Status | Used to determine if frame was received by the intended recipient(s). |

Each station downstream from the sending node looks at the destination address, and, if it is not for them, passes it on to the next station.

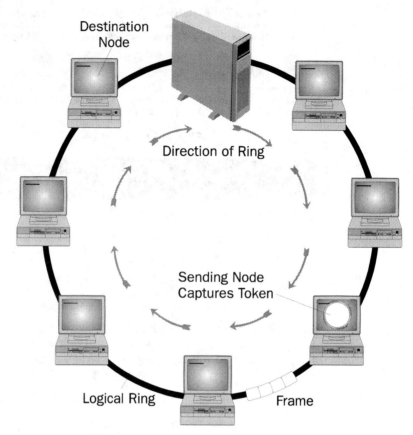

*Figure 3-14: 4 Mbps Token Ring Data Transmission*

When the frame reaches its destination, the recipient copies the data in the frame and adds a **stamp** or **flag** to the end of the frame to let the sender know that it was received. The frame then travels back to the originating node where it is removed from the network. The busy token is then changed to a free token and is passed downstream until it reaches the next waiting station.

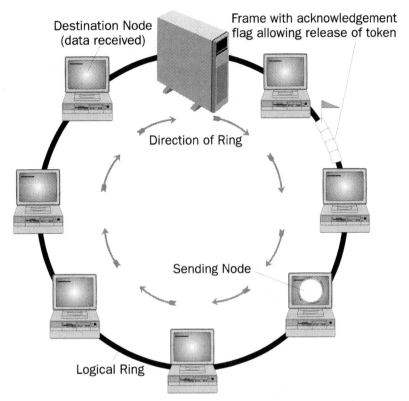

*Figure 3-15: Data Reception Complete on 4 Mbps Token Ring*

This method ensures that only one node at a time is sending data and gives each node in the network an even chance to transmit. This is a large contrast to the somewhat chaotic "transmit while its free" method used in Ethernet. Just as Ethernet is termed a contention-type Media Access Control protocol, Token Ring falls under the category of non-contention or collision-avoidance (see **page 44**). The main advantage is that there can be no collisions, and therefore collision-caused delays, on a Token Ring network. The main disadvantage is that each token and frame must pass through every node on the network, thus the more nodes that are added, the slower the network will become.

An optional 16 Mbps version of Token Ring uses a variation of the scheme explained above. With the 4 Mbps version, the network has to wait for a transmitted frame to make a circuit back to the originating node before the token becomes available again. In contrast, on a 16 Mbps Token Ring network the transmitting node can release the token just after transmitting the frame. This is known as "early token release."

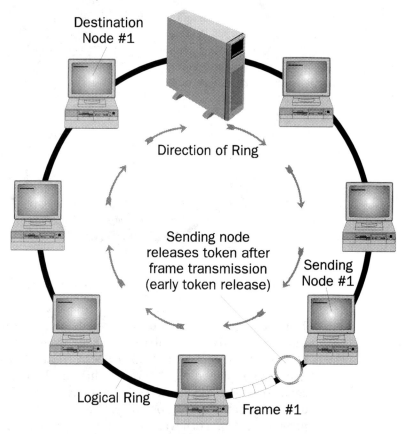

*Figure 3-16: First Frame Sent on 16 Mbps Token Ring*

The token then travels downstream on the heels of the transmitted frame, available for the next station that wishes to transmit. Every node on the ring must use the same setting, either the 4 Mbps or 16 Mbps protocol. Some 16 Mbps Token Ring NICs can automatically sense the speed that the network is operating at and adjust accordingly.

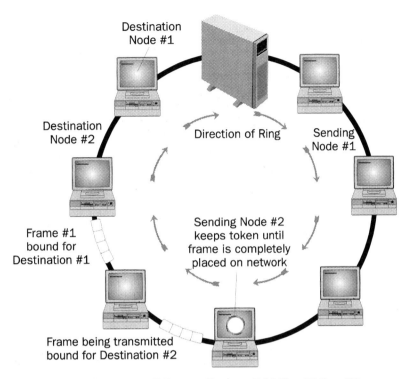

*Figure 3-17: Second Frame Sent on 16 Mbps Token Ring*

## Token Ring Multistation Access Units

Token Ring networks appear to be physically wired in a star topology, but the functional topology is really a ring. The **MAU** (**Multistation Access Unit**) is a type of concentrator that enables this logical ring. Each MAU has at least 8 **lobe ports** that are used to connect the various network devices. Each lobe port is connected to a ring in the MAU (shown below). The **Ring In** (**RI**) and **Ring Out** (**RO**) ports are used to connect other MAUs to the ring. If a single MAU is used, then the Ring In and Ring Out ports are not used and the signal is wrapped back into the other side of the same MAU.

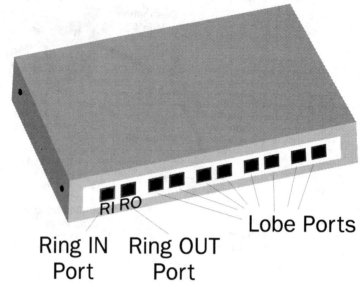

*Figure 3-18: Multistation Access Unit (MAU)*

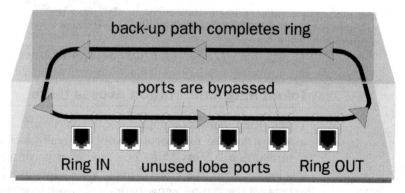

*Figure 3-19: MAU Detail*

When workstations are connected to the MAU, they must use a cable that has two twisted wire pairs. One of the pairs transmits to the node from the MAU, while the other pair transmits to the MAU from the node.

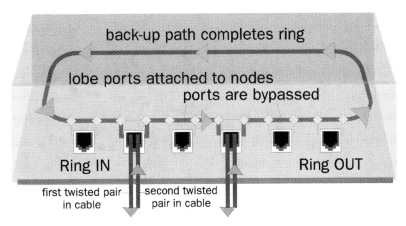

*All the signals between the MAU and the nodes are carried on two twisted pairs inside a single cable.*

*Figure 3-20: Workstations Attached to Lobe Ports*

When more than one MAU is used on the network, the Ring In and Ring Out ports are attached using a cable with two twisted pairs. The first pair is used for the normal network connection, while the second pair is used for a redundant link (see **Figure 3-21**).

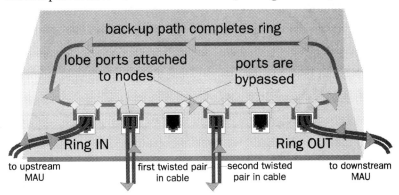

*The signals between two MAUs are also carried on two twisted pairs inside a single cable.*

*Figure 3-21: Two MAUs on the Token Ring*

Each Ring Out port must connect to the next MAUs Ring In port. If there are three or more MAUs linked together in this fashion, the last MAUs Ring Out port must be connected to the first MAUs Ring In port. This competes the ring and at the same time enables a fault-tolerant back-up path.

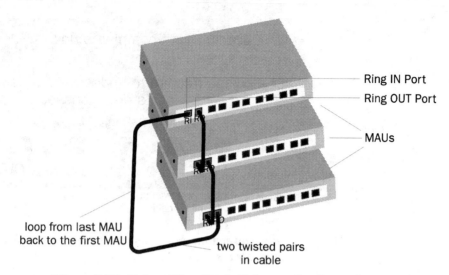

*Figure 3-22: Token Ring Fault-Tolerant Configuration (Realistic Depiction)*

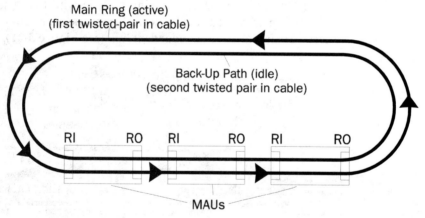

*Figure 3-23: Token Ring Fault Tolerant Configuration (Logical Depiction)*

The back-up path is normally idle, and flows in the opposite direction from the main ring. If a cable connection between MAUs is ever broken, the MAUs on both sides of the break will enable their internal "ring wrap" connection. This isolates the problem segment and enables the rest of the ring to continue to function using the back-up path.

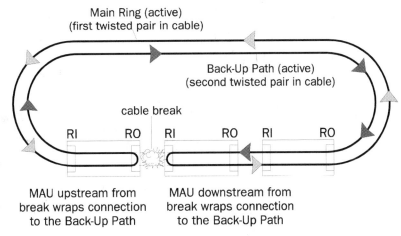

*Figure 3-24: Active Back-Up Path*

## ARCNET

**Attached Resource Computing Network (ARCNET)** was developed, trademarked, and licensed by the Datapoint Corporation. ARCNET is not an IEEE standard, although it is sponsored by ANSI as the ANSI 878 specification. ARCNET has generally been superseded by faster Ethernet versions and other more current LAN access standards, although it still has widely remaining installations and avid supporters.

> ARCNET specifications may be found in **Appendix C**.

ARCNET is capable of supporting up to 255 active devices (hubs do not count as an active device) per network segment. The standard data transfer rate is 2.5Mbps, although a 20Mbps version was introduced in the late 80s. While this may appear slow by current standards, in some situations ARCNET still has advantages over slower versions of Ethernet. Since ARCNET uses a collision-avoidance MAC (see **"Dedicated Bandwidth"** on **page 44**), increasing the number of nodes on a network has less effect on the transfer rate.

> For details on setting ARCNET node addresses, see **Appendix A**.

ARCNET is a modified token passing protocol that most often uses a star topology but occasionally may use a linear bus topology. Note that these are the physical wiring topologies used. ARCNET uses what is often called a **logical ring** topology. This means that the relationship between nodes is based on assigned **node addresses**, rather than on the physical wiring. Regardless of where nodes physically exist on the network, the token is passed between nodes of sequential addresses. See **Figure 3-25** for an example.

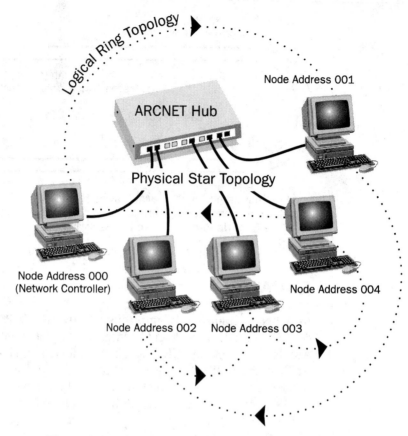

*Figure 3-25: ARCNET Physical and Logical Topologies*

All nodes on an ARCNET LAN receive all the messages that are broadcast onto the network at about the same time. Unlike the Ethernet and Token Ring networks, which have their unique node addresses pre-set at the factory, an ARCNET NIC node address must be set by a DIP switch or jumper on the card. This DIP switch usually has eight switches that correspond to a binary representation of a number between 0 and 255. When the network first becomes active, all NICs broadcast their node address and the NIC with the lowest node address becomes the **Network Controller**. This automatic network configuration is a key feature of ARCNET.

If an ARCNET NIC wants to originate a transmission, it must first wait for the Network Controller to send it the "**Permission Token**." Once this signal is received, the NIC broadcasts any message it may have, starting with the destinations' node address (**Figure 3-26**).

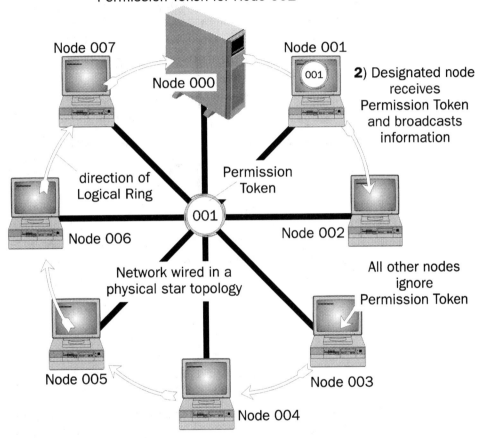

*Figure 3-26: ARCNET Token Passing, Part One*

Once the message has been removed from the network, the Network Controller sends the permission token to the NIC with the next highest node address (**Figure 3-27**).

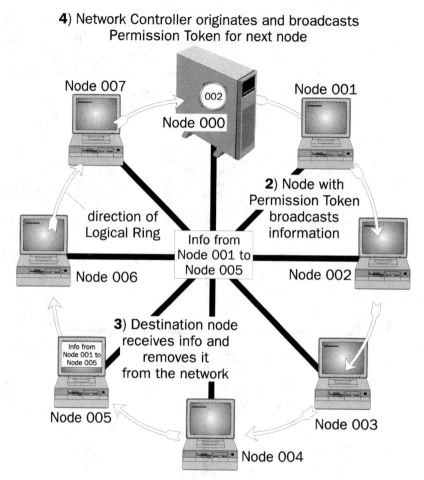

*Figure 3-27: ARCNET Token Passing, Part Two*

When an ARCNET network is being configured, the node addresses should be ordered sequentially in order to allow the Permission Token to be passed as quickly as possible. Because the Network Controller operation takes a very small amount of processing time, the lowest node address is normally given to the NIC installed in the most powerful system on the network.

When a new node enters the network, all NICs re-broadcast their node address on the network. This reconfiguration takes less than 65ms and should not interfere with the speed at which the network operates. It is vitally important that all ARCNET NICs have a unique node address – especially when adding a new node to the network. If the new addition has a node address set to an existing number and is powered on, it will fail to join the network. This is one of the most common installation errors, and therefore among the first things to check when an ARCNET NIC is not operating properly.

ARCNET NICs may use coaxial cable (standard) or unshielded twisted pair (UTP) cable. The major difference between the two is the allowable maximum distance between an active hub and a workstation. When coaxial cable is used, a workstation may be up to 2000ft. (606m) away from an active hub (on most implementations) and 100ft. (30m) away from a passive hub. When unshielded twisted pair cable is used, a workstation may be up to 400ft. (121m) away from a hub (all hubs in unshielded twisted pair ARCNET installations are active).

## FDDI

**FDDI** stands for **Fiber Distributed Data Interface**. FDDI is a set of standards established in the late 80s that deal mainly with the bottom two layers of the OSI model and are approved by ANSI and ISO. As the name implies, the primary feature of FDDI is the fiber-optic media. FDDI is a token-passing access method, currently capable of throughput up to 100Mbps. Enhanced versions are in the works. **FDDI-2**, adds sound and video channels in addition to data. **FFDT** (**FDDI Full-Duplex Technology**) is capable of up to 200Mbps data rates. Currently, FDDI is typically being applied as a high-speed **backbone**. FDDI connections down to the workstation level are rare due to cost and adequate performance of existing copper wiring.

Although FDDI has been available for several years, it has been slow in gaining acceptance due to the reluctance of potential customers in replacing existing twisted pair installations. **TP-PMD** (**Twisted Pair-Physical Media Dependent**), sometimes called **CDDI** (**Copper Distributed Data Interface**) has been fielded as a cost effective migration path to FDDI. Except for making use of existing twisted pair wiring, TP-PMD is functionally the same as FDDI.

FDDI uses a token-passing access method that is similar to Token Ring. The major difference between the two is that the FDDI token is appended to the end of a transmitted data frame, instead of remaining captured by the transmitting station as in 4 Mbps Token Ring.

FDDI token-passing works as follows:

1. When the network first becomes active, the first station on it sends a token downstream. The token is passed unaltered from station to station until it encounters one that needs to transmit.

2. When a station on the network wants to transmit, it waits for an available token to be passed to it. The station then captures the token and appends a frame to it that includes the destination address, source address, and data.

3. The frame is passed along the ring, with each station checking the addressing information.

4. When the addressee receives the frame and token, it extracts the data frame from the token.

5. The free token is then sent back downstream.

6. The next station downstream is now eligible to append a frame to the token to transmit.

See **Figure 3-28** on for a graphical depiction of this process.

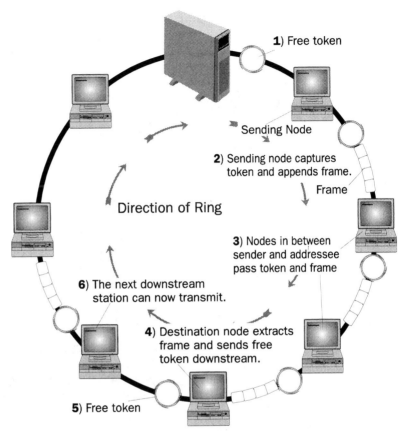

*Figure 3-28: FDDI Token-Passing*

Functionally, the big contrast between FDDI and 4 Mbps Token Ring token passing is that FDDI's method minimizes transmission delays. The frame does not have to make a complete circuit back to the sender before a free token is released back into the ring.

The FDDI data frame is similar in format to that used by Token Ring.

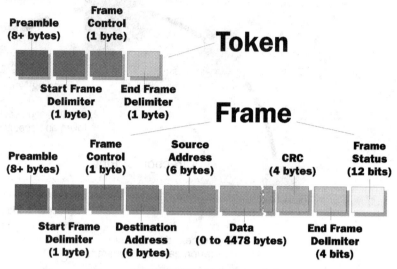

Figure 3-29: FDDI Token and Frame

| Table 3-4: FDDI Frame Format | |
|---|---|
| Preamble | Used for synchronization between nodes. |
| Start Frame Delimiter | Marks the start of the frame. |
| Frame Control | Defines frame type. A frame can be a Media Access Control (MAC) type, which are read by all stations, or Logical Link Control (LLC) type, which are read only by the destination station. |
| Destination Address | Originator of the packet. |
| Source Address | Intended recipient of the packet. |
| Information | Data being sent. |
| Frame Check Status | CRC-32 error checking. |
| End Frame Delimiter | Marks the end of the frame. |
| Frame Status | Used to determine if frame was received by the intended recipient(s). |

## SAS and DAS FDDI

The simpler form of FDDI, called **Single Attached Station (SAS)**, does not provide any protection against media failure. Any discontinuity in the ring brings down the network. **Dual Attached Station (DAS)** FDDI provides a high degree of fault tolerance through a redundant back-up ring. This **Dual Counter-Rotating Ring** is an option, but is one of the featured benefits of FDDI. **Figure 3-30** depicts DAS and SAS coverage existing in the same network.

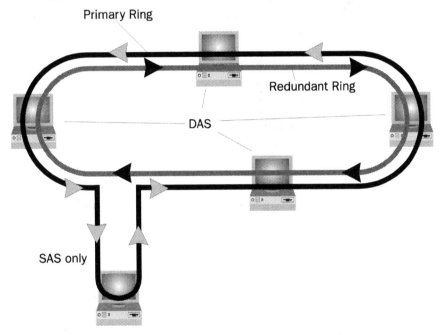

*Figure 3-30: SAS and DAS FDDI*

DAS uses a back-up ring in which data flows opposite to the direction of the primary ring. In the event a break occurs in the primary ring, the nodes on either side of the break sense the loss of the connection and close off the broken segments. The broken connection on the primary ring is then wrapped back to the redundant ring, effectively combining the two rings into one (see **Figure 3-31**).

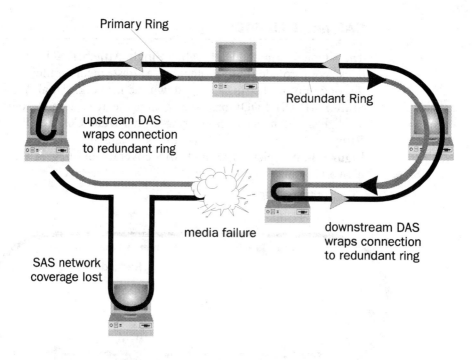

*Figure 3-31: FDDI Fault Tolerance*

## Emerging LAN Standards

The past five years have seen remarkable advances in computer technology, and networks have been no exception. Ethernet products are still the most popular, but newer technology is well on the way to replacing this standard. Newer standards typically take advantage high-bandwidth media, such as fiber-optics, and high-speed switching technology.

### ATM

**Asynchronous Transfer Mode (ATM)** is an attractive new technology that takes advantage of recent developments in the communications industry. ATM is a departure from traditional LANs in that ATM technology is used in both local and wide area connections. ATM was originally proposed for the bottom layers of the **Broadband ISDN (BISDN)** standard (see **Chapter Four**). The ITU is still working on the standard, but other organizations, notably the **ATM Forum**, have taken the technology and have come up with proposals for different uses, such as LANs.

Unlike Ethernet, Token Ring and FDDI, which basically broadcast information into shared-bandwidth media, ATM allots dedicated bandwidth to communicating nodes. At the heart of ATM is a high-speed switching technology made possible by small, fixed-length (53 bytes) data units called **cells** (instead of variable-length frames or packets). The fixed length of the cells requires less overhead in the switches, resulting in faster throughput.

Variable-length frames are better suited for data traffic that occurs in bursts – meaning a fairly small amount of information sent over a short period of time, such as text and database files. Modern applications often involve multimedia, such as sound, graphics, and even full-motion video. This type of data involves large, sustained data streams on a network. When being used to transmit this type of information, variable-length frames become larger to accommodate. The network can then become overloaded without any warning. Even a high-speed network (100 Mbps or greater) can quickly become saturated. This is an unacceptable situation when the network is being depended on to transmit real-time information.

The small and fixed size of ATM cells also aids in reducing network congestion. With the cell scheme, all types of data occupy the same amount of bandwidth for the duration of the transfer. This means that the bandwidth availability on the network is predictable, allowing for effective planning on the part of administrators.

Please see **Chapter Four** for information on how ATM fits into a WAN framework.

### 100VG-AnyLAN (100BaseVG)

**100VG-AnyLAN**, sometimes called 100VG, AnyLAN, or **100BaseVG**, is a high-speed LAN access standard developed by a group of companies led by Hewlett Packard. It has been approved by IEEE under the **802.12** standard.

The best way to understand this standard is to break down the components of the name.

- **100** represents the 100 Mbps maximum throughput.
- **VG** stands for the Voice Grade copper wiring used. 100VG can use category 3, 4, or 5 UTP cabling.
- **AnyLAN** means that it has built-in support for Ethernet or Token Ring frames. This means that network cabling and organization can remain the same, 100VG equipment simply replaces the old adapters and hubs for greater throughput. Note that this does not mean that Token Ring and Ethernet may coexist on the same network segment.

100VG is sometimes mistakenly called Fast Ethernet, which more accurately describes 100BaseT. 100VG does not use the CSMA/CD (IEEE 802.3) access method at all. Instead, IEEE 802.12 describes a non-contention access method known as **Demand Priority**.

Demand Priority is essentially an intelligent switching method. At the lowest level, 100VG switching hubs control access onto the network, allowing only one node at a time to transmit. Smaller networks may only need one hub. If more than one hub is necessary, they are in turn controlled by a higher level switching hub for overall network access. From this description, you can see that the network is basically structured as a tree, where each level depends on the level above it for access.

Demand Priority, as the name implies, allows data to be labeled with varying degrees of importance. Information that is labeled with a high priority, such as real-time voice and video, gets moved to the top of the queue at network switches. Because of this, Demand Priority is better suited to video conferencing and other time-sensitive transmissions than currently popular LAN access standards.

# Chapter Review

- There are several ways to classify LANs. Among these are topology, bandwidth distribution, and LAN access standards.

- Topology describes the relationship among network nodes. The physical topology is how nodes are actually wired together. Logical topology comes into play with token passing schemes where network addresses instead of physical locations define relationships.

- The dominant LAN standards are currently IEEE 802.3 CSMA/CD (Ethernet) and IEEE 802.5 Token Passing (Token Ring).

- The biggest issues driving the development of follow-on LAN standards are increased bandwidth and the ability to transmit real-time information reliably.

# CHAPTER FOUR

# LAN to LAN Connectivity

**Chapter Four Contents:**

Part 1: Inter-LAN Connectivity Overview ................... 77

Part 2: WANs ............................................................. 78

Part 3: Telecommunications Overview ........................ 79

Part 4: Telecommunications Standards and Services .... 87

Part 5: Inter-LAN Connectivity Devices ...................... 97

# Inter-LAN Connectivity Overview

In practice, LANs are rarely simple networks that can be wired from a single hub. Most often, networks experience growth along with the employee additions and increasing communications requirements that come with more sophisticated applications. At some point, the physical capabilities of a LAN are going to be exceeded, and the shared bandwidth becomes inadequate. The short term solution is to segment the network in order restore an adequate amount of bandwidth to the individual user. Some organizations may have separate departmental LANs and require wide area communications. In these cases, inter-LAN connectivity schemes are required.

## Segmenting LANs

Since most WANs use a shared bandwidth scheme, the logical solution is to only share bandwidth among nodes that communicate with each other the most. Dividing the network organizationally, such as by departments, limits broadcasts to areas that have a higher likelihood of containing the intended recipient. Physical segmenting a LAN by simply splitting them in two is inadequate since the purpose of the LAN is to provide comprehensive data communications across the entire organization. A link must still exist that allows communication between segments. These links are provided by devices such as bridges and routers. Please see **page 96** for more details on these devices.

Advances in networking technology have greatly enhanced this principle of selectively designated the nodes that share bandwidth. New switching techniques provide network administrators with the ability to create **logical LANs** by simply specifying the occupants of these domains. The physical locations of logical LAN members is irrelevant.

## Connecting Geographically Separate LANs

Connecting geographically separated networks involves the use of an intermediate data transport mechanism. These can range from campus backbones to long-distance telecommunications services. A **campus backbone** is used to join local networks in a building or office complex. A **Metropolitan Area Network (MAN)** is a group of interconnected LANs that typically exist within a relatively small geographic area, such as a few city blocks. A **Wide Area Network (WAN)** covers an unlimited distance. Depending on the source, the term "WAN" may sometimes be used to describe a MAN.

Campus backbones are usually private connections that use a high-speed fault-tolerant transmission method such as FDDI. MANs and WANs more than likely use public data transmission services, although larger corporations often use private wide area connections.

## WANs

The vast majority of wide area data links are made through public telecommunications networks. These are government-regulated companies that provide telecommunications services to the public, such as AT&T, MCI, and the Baby Bells. This following sections are devoted to wide area connections that are made through public telecommunications networks.

Proprietary WAN connections are normally used by larger corporations and government agencies to connect branch offices. Connections that do not go through public telecommunications providers often involve private telecommunications networks or sight-to-sight wireless links, such as microwave, infrared and satellites. Much of the technology used in private networks is identical to that provided by public services. Connections that involve proprietary technology will not be covered in this text.

# Telecommunications Overview

Wide area connections are generally achieved through the use of the services of one or more telecommunications agencies, such as local and long-distance telephone companies. These companies usually offer several WAN options scaled to fit the needs of the customer.

## Switching

There are basically two ways that telecommunications companies can allocate services to their customers: through leased lines or over switched networks.

Non-switched, or leased, lines are simply permanent end-to-end circuits that are reserved for the exclusive use between two end stations.

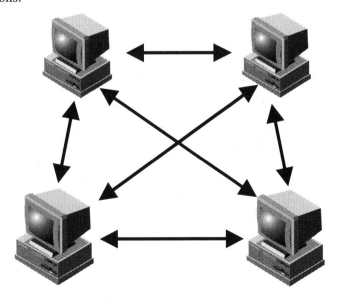

*Figure 4-1: Leased Lines*

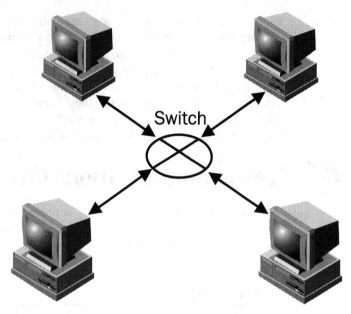

*Figure 4-2: Switched Lines*

Switching provides a means of distributing access to the transmission media among the nodes. This results in a more efficient utilization of the network. In comparing **Figure 4-1** with **Figure 4-2**, you can see that it takes half again as many lines in the former diagram to achieve the same network coverage as in the latter.

The three major switching services provided by international telecommunications companies are circuit-switched, packet-switched, and cell-switched.

**Disclaimer:**

The following sections outline the principles behind the different types of switching. The passages offer simplified explanations in order to help you understand the key points of the different technologies. In the real world, many networks combine elements of the different switching types.

## Non-Switched (Leased) Lines

These can range from analog data rates through DSx speeds. The costs of leased lines are usually not justified except by companies that go through enormous amounts of data. Most data transmissions occur in short, intense "bursts," with comparatively long intervals in between. Since a dedicated line is being paid for whether it is being used or not, this type of service is impractical for most applications. Many organizations find it more economical to pay for communications services on a usage basis. When this is the case, a switched networking scheme comes into play.

## Circuit Switching

> Circuit-switched networks temporarily establish a dedicated point-to-point channel for the exclusive use of two members.

Circuit-switched networks temporarily establish a dedicated point-to-point communications path for the exclusive use of two nodes. Before digital transmission, an analog switchboard was used to close the connection between two nodes. When the session was over, the circuits were opened, and the lines could be used by two other parties. Modern circuit switching works digitally, using highly complex logic to route information into the appropriate multiplexed synchronous (**STM**, see page **87**) channel. The result is the same, in that the channel is used exclusively between two nodes for the duration of the session.

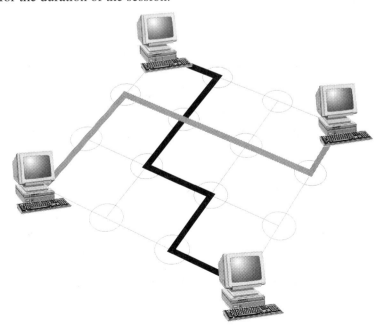

*Figure 4-3: Circuit Switching*

The main advantage of a circuit-switched system is that it provides guaranteed bandwidth, that is, the channel is unaffected by overall use of the network. This makes it especially useful for time-sensitive transmissions, such as voice and video. However, circuit switching is not practical for most data communications applications. In terms of digital communications, the time to set up a circuit-switched channel (10-20 seconds) is fairly long. Most data applications require better response times.

Common analog telephone lines are on switched circuits, collectively known as the **Public Switched Telephone Network (PSTN)**. **STM (Synchronous Transfer Mode)** and **ISDN (Integrated Services Digital Network)**, and **Switched 56** are examples of data communications systems that use circuit switching.

### Packet Switching

*Packet-switched networks share bandwidth by interleaving addressed data packets from different members into the channel.*

Packet-switched networks do not require a dedicated end-to-end connection between communicating nodes. Instead, data is broken up into small, numbered packets that contain addressing information. Individual packets are then routed to their destination, using channels that are shared by packets from other sources. **X.25**, **SMDS**, and **Frame-Relay** are examples of packet-switched services.

Packet switched networks can be connectionless (**datagrams**) or connection oriented (**virtual circuits**). Datagram is the more basic form of packet switching. It does not depend on a set physical path for the packets to follow. Since each packet contains complete addressing information, they may all be routed independently. Because of this, some of the packets may experience en route delays, thus arriving out of sequence. The numbered packets may then be reassembled in order at the receiving end.

A virtual circuit is a specific path that all packets follow through the network. Using this scheme, all packets always arrive in sequence.

Virtual circuits may be further divided into **Private Virtual Circuits (PVC)** and **Switched Virtual Circuits (SVC)**. A PVC is a fixed path, much like a leased line, that packets always take between two nodes. A PVC must be provisioned by the telecommunications service provider. An SVC is established only for the duration of a session.

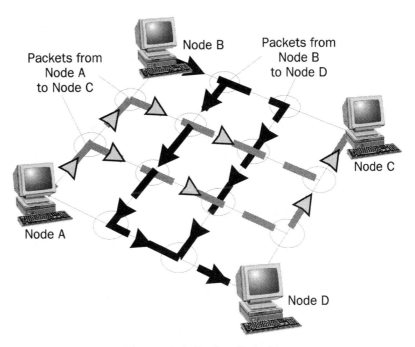

*Figure 4-4: Packet Switching*

Packet switching addresses the major shortcoming of leased and circuit-switched lines. No set-up delay is involved with a packet-switched connection, and it is typically cheaper to implement on a usage basis. The basic idea is that the cost of the aggregate bandwidth is spread across the network, rather than one company paying for dedicated bandwidth. A simple analogy would be taking the airport shuttle as opposed to renting a car. **Public Packet-Switched Data Network** (**PPSDN**) providers bill on a subscription basis, with an additional toll for the amount of data transmitted. Unlike a circuit-switched network, the only distance that is relevant in a PPSDN is that between the network member and the nearest switch.

The drawback to packet-switching is that bandwidth cannot be guaranteed. Since channels on the network are all shared, heavy traffic affects overall throughput. Non-sequential arrival of datagram packets becomes more significant with transmission delays. With either or a combination of these conditions, time-sensitive data, such as voice and video, is adversely affected. Packet switching was conceived for use in data networks that involved transmissions that are short with long intervals in between, such as file transfers. Today's data networks are increasingly burdened with sustained streams of data that includes real-time applications like voice and video.

## Cell Switching

Circuit and packet switching each have their benefits and drawbacks. Circuit-switching provides guaranteed bandwidth because the channel is dedicated to two nodes for the duration of session. The downside is that the switched circuit costs the user the same amount, regardless of the amount of data being transmitted, and is inefficient since surplus bandwidth is unavailable to other communicating nodes. Packet switching is more efficient, but is not practical for information that must be delivered in real time. With circuit and packet switching, it is necessary for telecommunications agencies to maintain more than one network to cover all types of information. A more practical approach is to use a single transport mechanism that is capable of carrying all types of data, which brings us to a third type of switching.

The term "packet" is normally used to describe a variable-length data unit. A **cell** is a small, fixed-length data unit optimized for transporting high-speed, time-sensitive information. A cell eliminates some of the processing overhead required by a variable-length packet during the switching process. The small size combines with the fixed-length to allow smaller, predictable delays between cells, making cells better suited to transporting voice and video.

The fixed length of cells means that transit times are predictable, allowing accurate algorithms to be implemented for multiplexing channels on statistical basis. Using a statistical multiplexing scheme allows flexible allocation of aggregate bandwidth in the transmission media. This means that the bandwidth of individual channels can be customized according to momentary need, rather than being fixed. This technique is sometimes referred to as **cell switching** or **cell relay**.

Cell switching is more efficient because a channel can be given as much bandwidth as it needs, and at the same time only as much as it needs (bandwidth on demand), rather than in fixed increments.

There is little doubt that cell switching will be the data transport mechanism of the future. **Asynchronous Transfer Mode (ATM)**, the international standard for cell relay, is receiving a considerable amount of attention because it has the potential of becoming a unifying standard for both local and wide area networks. Please see **page 87** for more information on ATM and its predecessor, STM.

## Telecommunications Media

### Guided Media (DSx Channel Hierarchy)

The foundation of digital telecommunications is the **Digital Signal (DS)** hierarchy of circuits. In the United States, the basic unit is the **DS1**, capable of up to 1.544 Mbps throughput. The DS1, better known as the **T1**, is time multiplexed (see **page 87**) into 24 **DS0** channels, with each channel being allocated 64 Kbps. The DS hierarchy extends up to **DS4 (T4)**, which has a maximum bandwidth of 274.176 Mbps. The T1 is the basis for several high-speed WANs, including PRI (primary rate) ISDN.

*Table 4-1: DSx Hierarchy*

| U.S. | | | Europe | | |
|---|---|---|---|---|---|
| DSX | BANDWIDTH | CHANNELS | DSX | BANDWIDTH | CHANNELS |
| DS0 | 64 Kbps | 1 | DS0 | 64 Kbps | 1 |
| DS1 (T1) | 1.544 Mbps | 24 | DS1 (E1) | 2.048 Mbps | 30 |
| DS1c (T1c) | 3.152 Mbps | 48 | DS2 (E2) | 8.448 Mbps | 120 |
| DS2 (T2) | 6.312 Mbps | 96 | DS3 (E3) | 34.368 Mbps | 480 |
| DS3 (T3) | 44.736 Mbps | 672 | DS4 (E4) | 139.264 Mbps | 1920 |
| DS4 (T4) | 274.176 Mbps | 4032 | DS5 (E5) | 565.148 Mbps | 7680 |

European telecommunications are based on the **E1**, which is a 2.048 Mbps channel. Note that the disparity between T1 and E1 speeds means that an intermediate translation is necessary for transcontinental data communications.

**Synchronous Optical NETwork (SONET)** is the name for the next generation international telecommunications backbone that is based on fiber-optic media. As in the DS system, SONET consists of a hierarchy of graduated bandwidth connections, from the STS-1 at 51.84 Mbps to the STS-192 at 9953.28 Mbps. The European counterpart to SONET, standardized under the ITU-T, is called **Synchronous Digital Hierarchy (SDH)**.

*Table 4-2: SONET/SDH Hierarchy*

| SONET (U.S.) | SDH (EUROPE) | BANDWIDTH |
|---|---|---|
| STS-1 | - | - |
| STS-3 | STM-1 | 155.52 Mbps |
| STS-9 | STM-3 | 466.56 Mbps |
| STS-12 | STM-4 | 622.08 Mbps |
| STS-18 | STM-6 | 933.12 Mbps |
| STS-24 | STM-8 | 1.244 Gbps |
| STS-36 | STM-12 | 1.866 Gbps |
| STS-48 | STM-16 | 2.488 Gbps |
| STS-192 | STM-64 | 9.953 Gbps |

## Unguided Media

Much of the telecommunications infrastructure involves unguided line-of-sight media, such as microwave satellite relays. Microwave relays use narrow-beam radio transmissions in the Gigahertz band range. They are typically used in areas where installing cabling may be prohibitive due to cost or physical obstacles (deserts, mountains, national parks, etc.).

Satellites are generally used for transcontinental links. Satellites introduce a significant problem in data communications. Geosynchronous satellites orbit the earth at around 20,000 miles above the surface. Even at the speed of light, a radio transmission takes about a quarter of a second between two transmitting/receiving stations on the surface. For a voice transmission, this delay may be distracting. Data transmissions that use a satellite need to compensate for this delay in order to avoid problems.

#  Telecommunications Standards and Services

At first glance, there seems to be a bewildering array of remote connection/networking options offered by public telecommunications companies. This is mostly because these providers offer scaled versions for most of the popular wide area data services. These scaled services typically are given appealing marketing names that may make them sound like something entirely new or different. Telecommunications companies, like all businesses, exist to make money. Providing scaled services is simply casting the widest possible net for customers.

The following sections outline the basic features of the most popular wide-area data transport technologies and services offered by public telecommunications companies. The data communications standards and services listed here are grouped by switching technology to maintain continuity with preceding sections. Please see Chapter Two for information on international standards organizations such as the ITU and ANSI.

## STM and ATM

Currently, telecommunications networks use **Synchronous Transfer Mode** (**STM**) as the predominant method of providing low-level data transport services across long-distance backbones. STM uses TDM to divide the physical media (such as a T1 line) among multiple channels. A channel is composed of individual data frames ("buckets") that occupy time slots that are interleaved at regular intervals. The regular intervals effectively divide the media into channels of fixed and equal bandwidth.

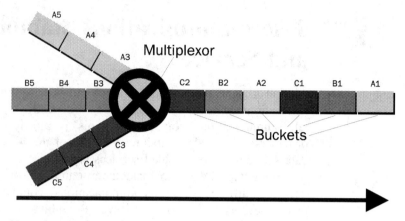

Note the synchronous order of cells being multiplexed.

*Figure 4-5: STM*

When a customer requests a connection to another location on the network, the telecommunications company establishes a fixed physical path for the channel (provisioning a channel). Transmitted information is divided into frames. Each frame must wait for a bucket that corresponds to its channel before it can be transmitted. When the session is over, the telecommunications company tears down the circuit. Only then is the bandwidth available to another customer.

STM is inherently inefficient in its use of bandwidth. If no information is waiting to be sent when an assigned bucket comes up, then that bucket is wasted, since it cannot be used by another channel. In addition, the dedicated channel also becomes a problem when more bandwidth is needed.

**Asynchronous Transfer Mode (ATM)** is a communications technology that is designed to make better use of available bandwidth by eliminating the need for fixed-bandwidth channels. At the heart of ATM is the high-speed cell switching technique that is described on **page 84**.

Like STM, ATM provides a fixed path for each channel. The fixed path provides a means that bandwidth can be dedicated, and ensures that time-sensitive packets arrive in order. The big advantage for ATM is that it customizes the amount of bandwidth a channel receives (bandwidth on demand).

STM channels have a fixed bandwidth because of the synchronous scheme used to divide the media. ATM channels are not fixed in size because cells are not sequenced by a common timing signal. This asynchronous transmission method allows much more flexibility in allocating bandwidth to individual channels. Each cell contains a label in its header, called the **Virtual Circuit Identifier** (**VCI**). The VCI lets the network switches know what channel each cell belongs to, much as STM frames are identified by the time slots they occupy. This allows a channel to occupy as little or as much space as it needs in the physical media. Bandwidth can be allocated according to the type of data, rather than a one-size-fits-all approach.

Contrast **Figure 4-5** with **Figure 4-6**. Notice that in the former, frames are interleaved at regular intervals. Since ATM cells are not allocated specific time slots, it is possible that a sequence of adjacent cells may belong to the same channel.

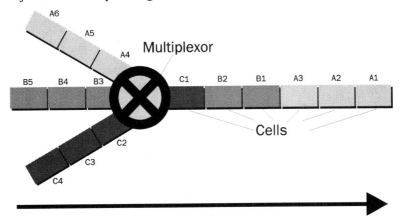

*Note the asynchronous order of cells being multiplexed.*

*Figure 4-6: ATM*

This highly scaleable approach to allocating bandwidth is attractive for service providers and customers alike. Bandwidth is guaranteed, and the amount is provided according to momentary needs, rather than on a fixed service.

ATM cells are 53 bytes long. The short length of the cells keeps transmission delays to an acceptable minimum. Delays introduced by long, variable-length packets hinder the transmission of time-sensitive information (sound and video).

In short, ATM combines the guaranteed bandwidth of circuit switching with the efficient bandwidth allocation of a packet-switched networks. With these features, ATM has the potential to unify all networks under a common transport service for all types of data.

ATM is a relatively new technology that is currently not widely available. This is because comprehensive standards have yet to be approved by the ITU and other international organizations. In the meantime, ATM developments have proceeded rapidly due to interest in its application as a LAN transport mechanism. An ad hoc group of computer and telecommunications companies known as the **ATM Forum** has taken up the task of lending some order to this process, as well as championing ATM as a universal networking standard.

## Leased Lines

See "**Switching**" on **page 79** for information on the difference between a switched and non-switched line. Leased lines are simply non-switched circuits or channels that are rented from a telecommunications company. This can be anything from analog to full DSx speeds (see **page 85**).

T1s are typically used by organizations to provide a high-bandwidth trunk from a building to the nearest phone company switching facility (central office). Many organizations provide separate internal phone lines to members through a Private Branch Exchange (PBX). The PBX and wide area networking devices, for example routers, are connected to a CSU/DSU (see **page 96**), that multiplexes the different channels into a T1 or **Fractional T1**. This leased line is then connected to the local phone company's central office, which in turn provides the appropriate switching to get the information on each channel to its intended destination.

Fractional T1 is just what it sounds like – a portion of T1 bandwidth in 64 Kbps (DS0) increments. Until recently, phone companies did not offer Fractional T1, preferring to sell bandwidth in full T1 increments only.

# Circuit-Switched Networks

## PSTN

One of the oldest methods of transporting information over wide areas is to simply use the universally available PSTN. Since computers communicate using digital information and most PSTN local loops are still analog, a translation is necessary. A modem modulates digital information into an analog carrier signal, and demodulates the analog signal back into digital information on the other end. Modems have enjoyed a tremendous boom in popularity in recent years in response to developments that have revolved around the increasing capabilities of inexpensive home computers. This includes a growing number of home offices and telecommuters, as well as the ongoing maturation of the Internet. Using a modem is the cheapest, easiest way to access remote computers and networks. All you need is a common residential account with the local phone company. The downside is that the connection is slow and subject to high error rates.

The fastest modems are currently rated at 33.6 Kbps (V.34+), which represents a departure from the trend of doubling throughput with a succeeding generation of hardware. This is because the bandwidth of analog lines is such that physical limits are being reached. Since analog telephone lines were only designed to convey a portion of the frequency range of the human voice, these lines are simply not capable of conveying distinct tones at much faster signaling rates. In addition, many modem manufacturers have seen the handwriting on the wall and are concentrating on ISDN.

## ISDN

**Integrated Services Digital Networks (ISDN)** is a set of ITU-T standards that define a comprehensive switched-circuit telecommunications network. ISDN is an all-digital network that is designed to transmit voice, video, and data. It is intended to eventually replace the analog portions of the PSTN, with availability already widespread in most parts of the U.S. To the end user, it will be a fairly transparent upgrade, using the existing copper telephone wiring found in most homes and offices. Depending on the provider, ISDN is billed in a similar manner to PSTN, with a usage fee added to the basic subscription.

There are two types of ISDN: baseband and broadband. The broadband standard is still in the works, but it will be based on ATM technology. Baseband ISDN is further divided into two classes. The first and most common option is the **Basic Rate Interface (BRI)**. BRI ISDN is composed of three channels: two 64 Kbps **B Channels** that bear the data, and one 16 Kbps **D channel** that is used as an out-of-band path for control signals. BRI is intended for the homes and small businesses.

The other type of baseband ISDN is the **Primary Rate Interface (PRI)**. PRI ISDN is built around the T1/E1 telecommunications line, being composed of twenty-three 64 Kbps B Channels and one 64 Kbps D Channel in the U.S. The European version uses thirty B Channels and one D Channel. PRI is intended for use as a trunk between an office and the phone company's local switching station.

Hooking into ISDN requires a terminal adapter (sometimes called an ISDN modem), and an NT1 terminator wired between the local switching station and the premises. For the end user, it basically works the same way as an analog line: dial up the remote node and wait for a circuit to be established. When the session is over, hanging up tears down the circuit.

One of the key selling points of ISDN is its ability to provide access to other types of networks. ISDN has or will have built in support for communicating with nodes on networks such as PSTN, X.25, Frame Relay, and ATM. ISDN service currently suffers from limited availability and high cost. Widespread subscription is expected to bring prices more in line with current rates for analog lines.

## Switched-56

Switched-56 is a circuit-switched telecommunications offering that provides scaled-down digital services to areas where ISDN is not yet offered, or as a lower-cost backup to primary leased or switched lines. Switched-56 is a data only dialed connection that provides one 56 Kbps data channel, as opposed to the two 64Kbps data channels for BRI ISDN.

Switched 56 is not a viable alternative to ISDN because it lacks many of ISDN's features, such as the ability to carry voice and out-of-band signaling (D channel).

# Packet-Switched WANs

## X.25

**X.25** is the ITU-T (formerly CCITT) standard for packet-switched networks, originally published in 1976.

An X.25 network consists of **Packet Assembler/Disassemblers** (**PADs**) that terminate a mesh of physical transmission media and packet switches. The PADs send packets to the nearest network switch. The switch then inspects each packet individually and routes it according to a predetermined logic. The network switches also perform flow control and error-checking at every stage, re-transmitting information when required.

X.25 was conceived in the era of mainframe computers and predominantly analog transmission media. This meant that remote terminals did not contain the logic to provide error-correction and flow control. In order for practical communications to be implemented, these functions had to reside with the network. Since each network switch is burdened with this overhead, most X.25 networks are limited to 64 Kbps throughput. Modern digital lines are very reliable, and even low-end terminal equipment is capable of performing error-correction and flow control. X.25 covers OSI Layers 1 to 3. More current transport technologies rely on relatively error-free digital circuits and logic-intensive terminal equipment. Eliminating the extra overhead of the OSI Layer 3 provides significant improvement in throughput.

The advantage of X.25 networks is that they use a mature technology and are almost universally available. X.25 provides virtually error-free transmission and is easy to obtain. X.25 provides bandwidth on demand (up to the 64 Kbps limit), so services are charged on the volume of data transmitted, rather than on the distance between connections.

On the downside, the line between mature and obsolete has already been crossed. Modern applications require much higher throughput. Channels are shared, so heavy use and over-subscription cause the network to slow down. X.25's variable-length packets on shared channels introduce unpredictable transmission delays between packets, so it is practical for data only (no voice or video).

X.25 networks are sometimes called **Public Packet-Switched Data Networks**, or **Public Data Networks** (**PDN**s), and are usually marketed under brand names such as TYMNET, Accunet, and SprintNet.

### Frame Relay

**Frame Relay** is a telecommunications service that is intended for use in metropolitan LAN to LAN (MAN) or other short-range connections as a cost effective replacement for older transmission methods. It maintains the cost effective bandwidth on demand scheme of X.25, while providing a tremendous increase in throughput. Another advantage of frame relay is that bandwidth can be guaranteed by reserving space on network buffers.

Like X.25, Frame Relay uses packet switching. Frame Relay, however, is good for speed in the DSx range. It essentially accomplishes this increase in speed by dumbing down network functions. Technological advances in terminal equipment and transmission media have made it possible to move control functions, such as error-correction and flow control, out of the network.

Like X.25, Frame Relay uses variable length packets. This means that Frame Relay packets are also subject to unpredictable transmission delays. This makes Frame Relay a data-only transmission mechanism, and presently unsuitable for voice and video. However, Frame Relay technology has not yet matured, and groups such as the **Frame Relay Forum** are proposing methods of working around this problem.

## Cell-Switched WANs

### ATM

Please see page 87 for general information on how ATM works. ATM is not yet generally available for the wide area connections for which was intended. Intense interest in the technology still exists, though, because of the potential of ATM to become a universal data connection. ATM can handle voice and video, and can be used as the transport mechanism all the from the workstation to long-distance backbone networks. The most dynamic area of ATM development is currently in its use for local area connections. Many hardware vendors are already shipping ATM products for use in LAN high-speed backbones.

## SMDS

**Switched Multi-Megabit Data Service (SMDS)** is a packet-switched data communications service developed by Bellcore that offers scaled throughput from Fractional T1 to DSx speeds. The primary purpose of SMDS is to provide cost-effective high-speed transport services that seamlessly extend LAN-like connectionless services over wide areas. A highly touted feature of SMDS is that it provides an easy migration path to ATM.

SMDS uses can be broken down into two segments. The first is the access point between a LAN and the SMDS network. This portion of the network. LAN frames are encapsulated in SMDS packets of up to 9188 bytes in length, then forwarded to the SMDS network.

The SMDS network takes the packets and breaks them down into individually addressed 53 byte cells that are similar to those used in ATM. These cells are then sent through an SMDS network of high-speed switches and transmission media. At the other end, the cells are reassembled into packets and forwarded on to the other LAN.

Currently, SMDS uses a variation on cell switching that is defined by the IEEE 802.6 MAN standard. The core network is well suited for replacement by ATM when it becomes available. When the change to ATM does occur, it will be transparent to end users, requiring no equipment upgrades on their part.

*Table 4-3: Wide Area Network Standards Comparison*

| Network | Switching Type | Throughput |
|---|---|---|
| ATM | Cell | scaleable to Gbps range |
| SMDS | Packet | scaleable to Mbps range |
| X.25 | Packet | 64 Kbps |
| Frame Relay | Cell | scaleable to Mbps range |
| ISDN | Circuit/Cell | 64 Kbps/D channel |
| Switched-56 | Circuit | 56 Kbps |
| PSTN | Circuit | 33.6 Kbps (standard modem connection) |

# Internet

The **Internet** is a nebulous worldwide collection of networks, linked together by the **Transmission Control Protocol/Internet Protocol (TCP/IP)** family of communications standards.

The Internet has its early origins in **ARPANET**, the original packet-switched wide area network that was developed in the 1960s. TCP/IP was run over ARPANET for the first time in the early 1980s, and has since gained widespread acceptance as a wide area transport protocol. Currently, the core of the Internet is formed by powerful switches and high speed links that are operated by large institutions. Access to this backbone is provided through telecommunications companies (Internet Service Providers) that typically distribute their available bandwidth on a subscription basis.

TCP/IP is intended to be independent of the hardware platform it is running on and ideally allows seamless cross-platform communication. It covers the mid to upper layers of the OSI model. **Transmission Control Protocol (TCP)** provides Network Layer services (see **Chapter Two** for more information), establishing a connection and error control between communicating nodes. **Internet Protocol (IP)** corresponds to the Transport Layer and contains addressing and routing control for transmitted information. The upper three layers are covered by specific TCP/IP applications, such as **File Transfer Protocol (FTP)**, **Simple Mail Transfer Protocol (SMTP)**, and **World Wide Web (WWW)**.

*Table 4-4: TCP/IP and OSI Layer Comparison*

| OSI LAYER | TCP/IP COMPONENT |
|---|---|
| Application<br>Presentation<br>Session | SMTP, FTP, WWW, TELNET, Gopher |
| Transport | Internet Protocol (IP) |
| Network | Transmission Control Protocol (TCP) |

The TCP part of TCP/IP provides transport protocol functions, which ensures that the total amount of bytes sent is received correctly at the other end. The IP part of TCP/IP provides the routing mechanism. TCP/IP is a routable protocol, which means that the messages transmitted contain the address of a destination network as well as a destination station. This allows TCP/IP messages to be sent to multiple networks within an organization or around the world, hence its use in the worldwide Internet (see Internet address).

# Inter-LAN Connectivity Devices

## Network Connecting/Segmenting Devices

The following are general categories of devices that provide inter-network connectivity. Some vendors may offer products that combine functions, and may also use their own terminology.

### Bridges and Switches (Low-Level Segmenting)

The primary purpose of **bridges** and **switches** is to connect separate LAN segments or separate LANs. They are also commonly used to divide a large LAN into smaller segments. Segmenting a network may be desirable for a variety of reasons. The most common reason is to divide a network in order to keep data traffic more manageable. In a uniform LAN, the available bandwidth is the same, whether it has 10 nodes or 100. Obviously, if more nodes are vying for the same overall bandwidth, each node will have a smaller share. The result is slower data transfers for each node.

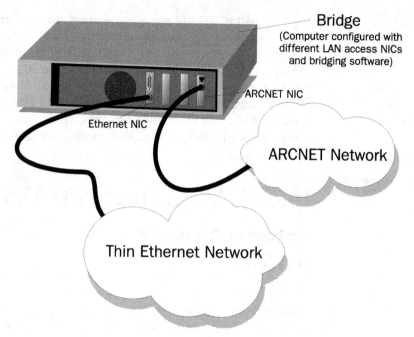

*Figure 4-7: Ethernet to ARCNET Bridge*

Although bridges and switches have a similar functions, they operate very differently. Bridges can provide connections for both similar and dissimilar LANs. For example, a bridge could be used to connect two Ethernet segments or an Ethernet segment to a ARCNET segment (**Figure 4-7**).

A switch takes the segmenting approach to its logical conclusion by providing each node with a dedicated connection to the network. In a non-switched network environment, each network node simultaneously shares an equal portion of the available bandwidth with other nodes. Switches provide each node with a short turn at the entire bandwidth of the network segment.

Switching is a newer technology that is becoming widely accepted in high-speed networks. **ATM (Asynchronous Transfer Mode)** is a recent LAN/WAN category that is dependent on high-speed switching technology.

Switching can segment networks logically, rather than physically. This means that LAN segments can be administered according to organizational requirements, rather than the physical locations of individual users on the network. A logical segment is known as a **Virtual LAN**, or **VLAN**.

### Gateways and Routers (High-Level Segmenting)

A **gateway** is similar to a bridge, in that it is used to allow communication between dissimilar networks. Gateways, however, operate on a higher level, meaning they are protocol and application specific, rather than physical and data-link specific (see Chapter Two for details on the OSI Layers). Typical uses of gateways include connecting a LAN to a mainframe system and providing conversions between different messaging (such as e-mail) formats.

A **router** is similar to a switch, in that it provides a connection between LANs or LAN segments. Routers provide their services at a fairly high communications level that involves addressing protocols, rather than the low-levels that switches work at. In addition to inter-network connectivity, routers can provide a measure of redundancy and/or added efficiency for getting data traffic to the intended destination. They do so by intelligently determining the best path for the data.

For a more detailed illustration on how inter-network connectivity devices fit into the scheme of a network architecture, please consult **Chapter Two**.

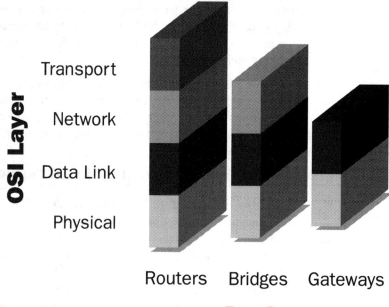

*Figure 4-8: Inter-LAN Connectivity Devices*

## Remote Access Adapters

The devices listed in this section provide access to the physical media as well as the low-level signaling protocols that are specific to various wide area connections.

### Modems

A modem is used to allow digital devices to communicate over common analog media, namely, the PSTN. **Modem** is a contraction of **MODulate/DEModulate**, which is basically what a modem does. It modulates square digital waves into continuously variable analog waves (sine waves) that correspond to distinct tones, then performs the conversion in reverse on the other end.

In addition to digital/analog signal conversion, modems control other aspects of establishing, maintaining, and terminating data communications between host computers such as dialing, negotiating protocols with the remote modem, and error detection/correction.

The most current modems are capable of speeds of up to 33.6 Kbps. Transmission speeds are limited by the bandwidth of analog lines. These lines are only designed to handle a small part of the range of the human voice, and are highly prone to interference.

## ISDN Terminal Adapters

ISDN terminal adapters are sometimes called ISDN modems, although they do not perform an analog to digital conversion. The other major difference lies in ISDN's three transmission lines, versus the PSTN's single line. ISDN modems use out-of-band control signals that are carried over the D Channel (see page 91), versus the in-band signaling that is necessitated by an analog serial line. The two data (B Channel) lines enables simultaneous transmission of voice and data, or the combining of data lines for doubling of throughput.

## T1/E1 Adapters (CSU/DSU)

A **Digital Service Unit** (**DSU**) is the equipment that translates signals from the local terminal equipment into a format that can be transmitted over high-speed lines such as a T1. A DSU also adapts received signals back into the format being used by the terminal equipment.

Some terminal equipment, such as a WAN router or multiplexor, may incorporate a DSU.

Telecommunications lines must be terminated at the customers premises. This function is provided by the **Channel Service Unit** (**CSU**). The CSU also provides diagnostic features such local and remote loopback testing. Many DSUs combine CSU features, which is why they usually are mentioned as a unit (DSU/CSU) rather than individually.

## Chapter Review

- LAN to LAN connectivity can involve anything from departmental links all the way to international telecommunications.

- Most metropolitan and wide area networks involve the services of one or more telecommunications agencies.

- Telecommunications companies can allocate transmission channels to their customers through dedicated (leased) or switched lines.

- There are three methods of switching. Circuit switching establishes a temporary end to end channel. Packet switching breaks up information into individually addressed data units (packets) that depend on network switches to route them to their intended destination, where they are reconstituted into their original form. Cell switching, sometimes called fast packet switching, is a hybrid of circuit and cell switching that enables more efficient allocation of network resources.

- Inter-LAN connectivity can involve simple devices such as bridges to more sophisticated, logic-intensive routers and gateways.

**CHAPTER FIVE**

# Cable Guide

## Chapter Five Contents:

Part 1: Overview .......................................................... 105
Part 2: Cable Types ...................................................... 105
Part 3: Cabling Considerations ..................................... 111

# Overview

The common cable types used in network installations are listed below. These are only presented to give you a basic understanding of the cable types. Please refer to Appendix D for details on specific recommended cable types and maximum cable lengths.

# Cable Types

## Twisted-Pair

For specifications on the various cable types, please refer to Appendix D.

Shielded Twisted Pair

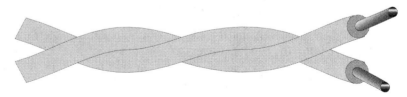

Unshielded Twisted Pair

*Figure 5-1: Twisted-Pair Cables*

**Twisted-pair** cable is the least expensive type of wiring that can be used for network installations. This type of cable consists of two insulated wires twisted together a specified number of times per foot. The purpose of the twisting is to reduce the effect of outside electrical interference, or "**noise**." The more twists per foot, the less the interference. Twisted-pair cables come in two general categories: unshielded and shielded. The difference is that **shielded twisted-pair** (**STP**) features a copper shield that covers the wire pairs. The copper shielding further reduces the interference, but makes the cable more expensive than **unshielded twisted-pair**.

Twisted-pair wires come in many different sizes. These sizes are rated by an **American Wire Gauge** (**AWG**) number. The standard AWG wire sizes for network installations are 22- and 24-AWG. Twisted-pair cabling is categorized for communications applications by an organization called the **Electronics Industry Association/Telecommunications Industry Association** (**EIA/TIA**). There are five EIA/TIA twisted-pair categories; these are outlined in **Table 5-1**.

*Table 5-1: EIA/TIA 568 Specification for Twisted-Pair Categories*

| Category | Specification |
|---|---|
| 1 | Not used for LAN cabling |
| 2 | UTP cable for LANs having a throughput of up to 4Mbps. |
| 3 | UTP cable for LANs having a throughput of up to 10Mbps. |
| 4 | UTP cable for LANs having a throughput of up to 16Mbps. |
| 5 | STP cable for LANs having a throughput of up to 100Mbps. |

Twisted-pair networks were initially only able to handle speeds of up to 1 Mbps over a few hundred feet. Current twisted-pair technology can transfer data at speeds of up to 100 Mbps, although lower speeds of 10Mbps and below are much more common.

Twisted-pair cables are bundled in pairs within a protective cover. LANs use anywhere between 2 and 25 wire pairs per cable. Enough pairs should be used to allow for current needs as well as future expansion. If your LAN only requires 1 twisted-pair, but 4 or 6 pairs are installed, adding stations will be simplified in the future because you will not have to run a new cable beside the existing cable.

## Coaxial

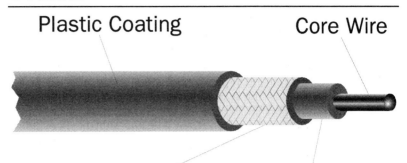

*Figure 5-2: Coaxial Cable*

Coaxial cable or "coax" is probably the most widely used cable for network installations. It consists of a copper core surrounded by insulation, a copper or aluminum wire mesh, all covered by a plastic insulator. The wire mesh serves as one of its conductors and as a shield to outside interference. You will also find this type of cable used with cable television. Coaxial cable may be used in two distinctly different modes, baseband and broadband.

### Baseband Coaxial

This type of coaxial cable has one channel, therefore it can only transfer a single signal at a time. It uses the core wire as the carrier and the wire mesh as a shield and ground. The digital information is sent down the cable in a **serial** fashion, one bit at a time. Because of this limitation, it isn't possible to send multiple signals composed of data, voice and video down a baseband cable. Baseband coaxial cable can handle data between 10 and 80 Mbps, although the actual speed depends on the type of LAN used.

Ethernet is one type of LAN that uses baseband coax. Ethernet normally transfers data at 10 Mbps over the this type of media. This is sometimes a misleading figure. Due to the serial transmission nature of baseband, the more workstations that are attached, the slower the network becomes. Not only does the network become more active, but more cable is used with every node added, in turn slowing the over-all data transfer rate.

## Broadband Coaxial

In contrast to baseband, broadband coax can carry multiple signals simultaneously. Each signal is broadcast at a different frequency over the cable at the same time. This transmission technique is also used by cable television, with each channel being broadcast over the coax at a different frequency.

Broadband network installation requires far more planning than a baseband LAN. Since multiple signals share the cable, amplifiers must be installed to maintain the signal strength over the entire network. The broadband cable must be tapped by each node. The broadband cable is usually kept above the ceiling and has tap lines that consist of smaller coax that drop down to the workstations.

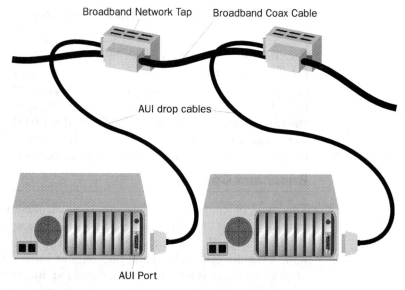

*Figure 5-3: Broadband Coaxial Backbone and Drop Cables*

## Fiber-Optic

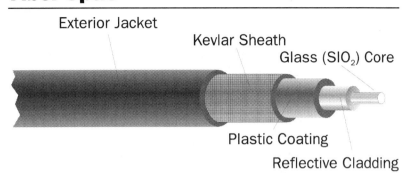

*Figure 5-4: Fiber-Optic Cable*

Fiber-optic cables use light, rather than electricity, as the transmission medium along the network. Electrical signals are converted to light signals, then transmitted by a laser or LED (Light-Emitting Diode) through the core section of the cable. Some loss of light occurs along the cable. This attenuation can be compensated for by optical repeaters. At the destination end of the cable, the light is translated back into a digital or analog signal by a photo-diode (a device that is able to read light fluctuations).

The core of fiber-optic cable is composed of pure glass pulled into a very thin fiber. The core is surrounded by another layer of glass called the cladding. The cladding has a lower refraction index than the core, so light signals are contained (reflected) back into the core with minimal attenuation.

Fiber-optic cable is categorized as **monomode** or **multimode**. Monomode fiber is the better of the two. It uses a small diameter fiber to transmit a cohesive (monomode) light ray. Monomode fiber allows the light to travel a greater distance before becoming excessively distorted. Lasers must be used as the signaling source in monomode cable – LEDs cannot produce the quality of light needed. Unfortunately, laser emitters are much more expensive than LEDs, and the thin fiber is difficult to splice.

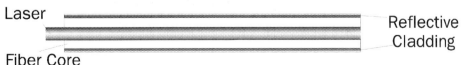

*Figure 5-5: Monomode Fiber Optic Cable*

Multimode fiber has a smaller bandwidth but is much easier to splice. Multimode fiber has a wider diameter than monomode. Standard (**step index**) multimode fiber is currently the most widely used type. Step index fiber uses a constant index of refraction through the radius of the core. This type of fiber is easier to manufacture, but cannot be used over as great a distance as others. This is because the light rays tend to scatter until the signal is no longer reliable (see **Figure 5-6**). There are three popular sizes of multimode fiber-optic cables used in network installations; 100 x 140µm, 62.5 x 125µm, 50 x 125µm (µm = microns/millionths of a millimeter). The numbers refer to the core (first number) and cladding (second number) diameters.

The other type of multimode fiber, **graded-index**, is generally more expensive, but provides a higher rate of data transmission over a greater distance. This type of fiber uses a core that is composed of several layers. The layers have graduated indexes of refraction, dropping slowly from the center of the fiber towards the outside (see **Figure 5-6**). The result is a lens effect that keeps multiple light rays focused over a greater distance.

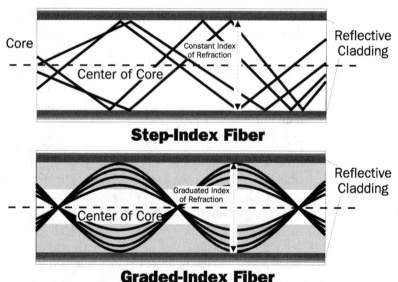

*Figure 5-6: Step-Index vs. Graded-Index Multimode Fibers*

Fiber-optic LANs can only transfer data in one direction at a time, light can not be sent from both directions. Most fiber-optic cables have two separate fibers contained in one sheath for duplex communications.

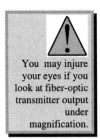
You may injure your eyes if you look at fiber-optic transmitter output under magnification.

Some words of caution concerning the handling of fiber-optic cable should be noted. Remember, this type of cable is made from glass, bending it excessively (in more than a 1 cm radius) can cause it to break. You may injure your eyes if you look at the fiber-optic transmitter output under magnification. Looking at it with the naked eye is usually safe but is not recommended. An Optical Power Meter may be used to check the output on a transmitter If this equipment is not available, then you may hold a piece of paper in front of the port in a darkened room and see if the light is being transmitted to the paper.

## Wireless

When implemented in a LAN application, wireless media involves radio, such as FM or microwave, or light, such as laser or infrared links. Light connections are line of sight, meaning that no obstructions can exist between the transmitter and receiver. Radio connections are usually in the FM band, and do not require a line of sight. This means that FM radio links can be made through walls or floors. High-power microwave (in the GHz range) links, much like infrared, require a line of sight. Microwaves are typically used to connect LANs that are within a campus or metropolitan area.

Wireless links are much more exposed to environmental impairments than cables. Physical and electromagnetic obstructions are much more likely to affect a signal carried by radio or infrared than in an insulated and/or shielded cable.

Wireless links are inherently more of a security risk – anyone desiring to "listen in" only has to have the appropriate receiver. Encrypting data traffic and narrow beam media (such as laser) are two ways of limiting this risk.

For the reasons listed above, wireless links normally are used only in places where cables are impractical. Wireless FM links are very flexible and easy to implement, making them an attractive short-term solution. Metropolitan or campus LAN links may cross property that is owned by someone else. If this is the case, an organization is in the position of having to lease the right-of-way that its cable must run across, or use a telecommunications service to provide the link. Some companies may find a wireless link is more economical than either of these options.

# Cabling Considerations

When designing a network, all the different cable types should be considered, and the benefits and downfalls of each should be carefully examined. Some important factors to look at are:

- Ability to easily upgrade
- Cost
- Flexibility
- Performance
- Safety
- Security
- Shielding and environmental interference

## Upgrading

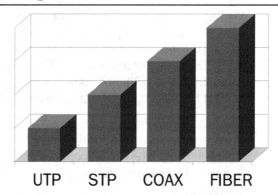

*Figure 5-7: Upgrade Potential Comparison*

The networking industry is moving so fast, that it is hard to predict where the next best, fastest, or better standard will come from. If a network is designed properly, upgrading sometime in the future does not have to be an ordeal. The cable type that has the highest upgrade potential currently is fiber-optic. Networks that currently utilize fiber-optic cable are nowhere close to the upper limits of the cable.

## Cost

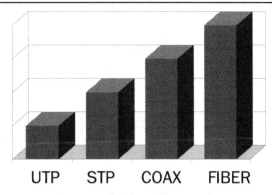

*Figure 5-8: Cost Comparison*

This can be one of the most frustrating factors while planning a network. Often, the cost of cable and the expense in running the cable can limit the overall network. If price is the only concern, then use Unshielded Twisted-pair; but remember the limitations of UTP wire. If price is no object, then install an entire Fiber-optic network. Cost should be just one of the factors involved in planning a network, not the main concern.

## Flexibility

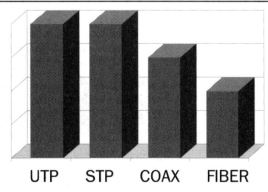

*Figure 5-9: Flexibility Comparison*

Should the network ever change protocols (i.e., ARCnet to Ethernet) the old cable should still be usable. The top cable for across-the-board compatibility and flexibility is twisted-pair cable. Most protocols can use it, and if the network is ever removed, telephone systems can use the same cable. Fiber is at the bottom only because it's the newest and largely used in limited applications (such as a fiber-only backbone).

## Performance

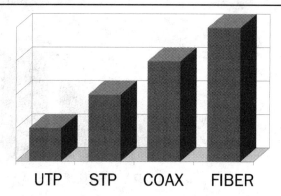

*Figure 5-10: Performance Comparison*

Performance is one of the most complex factors to consider because of the development of new standards. For the foreseeable future, nothing will be able to beat the capacity of fiber-optic. Next in line is coax, since it is used for broadband networks that can carry a number of separate channels. Shielded twisted-pair bests unshielded twisted-pair since STP is less prone to interference.

## Safety

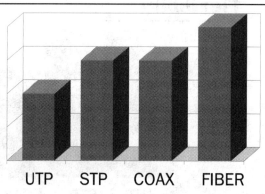

*Figure 5-11: Safety Comparison*

If the network cable will be traversing hazardous areas where a spark might cause a fire or explosion, or if the environment is corrosive to metals, then fiber-optic cable is by far the best choice. Coax and shielded twisted-pair are tied, due to the heavy amount of shielding available on both these cable types. Unshielded twisted-pair is at the bottom, because of its susceptibility to wear and breakage. For most applications, these cables types are equally safe.

## Security

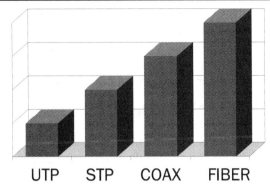

*Figure 5-12: Security Comparison*

If the network will be transmitting sensitive or extremely private information, then security is a major concern. The security of the network could be compromised by someone "tuning-in" to the radiated emissions from poorly shielded cable or connections. Another method is to tap directly into the network cable itself. Fiber-optic cable is immune to both of these snooping methods and is the most secure cable type.

## Shielding and Interference

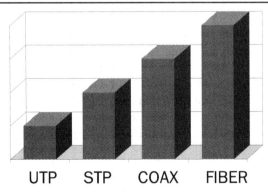

*Figure 5-13: Shielding Comparison*

If the network media must be placed in electrically noisy environments (i.e., near air conditioners, fluorescent lights, etc.) or near sensitive equipment (i.e., lab analyzers, high frequency radios, etc.) then the amount of shielding on the cable is of major importance. The more shielding, the less interference with network transmissions, and the less unwanted emissions from the cable itself. Due to the nature of fiber-optic cable, it is the best choice.

**Crosstalk** is the leakage of electronic signals from one channel to another, and often happens when the cables cross over each other. This type of interference can be very disruptive to data transmission. A cable with even light shielding has better crosstalk isolation than a cable with no electrical shielding at all. Crosstalk is of major concern to twisted-pair networks, and should be carefully considered when planning cable installation.

## Conclusions

The type of cable to be installed depends on what type of network is being planned. Fiber-optic cable for a five node network that will be used for mail and some file sharing would be overkill. For a 150 node network that must routinely send high-speed sensitive data, twisted-pair would be a poor choice.

Coaxial cable is by far the most commonly used cable for network installations, and is a good choice for most network installations. Twisted-pair installations take advantage of existing telephone cable, thus saving the cost of purchasing and laying new cable. Twisted-pair is the least expensive network cable and the simplest to install. Fiber-optic cable is growing in popularity and will likely become the standard for future networks. The highest transmission speeds and security are available with fiber. Among the limiting factors for fiber-optic cable are (presently) price and complexity, which has largely limited its use to high-speed backbone applications. The devices used with fiber-optic networks can be several times more expensive than those used with twisted-pair. Most networks support the use of adapters that allow a twisted-pair cable to connect to a fiber or coaxial cable length and vice versa. Keep in mind that sometime in the future, the necessity to upgrade will exist. The cable that you choose today will affect how much cable you might have to install in the future.

# Chapter Review:

- Twisted pair, coaxial, and fiber-optic are the three most common cable types used in network applications.

- Cost vs. performance is the major consideration in choosing among the different cable types. Most networks do not exclusively use one type, instead applying a mix of cabling that maximizes cost-effectiveness and performance where needed.

**CHAPTER SIX**

# Network Operating Systems

## Chapter Six Contents:

Part 1: Overview .......................................................... 121

Part 2: Popular NOSs .................................................. 125

Part 3: Network Operating System Trends .................... 129

# Overview

*A Network Operating System is the logic used to control access to the network and its services.*

Network Operating Systems, or NOSs, are designed to orchestrate the links between computers and their shared information so that individual users are able access or transfer information from one system to another without exchanging physical media such as floppy disks or tapes. The first practical networks and network operating systems originated with very large companies like Xerox and IBM, who needed relatively quick and reliable methods for exchanging computer-based data between individuals, departments or offices that were often separated by large distances.

## Background

The need for collaboration on a large scale all but necessitated such technological developments, but in many cases the companies that pioneered the use of such technologies were too short-sighted to realize the commercial applications of their research. For instance, Xerox, during the early to mid 1970's, created and used local area networks (in addition to developing the first graphical user interface and object-oriented programming language) that were functionally equivalent to those in use today. But in the absence of a pre-existing market, Xerox was unable to recognize the inherent commercial worth of their forward-thinking developments. Their failure to capitalize on technological innovation left the door open for many smaller, more commercially astute companies (like Novell, Microsoft, Apple, and others) to create a booming industry. But the collaborative revolution unleashed by the birth of the network would not have been possible without other developments that were occurring at an equally frantic pace – namely, the creation and proliferation of the personal computer.

Networks and their operating systems would have had little use without the advent of the desktop PC. Prior to the invention of the personal computer, mainframes were the sole source of computing power. Access to the mainframe's resources, both data and CPU time, was funneled through a series of so-called dumb terminals – essentially a keyboard and display linked directly to the mainframe with little or no computing power residing with the user. These dumb terminals were akin to modern networks only in the most basic sense – they offered distributed access to computing resources, but did not distribute the resources themselves (unlike the client/server or peer-to-peer models, in which desktop PCs, with their on-board processors, memory, and storage media, decentralize the computing muscle of the network). The mainframe world had little use for network operating systems as we know them today.

But the advent and explosive growth of the desktop PC all but mandated the development of cheaper, more efficient networks. Initially, this was for the purpose of sharing (then) expensive and sporadically used equipment such as printers, plotters, and mass storage devices. Eventually, it became a means of sharing the most valuable business resource of all – information. It is the quest for more reliable transfer of data through myriad types of personal computers and peripherals that continues to spur the standardization of the network and its operating system.

## Common NOS Functions

> Common NOS functions include controlling network access, file handling, and allowing the sharing of peripherals.

Despite the varied strategies used by NOS vendors for networking shared resources, most network operating systems are in fact doing many of the same things. These functions, common to virtually all networks, include the following.

### Control Network Access

All networks must include effective methods for controlling and customizing user access to their information resources. When a new network user is added, he or she is typically given an account, a password, and a set of physical or logical network volumes that he or she will be able to access (with corresponding read/write rights for each network drive).

### Network File Handling

In order to maintain data integrity, only one user at a time can be allowed to alter the contents of a file. Ideally, network file access is a transparent process to a user on a workstation. The NOS is relied on to manage the access of files from multiple users. If more than one user is attempting to access the same file, the NOS resolves the conflict. Other file handling functions include error detection and correction, file encryption, and compression. Some network operating systems also include regular backups as part of their file handling regimen.

### Shared Applications

Most useful and effective networks also allow for shared applications to be run across the network. Common reasons for running applications over a network are to maintain licensing control over the number of concurrent users of a given application (since it is possible to limit the number of users accessing a given network file), or the need to keep source apps centralized for updating purposes.

### Sharing Peripherals

One of the major impetuses for the development of networks was to enable the sharing of very expensive peripherals. The NOS controls access and maintains a queue if multiple users request the services of a peripheral.

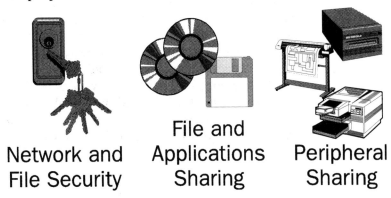

Network and File Security    File and Applications Sharing    Peripheral Sharing

*Figure 6-1: NOS Functions*

# Network Operating Systems and the OSI Reference Model

The OSI Reference Model is a useful tool to illustrate the interaction of network hardware and software functions. Please see **Chapter Two** for details on the OSI model. The manner in which Network Operating Systems orchestrate the links between diverse pieces of computer and communications software is especially well suited to comparison against the OSI model.

Rather than being a literal interdependence, the OSI model provides designers with an overall context for understanding the interaction between the logical and physical parts of a network. By understanding the distinctions outlined in the OSI model, network designers are able to break down the myriad functions of the network into manageable portions. This division of labor, when effectively implemented, means that operations that are performed at the hardware (physical) level, or in the layers immediately above it are largely taken for granted by the network operating system. Consider that managing functions, such as adherence to media-specific throughput standards, have little business residing near applications-level software, the chief domain of the NOS. And thanks to the general agreement on the OSI standard, or at least on the need for the kind of distinctions that it outlines, the NOS can leave many such monitoring functions to board-level firmware.

Adherence to the OSI model varies among NOS vendors. Some have gone so far as to virtually ignore the OSI model, since it is only one way that a protocol stack may be implemented. The OSI model is most useful as a common basis of comparison among the various networking products on the market. NOS vendors give a bewildering array of names to the various levels in their product's protocol stacks. Using the OSI model will help you sort out what functions the names are actually referring to. **Table 6-1** shows a comparison of two NOS protocol stacks and how the components correspond to in the OSI model.

## Table 6-1: Network Operating Systems and the OSI Model

| OSI Layer | Novell NetWare v3 | Banyan Vines |
|---|---|---|
| Application | NetWare Shell | Redirector |
| Presentation | Network Core Protocols (NCP) | VINES Remote Procedure Calls |
|  |  | Server Message Block (SMB) |
| Session | NetBIOS Emulator | NetBIOS Services |
| Transport | Sequenced Packet Exchange (SPX) | VINES Interprocess Communications Protocol (VIPC) |
| Network | Internet Packet Exchange (IPX) | VINES Internet Protocol |
| Data Link | Open Data-Link Interface (ODI) | Network Driver Interface Specification (NDIS) – Product Specific Driver |

## Local and Network OS Interaction

There has traditionally been a clear distinction between local and network operating systems. Local operating systems provided a common interface between applications and the workstation, and network operating systems provided access to shared services. This is the client/server model described in **Chapter One**. The goal was to make network service as transparent as possible to the end user. Therefore the interaction of the local with the network OS occurred at mid level communications layers.

Recent developments in computer hardware, software, and communications have moved the spotlight to the network itself. The trend in newer local operating systems has been to incorporate network functions, resulting in even less dependency on the network OS. This shift has put more of an emphasis on peer-to-peer model communications (also covered in **Chapter One**). In addition, more applications are incorporating networked features, meaning the communication is occurring at the higher layers.

## Popular NOSs

Many companies have tried their hand in the NOS arena, but only a few have succeeded in the long term. The companies cited below are among those that have retained a significant share of the market.

## Novell NetWare

Founded in Orem, Utah in 1983, Novell has been a dominant force in the Network Operating Systems arena since its inception. One of the first companies to bring NOS technology to small and medium sized businesses, Novell experienced explosive growth through distribution of its NetWare products during the mid-80s and early 90s. NetWare, with its client-server networking model, proved to be a symbiotic addition to the emerging PC market. But for all of its technical innovation and market savvy, Novell simply built on a pre-existing foundation. Like the networks or mainframe terminal systems that preceded it, NetWare-based networks needed to do a number of things common to all distributed computing environments: restrict and personalize access to network resources, ensure the integrity of the data being created and stored, and effectively manage the resources of the network in a way that makes it as transparent to the user as possible.

Initial releases of NetWare did all of these things, and did them better than any competing products being offered for the personal computer at the time (many of which were little more than enhanced serial links). And where previous networking schemes often demanded rigid installation guidelines, NetWare offered many options. It supported a wide variety of cabling types, network types, topologies, and system architectures. With this innovation in flexibility and relative affordability, Novell enjoyed an almost insurmountable lead in the NOS market for the better part of a decade.

## Microsoft Windows NT

In recent years, Microsoft and its Windows NT operating system have been slowly eating away at Novell's still massive market share (as of this writing Novell still commands a market share slightly greater than 60%) in the NOS arena. In a strategy that mirrors their move into the operating system (text-based and GUI), net browser, and applications arenas, Microsoft has once again waited for a settling of standards and user expectations in order to make their entry into the marketplace.

Early projections of Windows NT's success soon after its release have yet to be realized, thanks in part to missteps by Microsoft and aggressive competition from Novell and others. But by 1994, Windows NT presented network administrators with an attractive choice for their local area networks—a network OS that was more or less fully integrated with a desktop operating system. And the ability for NT to mimic its less capable cousins, Windows for Workgroups (WFW 3.11) and later Windows 95, meant that Windows NT had the dual advantages of interface familiarity and a wide applications base. These advantages, combined with the relative ease of set-up that comes from having designed virtually all of the pieces in the networking puzzle, made it easy to see why some observers were quick to declare Microsoft the de facto winner in the NOS war long before the first big battle had even been fought.

Recent releases of Windows NT have shown that Microsoft is every bit as vulnerable to competitive pressures as the next player in this ever-changing arena. After being lauded for improvements in performance and speed after the release of Windows NT version 4.0, Microsoft has since become the target of criticism by those who claim Microsoft has traded speed for stability. Specifically, users have questioned Microsoft's decision to relocate certain network interface functions to the inner-most layer of Intel's protected-mode CPU design. By removing many of the insulative layers between the network operating and the CPU that drives it, Microsoft may be gambling that the benefits of increased performance will offset customer dissatisfaction with system instability.

## Banyan VINES

Banyan VINES has been a widely used NOS for many years, finding its niche at first with large companies and universities.

Unlike NetWare, which focused largely on LAN functionality (at least in initial releases), VINES has from the onset dealt with the question of networking from a more wide-area perspective.

As companies like Novell and Microsoft later began to turn their attention to wide-area networking, many observers predicted the demise of VINES. Such predictions were premature, especially given the loyalty which VINES and its StreetTalk file access system have instilled in network administrators charged with linking far-flung, heterogeneous network resources.

## UNIX

Like DOS, most flavors of UNIX take a command line approach to the operating system/network operating system. Unlike DOS, UNIX was built with the networking power and functionality needed to help it survive the GUI revolution. UNIX has been a workhorse operating system, used at first in corporate workstation environments for the tasks that DOS was too under-powered to handle, and more recently its components have served as the organizational underpinning of the Internet.

The first versions of UNIX were created almost thirty years ago, in 1969, by researchers at Bell Telephone Laboratories. Work on UNIX began after Bell pulled out from further development of another ground-breaking operating system called Multics (Multiplexed Information and Computing Service), which was first released in 1965. The name UNIX was essentially meant as a pun, derived from the name Multics, with 'Uni' replacing 'Multi' and 'X' replacing 'cs'. Much of UNIX's resilience in the Microsoft-dominated marketplace of the 90's can be attributed to its Multics-based lineage. Built for military and commercial use over thirty years ago, Multics could support parallel processing, segmented memory blocks, virtual memory, and was also the first operating system to offer a true relational database capability. When Bell pulled out of the Multics project, their best and brightest took the lessons learned there and applied them to the fledgling UNIX.

Unlike DOS, UNIX has been slow to gain widespread acceptance with end-users. One reason is a somewhat cryptic command line structure. Another is that UNIX is in the public domain, which has encouraged vendors to produce several distinct versions of UNIX. Offshoots such as AIX, XENIX, and LINUX have had the effect of dividing the UNIX user base across several incompatible platforms.

From the onset, UNIX has been a less commercially-minded operating system than its competitors from Microsoft (Windows) and IBM (OS/2), and from all appearances this is at least in part by choice. Despite the release of products like Windows NT and Windows 95, which have begun to narrow the performance differential between UNIX and the rest of the OS/NOS world, UNIX users are still to be found everywhere. And the widespread availability of UNIX versions like LINUX, which is available for free download on the World Wide Web, helps to ensure this longevity.

### Artisoft LANtastic

An often overlooked but still resilient player in the NOS market, Artisoft's LANtastic has proven itself a durable competitor in the network operating systems arena. Gaining wide acceptance initially on the basis of sheer affordability, users soon discovered that LANtastic did most of the things that they had come to expect from more sophisticated NOSs. LANtastic could recognize a number of different LAN protocols, and offered support some forms of wide area networking. It also had a degree of programmable customization that made it an attractive choice for small NOS users who wanted their NOS to grow with their business.

With its latest release on LANtastic, which offers greater support for NetWare and NT clients, Artisoft is hoping to increase its market share in what has become an extremely competitive and volatile industry.

# Network Operating System Trends

## OS/NOS Integration

Networking functions have become almost inseparably related with the functions of the local desktop operating system. As UNIX, Windows NT, and OS/2 become more feature-laden and refined, we see the OS and NOS moving toward the seamless integration of local and network resources. Mainstream PC networking was once little more than a third party add-on capability to an existing OS. It is now factored into almost every facet of local operating system, hardware, and ancillary software development.

In addition to recognizing the need for tighter integration between OS and NOS functionality, operating systems developers have also taken steps to support many different networking protocols. NOSs such as UNIX have included such protocol flexibility for some time, mainly due to a largely technical, programming-literate user-base. However, this flexibility has been slow to appear in more consumer-oriented operating systems. NOSs like Windows 3.11 and Windows 95 have only recently begun to reach the levels of functionality and flexibility that more demanding network users have come to expect.

Since late 1991, Microsoft has pursued a popular strategy in addressing interface flexibility issues (whether applications or OS related): the use of a modular plug-in facilities that allow developers to enhance the ability of the NOS to recognize different protocols. Through the use of modules such as Winsock, which is used for TCP/IP support, Microsoft has been able to pass some of their development burden on to third-party developers and end users. This strategy has generally been welcomed because it offers the opportunity to customize network environments for a wide variety of needs.

## Intranet

In recent years, Novell, Microsoft and others have also turned their attention to one of the most lucrative and attractive (at least from a pure marketing perspective) sectors of the networking marketplace – corporate intranets. Much of the attractiveness of this segment of the networking market comes from the fact that it is largely a repackaging of pre-existing Internet technologies.

In a nutshell, intranet technology seeks to do for companies and organizations what the Internet did for computer users across the globe – offer a means of linking users with like interests and backgrounds to a common point of reference. Where the Internet must accomplish this feat over vast geographical distances, intranets often perform it within a single building.

Corporate intranets are often simply web-pages designed for the sole use of persons within an organization. They may contain information on employee benefits, or serve as a bulletin board for posting messages, or even offer a means of direct data entry via HTML-based templates. Intranet software, non-existent until a couple of years ago, has ballooned into an industry of its own. Companies seeking to exert more control over distribution of their corporate policies, use of their applications environments, and dissemination of information through their organizations, found in the intranet a tool for single-source control of corporate information.

Consider the all-too-familiar question of application upgrades. In a large organization, upgrading a single piece of client-based software – for instance a sales tracking program – may take weeks or months to complete if hundreds of systems are involved. The same task performed on an intranet-based application can be done in a single installation. Perhaps even more importantly, extensions to HTML standards have made it possible to create and modify web-base applications with surprising ease and speed. This capability makes intranet applications even more attractive than server-based applications that typically cannot be modified. Web-based programs have already come close to matching the level of functionality of traditional applications, and their no-hassle installation and ease of upgrading will no doubt lend momentum to their industry acceptance.

Intranet integration with existing network operating systems is currently generating a great deal of activity among vendors. With its recently released intranet package, Novell has taken an early lead in the intranet integration arena, but Microsoft and others are in close pursuit. Whether intranets will prove to be a useful and durable form of corporate networking has yet to be determined – at this point it is difficult to separate the marketing bluster from the tangible benefits.

## Chapter Review

- An NOS is the intelligence behind the network. It is used to provide orderly access to shared services, such as file handling and use of printers.

- The first network operating systems were developed for a client/server environment and had simple features such as network security and enabling shared access to expensive peripherals. Modern NOSs are being developed in an age of highly capable workstations that often work in a peer-to-peer environment and are better integrated with the features of local (workstation) operating systems.

- Popular NOS packages include Novell's NetWare, Microsoft's Windows NT, Artisoft's LANtastic, and UNIX. The selection of one NOS over another is determined on a case-by-case basis, depending on the needs of the customer.

- Intranets are web-based communications systems designed for exclusive use within an organization. While intranets initially served as little more than company bulletin boards, advances in web-based programming have made it possible for intranets to feature custom-built applications suites.

**CHAPTER SEVEN**

# Hardware Settings

## Chapter Seven Contents:

Part 1: Overview ............................................................ 135

Part 2: Jumper and Switch Position Indicators .............. 135

Part 3: Board Element Naming Conventions................. 136

Part 4: Diagnostic LEDs ................................................ 139

Part 5: Jumper Settings ................................................. 140

## Overview

Network Interface Cards that do not have any jumpers or switches to configure are not included in this book. These NICs are usually set by software provided by the manufacturer, or they are not user configurable.

Network interface adapters may have numerous switches or jumpers used to configure the hardware as desired. This chapter provides hardware settings for many of the most popular network adapters produced. Plug and play (software configured) devices obviate the need for hardware setup information, therefore are not included in this chapter.

The CD-ROM included with this text contains a fully operational *Micro Technical Library of Network Interface Cards*. The MTL contains the information found in this chapter and much more. Its powerful search engine will help you find the right settings – saving you time and money.

With the proliferation of various types of peripherals for personal computers, the potential for encountering resource conflicts has become more pronounced. The following sections explain the purpose of commonly encountered switches and jumpers, and their recommended settings.

In order to avoid resource conflicts and aid in troubleshooting problems, be sure to keep a record of the current hardware settings of all peripheral devices installed in a system.

## Jumper and Switch Position Indicators

The default or most common jumper and switch positions are denoted by the symbol "➪." Pin 1, Jumper 1, and Switch 1 are indicated by a white pin or switch.

Pin 1 and Jumper 1          ON position indicator dot

  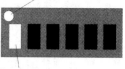

Switch 1

**Jumper Blocks**          **Switch Block**

*Figure 7-1: Jumper and Switch Block Conventions*

The switch's ON position is indicated by the white dot. Always verify the ON position by examining the switch on the actual hardware. When the ON position is not known to us there will be no white dot.

# Board Element Naming Conventions

Whenever possible, diagrams contained in this chapter use the same jumper and switch labels as the manufacturer. If the labeling information provided by the manufacturer is unclear or not available, an alternate label standard is used. Under the alternate standard, all jumpers are labeled with the prefix "JP," while all connectors are labeled with the prefix "CN." In most cases, the jumper *numbers* remain the same as the manufacturer's labels. In the cases where the jumper numbers in this book differ from those silk-screened on the card, the settings listed are consistent with this book's diagrams. Where a whole set of jumpers are labeled with one number, we label them with that same number followed by a letter.

In **Figure 7-2**, the top diagram is as typically found in the manufacturer's documentation. The diagram on the bottom is the same device using the conventions outlined in this chapter.

The Network Technical Guide **137**

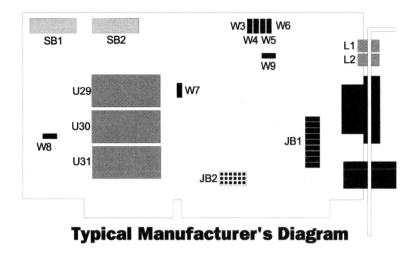

**Typical Manufacturer's Diagram**

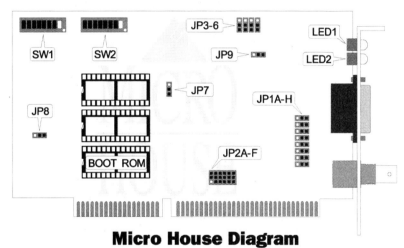

**Micro House Diagram**

*Figure 7-2: Diagram Comparison*

 # Additional Hardware Notes

## System Resources Dedicated to the Serial Port

I/O devices need a standardized way to access the host computer's system resources. For network interface devices, this usually involves assigning an **I/O address**, a hardware **Interrupt Request (IRQ)**, and possibly a boot ROM address.

Default settings depend on the type of device being used. Prior to installing any new device, check the system resource settings of all devices installed. If a conflict exists, change the hardware settings as necessary.

## I/O Address

The I/O address is a location in the computer's memory assigned to store information coming from or going to a particular device.

## Interrupt Request (IRQ)

Peripherals need an orderly way to share CPU resources. PCs use IRQ lines to gain the CPU's attention when needed. Recent model PCs use 16 IRQ lines. A hierarchy exists among the IRQ lines to give some peripherals priority over others. Lower IRQ numbers have a higher priority.

## Boot ROMs

In some cases, it may be desirable to connect what is known as a diskless workstation, or a computer that has no floppy or hard drives installed, to the network. This way, the workstation cannot be used to copy files or secure data to the floppy or hard drive. Most NICs include an empty socket used for the installation of an optional Boot ROM. The Boot ROM forces the workstation to use the DOS startup files from the network server, instead of booting from floppy or a hard drive. This procedure is often referred to as a **Remote Program Load** (**RPL**). The ROM can be purchased separately and must be designed for the specific NIC that you will be using.

Before installing an optional Boot ROM, check to see that all the pins are straight, perpendicular to the body of the chip. Insert the ROM carefully into its socket on the NIC. The notches on both ROM and socket must line-up (see diagram below). The ROM and possibly the NIC may be destroyed if it is inserted the wrong way. Make sure that all the pins are inserted into the socket correctly and that none are bent underneath. Pins that are bent under the ROM are difficult to detect, and is one of the most common installation errors.

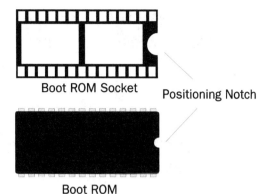

*Figure 7-3: Boot ROM Installation*

In addition to physically installing the chip, it may also be necessary to change one or more jumpers on the card. Look for jumpers that enable/disable Boot ROM, as well as jumpers that assign a Boot ROM address or Base Memory Address.

# Ethernet AUI Fuse

Some Ethernet cards include a fuse that protects the AUI port from transceiver cabling faults that could subject the adapter to a power surge. The Thin-Ethernet BNC or 10BaseT modular connector on the NIC may still be used even if the AUI fuse is defective or blown. To replace the AUI port fuse, use an 800 mA, 250V type fuse.

# Diagnostic LEDs

Many interface cards include diagnostic LEDs to provide "at-a-glance" status indications. Most of the color and on/off state conventions (such as blinking) are particular to the manufacturer of the hardware. The table below gives verbose descriptions of common LED indications.

*Table 7-1: Diagnostic LED Description – General*

| LED | Status | Condition |
|-----|--------|-----------|
| RX  | On     | Data is being received |
| RX  | Off    | Data is not being received |
| TX  | On     | Data is being transmitted |
| TX  | Off    | Data is not being transmitted |
| NC  | On     | Network connection is good |
| NC  | Off    | Network connection is broken |

*Table 7-2: Diagnostic LED Description – General (Continued)*

| LED | Status | Condition |
|---|---|---|
| DRV | On | Device driver is loaded |
| DRV | Off | Device driver is not loaded |
| PWR | On | Power is on |
| PWR | Off | Power is off |

*Table 7-2: Diagnostic LED Description – Ethernet*

| LED | Status | Condition |
|---|---|---|
| COL | On | A data collision is detected on the network |
| COL | Off | No data collision detected on the network |
| POL | On | Polarity is normal |
| POL | Off | Polarity is reversed (twisted pair receive or transmit wires are reversed) |
| JAB | On | Network jabber is detected (Continuous defective transmissions due to faulty wiring or node connections) |
| JAB | Off | Network jabber is not detected |

*Table 7-3: Diagnostic LED Description – Token Ring*

| LED | Status | Condition |
|---|---|---|
| NET | On | Adapter successfully joined the ring – network connection is good |
| NET | Off | Adapter did not successfully join the ring – network connection is not available |
| 16/4 | On | 16 Mbps transmission speed is enabled |
| 16/4 | Off | 4 Mbps transmission speed is not enabled |

*Table 7-1: Diagnostic LED Description – ARCNET*

| LED | Status | Condition |
|---|---|---|
| NET | On | Network configuration is good |
| NET | On | Card is reconfiguring to the network |
| NET | Off | Network configuration is bad |

## Jumper Settings

The remainder of this chapter contains set-up information on selected network interface devices.

## 3COM CORPORATION
## ETHERLINK (3C501; ASSY.#1221)

**NIC Type** Ethernet
**Transfer Rate** 10Mbps
**Data Bus** 8-bit ISA
**Topology** Linear Bus
**Wiring Type** RG-58A/U 50ohm coaxial
AUI transceiver via DB-15 port
**Boot ROM** Available

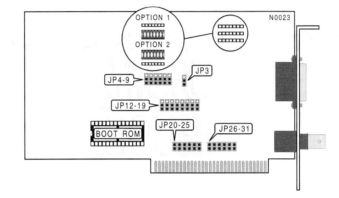

| CABLE TYPE SELECTION ||
|---|---|
| Type | Setting |
| ⇨ RG-58A/U 50ohm coaxial | Option 1 |
| AUI transceiver via DB-15 port | Option 2 |

| BOOT ROM CONFIGURATION ||
|---|---|
| Setting | JP3 |
| ⇨ Disabled | Pins 2 & 3 Closed |
| Enabled | Pins 1 & 2 Closed |

| I/O BASE ADDRESS SELECTION ||||||| 
|---|---|---|---|---|---|---|
| Address | JP4 | JP5 | JP6 | JP7 | JP8 | JP9 |
| 300h | Pins 2 & 3 | Pins 2 & 3 | Pins 2 & 3 | Pins 2 & 3 | Pins 1 & 2 | Pins 1 & 2 |
| 310h | Pins 1 & 2 | Pins 2 & 3 | Pins 2 & 3 | Pins 2 & 3 | Pins 1 & 2 | Pins 1 & 2 |

Note: Pins designated should be in the closed position.

| BOOT ROM ADDRESS SELECTION |||||||||
|---|---|---|---|---|---|---|---|---|
| Address | JP12 | JP13 | JP14 | JP15 | JP16 | JP17 | JP18 | JP19 |
| CC000h | Pins 2&3 | Pins 2&3 | Pins 1&2 | Pins 1&2 | Pins 2&3 | Pins 2&3 | Pins 1&2 | Pins 1&2 |
| EC000h | Pins 2&3 | Pins 2&3 | Pins 1&2 | Pins 1&2 | Pins 2&3 | Pins 1&2 | Pins 1&2 | Pins 1&2 |

Note: Pins designated should be in the closed position.

## 3COM CORPORATION
## ETHERLINK (3C501; ASSY.#1221)

| | | | INTERRUPT REQUEST SELECTION | | | |
|---|---|---|---|---|---|---|
| IRQ | JP20 | JP21 | JP22 | JP23 | JP24 | JP25 |
| 2 | Closed | Open | Open | Open | Open | Open |
| 3 | Open | Closed | Open | Open | Open | Open |
| 4 | Open | Open | Closed | Open | Open | Open |
| 5 | Open | Open | Open | Closed | Open | Open |
| 6 | Open | Open | Open | Open | Closed | Open |
| 7 | Open | Open | Open | Open | Open | Closed |

| | | | DMA CHANNEL SELECTION | | | |
|---|---|---|---|---|---|---|
| Channel | JP26 | JP27 | JP28 | JP29 | JP30 | JP31 |
| DMA1 | Closed | Open | Open | Closed | Open | Open |
| DMA2 | Open | Closed | Open | Open | Closed | Open |
| DMA3 | Open | Open | Closed | Open | Open | Closed |

Note: Setting the DMA Channel on this card is required even though "No Channel" is listed when setting up the LAN configuration software.

# 3COM CORPORATION
## ETHERLINK II/16 (3C503)

**NIC Type**        Ethernet
**Transfer Rate**   10Mbps
**Data Bus**        8/16-bit ISA
**Topology**        Linear Bus
**Wiring Type**     RG-58A/U 50ohm coaxial
                    AUI transceiver via DB-15 port
**Boot ROM**        Available

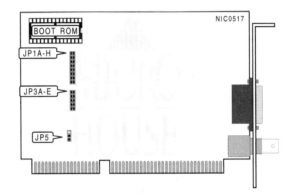

| I/O BASE ADDRESS ||||||||| 
|---|---|---|---|---|---|---|---|---|
| Address | JP1A | JP1B | JP1C | JP1D | JP1E | JP1F | JP1G | JP1H |
| 250h | Open | Open | Open | Open | Closed | Open | Open | Open |
| 280h | Open | Open | Open | Open | Open | Closed | Open | Open |
| 2A0h | Open | Open | Open | Open | Open | Open | Closed | Open |
| 2E0h | Open | Open | Open | Open | Open | Open | Open | Closed |
| ⇨300h | Closed | Open | Open | Open | Open | Open | Open | Open |
| 310h | Open | Closed | Open | Open | Open | Open | Open | Open |
| 330h | Open | Open | Closed | Open | Open | Open | Open | Open |
| 350h | Open | Open | Open | Closed | Open | Open | Open | Open |

| BOOT ROM ADDRESS |||||| 
|---|---|---|---|---|---|
| Address | JP3A | JP3B | JP3C | JP3D | JP3E |
| ⇨Disabled | Closed | Open | Open | Open | Open |
| C8000h | Open | Closed | Open | Open | Open |
| CC000h | Open | Open | Closed | Open | Open |
| D8000h | Open | Open | Open | Closed | Open |
| DC000h | Open | Open | Open | Open | Closed |

| DATA BUS SIZE ||
|---|---|
| Setting | JP5 |
| ⇨16-bit | Pins 2 & 3 closed |
| 8-bit | Pins 1 & 2 closed |

## 3COM CORPORATION
## ETHERLINK 16 TP (3C507-TP)

| | |
|---|---|
| **NIC Type** | Ethernet |
| **Transfer Rate** | 10Mbps |
| **Data Bus** | 16-bit ISA |
| **Topology** | Star |
| **Wiring Type** | Unshielded twisted pair |
| **Boot ROM** | Available |

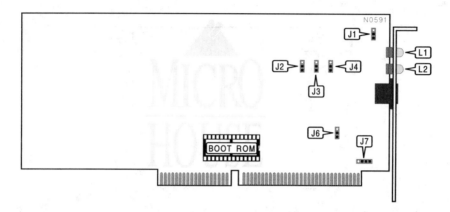

| 10BASE-T STANDARD COMPATIBILITY CONFIGURATION | | |
|---|---|---|
| **Setting** | **Jumper** | **Setting** |
| ▷ DC signal disabled[2] | J1 | Pins 2 & 3 closed |
| DC signal enabled (On receive wires)[3] | J1 | Pins 1 & 2 closed |
| ▷ Link beat enabled | J2 | Pins 1 & 2 closed |
| Link beat disabled | J2 | Pins 2 & 3 closed |
| ▷ Receive threshold select normal | J3 | Pins 1 & 2 closed |
| Receive threshold select low[1] | J3 | Pins 2 & 3 closed |
| ▷ Equalization disabled[2] | J4 | Pins 2 & 3 closed |
| Equalization enabled[3] | J4 | Pins 1 & 2 closed |
| ▷ Transmit level select normal (±5V) | J6 | Pins 1 & 2 closed |
| Transmit level select low (±2V) | J6 | Pins 2 & 3 closed |
| ▷ Cable Impedance select 100ohms | J7 | Pins 2 & 3 closed |
| Cable Impedance select 150ohms | J7 | Pins 3 & 4 closed |
| Cable Impedance select 75ohms | J7 | Pins 1 & 2 closed |

Note [1]: Use this setting with the 3Com MultiConnect TP Module. Do not use this setting with the 3Com LinkBuilder, or the SynOptics LattisNet Model 3308.
Note [2]: This is the 10BASE-T standard and should be used with all 10BASE-T standard hubs.
Note [3]: This setting is required when using any Pre-10BASE-T SynOptics hub.

## 3COM CORPORATION
## ETHERLINK 16 TP (3C507-TP)

| \multicolumn{4}{c}{DIAGNOSTIC LED(S)} |
|---|---|---|---|
| **LED** | **Color** | **Status** | **Condition** |
| L1 | Yellow | Blinking | Data is being transmitted or received |
| L1 | Yellow | Off | Data is not being transmitted or received |
| L2 | Green | On | Twisted pair network connection is good |
| L2 | Green | Blinking | Twisted pair wire polarity reversed |
| L2 | Green | Off | Twisted pair network connection is broken |

Note: Reversed polarity is automatically corrected for but this may impede performance. It is suggested that reversed polarity wires be corrected to attain maximum performance.

## 3COM CORPORATION
## TOKENLINK

| | |
|---|---|
| NIC Type | Token-Ring |
| Transfer Rate | 4Mbps |
| Data Bus | 16-bit ISA |
| Topology | Ring |
| Wiring Type | Unshielded twisted pair |
| | Shielded twisted pair (DB-9 port) |
| Boot ROM | Available |

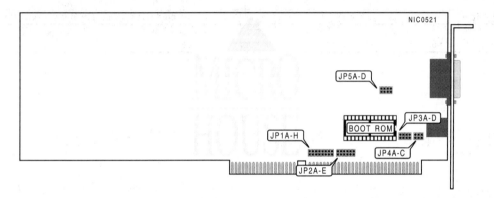

| DMA CHANNEL | | | | | | | | |
|---|---|---|---|---|---|---|---|---|
| Channel | JP1A | JP1B | JP1C | JP1D | JP1E | JP1F | JP1G | JP1H |
| ⇨DMA3 | Open | Open | Open | Open | Open | Open | Closed | Closed |
| DMA5 | Open | Open | Open | Open | Closed | Closed | Open | Open |
| DMA6 | Open | Open | Closed | Closed | Open | Open | Open | Open |
| DMA7 | Closed | Closed | Open | Open | Open | Open | Open | Open |

| INTERRUPT REQUEST | | | | | | |
|---|---|---|---|---|---|---|
| IRQ | JP2A | JP2B | JP2C | JP2D | JP2E | JP2F |
| ⇨5 | Open | Open | Open | Open | Open | Closed |
| 10 | Open | Open | Open | Open | Closed | Open |
| 11 | Open | Open | Open | Closed | Open | Open |
| 12 | Open | Open | Closed | Open | Open | Open |
| 14 | Open | Closed | Open | Open | Open | Open |
| 15 | Closed | Open | Open | Open | Open | Open |

## 3COM CORPORATION
### TOKENLINK

| I/O BASE ADDRESS | | | | |
|---|---|---|---|---|
| **Address** | **JP3A** | **JP3B** | **JP3C** | **JP3D** |
| ⇨300h | Closed | Open | Open | Open |
| 310h | Open | Closed | Open | Open |
| 320h | Open | Open | Closed | Open |
| 350h | Open | Open | Open | Closed |

| PRIMARY/ALTERNATE MODE | | | |
|---|---|---|---|
| **Setting** | **JP4A** | **JP4B** | **JP4C** |
| ⇨Single card | Open | Open | Closed |
| Primary | Closed | Open | Open |
| Alternate | Open | Closed | Open |

Note: If two cards are to be used in a single system, one of the cards must be set to primary and the other one must be set to alternate. If only one card is used this option must be set to single.

| BOOT ROM ADDRESS | | | | |
|---|---|---|---|---|
| **Address** | **JP5A** | **JP5B** | **JP5C** | **JP5D** |
| ⇨Disabled | Closed | Open | Open | Open |
| C4000h | Open | Closed | Open | Open |
| D4000h | Open | Open | Open | Closed |
| D6000h | Open | Open | Closed | Open |

## ACCTON TECHNOLOGY CORPORATION
## ETHERPAIR-16 (EN1641)

| | |
|---|---|
| NIC Type | Ethernet |
| Transfer Rate | 10Mbps |
| Data Bus | 8/16-bit ISA |
| Topology | Star |
| Wiring Type | Unshielded twisted pair |
| | AUI transceiver via DB-15 port |
| Boot ROM | Available |

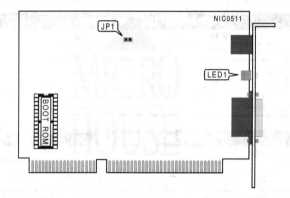

| CARD CONFIGURATION | | | | | |
|---|---|---|---|---|---|
| Mode | I/O Address | IRQ | Cable Type | Boot ROM | JP1 |
| ➪Software | Software | Software | Software | Software | Open |
| NE1000 | 280h | Disabled | AUI transceiver | Disabled | Closed |

Note: The card is normally configured using the included software utility. The second configuration is provided to allow easier troubleshooting if the system will not boot or locks-up.

| DIAGNOSTIC LED(S) | | | |
|---|---|---|---|
| LED | Color | Status | Condition |
| LED1 | Green | On | Twisted pair network connection is good |
| LED1 | Green | Off | Twisted pair network connection is broken |

## ACCTON TECHNOLOGY CORPORATION
### ETHERPAIR-8 (EN1801)

| | |
|---|---|
| **NIC Type** | Ethernet |
| **Transfer Rate** | 10Mbps |
| **Data Bus** | 8-bit ISA |
| **Topology** | Linear Bus |
| **Wiring Type** | Unshielded twisted pair |
| | AUI transceiver via DB-15 port |
| **Boot ROM** | Available |

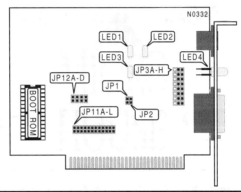

| SIGNAL QUALITY ERROR (SQE) TEST ||
|---|---|
| **Setting** | **JP1** |
| ⇨Enabled | Closed |
| Disabled | Open |

Note: Signal Quality Error (SQE) is a test of the collision circuitry and path.

| LINK INTEGRITY TEST ||
|---|---|
| **Mode** | **JP2** |
| ⇨Enabled | Closed |
| Disabled | Open |

Note: Link integrity is only valid when the cable type is twisted pair.

| CABLE TYPE ||
|---|---|
| **Type** | **JP3A-H** |
| ⇨Unshielded twisted pair | Pins 1 & 2 Closed |
| AUI transceiver via DB-15 port | Pins 2 & 3 Closed |

| INTERRUPT REQUEST ||||||| 
|---|---|---|---|---|---|---|
| **IRQ** | **JP11A** | **JP11B** | **JP11C** | **JP11D** | **JP11E** | **JP11F** |
| 2 | Open | Open | Open | Closed | Open | Open |
| ⇨3 | Open | Open | Open | Open | Closed | Open |
| 4 | Open | Open | Open | Open | Open | Closed |
| 5 | Closed | Open | Open | Open | Open | Open |
| 6 | Open | Closed | Open | Open | Open | Open |
| 7 | Open | Open | Closed | Open | Open | Open |

## ACCTON TECHNOLOGY CORPORATION
## ETHERPAIR-8 (EN1801)

| DMA CHANNEL ||||| 
|---|---|---|---|---|
| Channel | JP11G | JP11H | JP11I | JP11J |
| ⇨Disabled | Open | Open | Open | Open |
| DMA1 | Closed | Open | Closed | Open |
| DMA3 | Open | Closed | Open | Closed |

| I/O BASE ADDRESS |||
|---|---|---|
| Address | JP11K | JP11L |
| ⇨300-31Fh | Closed | Closed |
| 320-33Fh | Open | Closed |
| 340-35Fh | Closed | Open |
| 360-37Fh | Open | Open |

| BOOT ROM ||
|---|---|
| Setting | JP12A |
| ⇨Disabled | Open |
| Enabled | Closed |

| BOOT ROM ADDRESS ||||
|---|---|---|---|
| Address | JP12B | JP12C | JP12D |
| ⇨C8000h | Closed | Closed | Closed |
| CC000h | Open | Closed | Closed |
| D8000h | Closed | Open | Closed |
| DC000h | Open | Open | Closed |
| E8000h | Closed | Closed | Open |
| EC000h | Open | Open | Open |

| DIAGNOSTIC LED(S) ||||
|---|---|---|---|
| Led | Color | Status | Condition |
| LED1 | Red | Blinking | Collision detected on network |
| LED1 | Red | Off | Normal operation |
| LED2 | Green | Blinking | Data is being transmitted |
| LED2 | Green | On | Card is receiving power |
| LED2 | Green | Off | Card is not receiving power |
| LED3 | Red | On | Card is continuously transmitting, jabbering |
| LED3 | Red | Off | Normal operation |
| LED4 | Green | On | Twisted pair network connection is good |
| LED4 | Green | Off | Twisted pair network connection is broken |
| LED4 | Green | Blinking | Data is being received |

Note: If LED3 lights the twisted pair driver will be disabled.

## ACCTON TECHNOLOGY CORPORATION
## RINGPAIR-4/16T (TR1615 REV. 2)

**NIC Type**      Token Ring
**Transfer Rate**      4/16Mbps
**Data Bus**      16-bit ISA
**Topology**      Ring
**Wiring Type**      Shielded twisted pair
     Unshielded twisted pair
**Boot ROM**      Not available

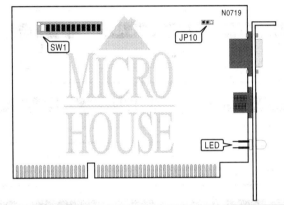

| BASE MEMORY ADDRESS | | | | | | |
|---|---|---|---|---|---|---|
| Address | SW1/1 | SW1/2 | SW1/3 | SW1/4 | SW1/5 | SW1/6 |
| ⇨ CC000h | On | Off | Off | On | On | Off |
| C0000h | On | Off | Off | Off | Off | Off |
| C2000h | On | Off | Off | Off | Off | On |
| C4000h | On | Off | Off | Off | On | Off |
| C6000h | On | Off | Off | Off | On | On |
| C8000h | On | Off | Off | On | Off | Off |
| CA000h | On | Off | Off | On | Off | On |
| CE000h | On | Off | Off | On | On | On |
| DC000h | On | Off | On | On | On | Off |
| D0000h | On | Off | On | Off | Off | Off |
| D2000h | On | Off | On | Off | Off | On |
| D4000h | On | Off | On | Off | On | Off |
| D6000h | On | Off | On | Off | On | On |
| D8000h | On | Off | On | On | Off | Off |
| DA000h | On | Off | On | On | Off | On |
| DE000h | On | Off | On | On | On | On |

## ACCTON TECHNOLOGY CORPORATION
## RINGPAIR-4/16T (TR1615 REV. 2)

### INTERRUPT SELECT

| IRQ | SW1/7 | SW1/8 |
|---|---|---|
| ▷ IRQ3 | Off | On |
| IRQ2 | Off | Off |
| IRQ6 | On | Off |
| IRQ7 | On | On |

### PRIMARY/ALTERNATE ADAPTER

| Setting | SW1/9 |
|---|---|
| ▷ 4/16T as primary or sole LAN adapter | Off |
| 4/16T as alternate LAN adapter | On |

### SHARED RAM

| Setting | SW1/10 | SW1/11 |
|---|---|---|
| ▷ 16KB shared RAM | Off | On |
| 8KB shared RAM | Off | Off |
| 32KB shared RAM | On | Off |
| 64KB shared RAM | On | On |

### DATA BUS TRANSFER RATE

| Setting | SW1/12 |
|---|---|
| 16Mbps | Off |
| 4Mbps | On |

### BUS WIDTH

| Setting | JP10 |
|---|---|
| 8-bit bus width enabled | Pins 2 & 3 closed |
| 16-bit bus width enabled | Pins 1 & 2 closed |

### DIAGNOSTIC LED

| LED | Color | Status | Condition |
|---|---|---|---|
| LED1 | Green | On | 16Mbps transmission speed |
| LED1 | Red | On | 4Mbps transmission speed |

## ADDTRON TECHNOLOGY COMPANY, LTD.
## AE-200LC

| | |
|---|---|
| **NIC Type** | Ethernet |
| **Transfer Rate** | 10Mbps |
| **Data Bus** | 16-bit ISA |
| **Topology** | Linear Bus |
| **Wiring Type** | Unshielded twisted pair |
| | RG-58A/U 50ohm coaxial |
| **Boot ROM** | Available |

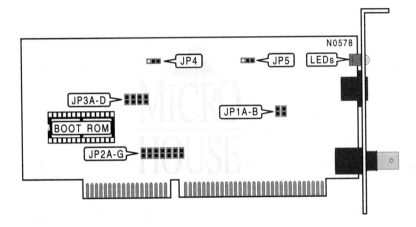

| I/O BASE ADDRESS | | |
|---|---|---|
| Address | JP1A | JP1B |
| ⇨300h | Open | Open |
| 320h | Closed | Open |
| 340h | Open | Closed |
| 360h | Closed | Closed |

| INTERRUPT REQUEST | | | | | | | |
|---|---|---|---|---|---|---|---|
| IRQ | JP2A | JP2B | JP2C | JP2D | JP2E | JP2F | JP2G |
| 2 | Closed | Open | Open | Open | Open | Open | Open |
| 3 | Open | Closed | Open | Open | Open | Open | Open |
| 4 | Open | Open | Closed | Open | Open | Open | Open |
| 5 | Open | Open | Open | Closed | Open | Open | Open |
| 10 | Open | Open | Open | Open | Closed | Open | Open |
| 11 | Open | Open | Open | Open | Open | Closed | Open |
| 12 | Open | Open | Open | Open | Open | Open | Closed |

# 154 Hardware Settings

## ADDTRON TECHNOLOGY COMPANY,, LTD.
## AE-200LC

| \ BOOT ROM ADDRESS | | | | |
|---|---|---|---|---|
| Address | JP3A | JP3B | JP3C | JP3D |
| ⇨Disabled | Open | Open | Open | Open |
| C8000h | Closed | Open | Closed | Open |
| CC000h | Closed | Closed | Closed | Open |
| D0000h | Closed | Open | Open | Closed |
| D4000h | Closed | Closed | Open | Closed |
| D8000h | Closed | Open | Closed | Closed |
| DC000h | Closed | Closed | Closed | Closed |

| COMPATIBILITY MODE | |
|---|---|
| Setting | JP4 |
| ⇨Enabled | Pins 1 & 2 closed |
| Disabled | Pins 2 & 3 closed |

Note: On some systems the data bus timing is not truly IBM PC/AT compatible. If the card does not initialize or you receive an "Error Reading Address PROM" message, compatibility mode should be enabled.

| CABLE TYPE | |
|---|---|
| Type | JP5 |
| ⇨RG-58A/U 50ohm coaxial | Pins 2 & 3 closed |
| Unshielded twisted pair | Pins 1 & 2 closed |

| DIAGNOSTIC LED(S) | | | |
|---|---|---|---|
| LED | Color | Status | Condition |
| LED1 | Red | Off | data is not being transmitted/received |
| LED1 | Red | On | data is being transmitted/received |
| LED2 | Green | On | 10BaseT link integrity exists |
| LED2 | Green | Off | 10BaseT link integrity broken |

Note: LED2 is only functional when link integrity is enabled, and when installed on a 10BaseT network.

# ADDTRON TECHNOLOGY COMPANY, LTD.
## AE-220ST

**NIC Type**  Ethernet
**Transfer Rate**  10Mbps
**Data Bus**  16-bit ISA
**Topology**  Linear bus
**Wiring Type**  RG58A/U 50ohm coaxial
 DIX transceiver via DB-15

**Boot ROM**  Available

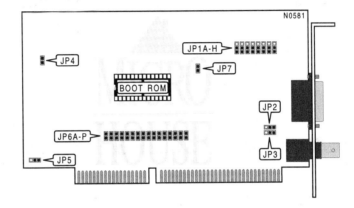

| CABLE TYPE ||
| --- | --- |
| Type | JP1A-H |
| RG58A/U 50ohm coaxial | Pins 1 & 2 closed |
| DIX transceiver via DB-15 port | Pins 2 & 3 closed |

| SEGMENT LENGTH ||
| --- | --- |
| Length | JP2 |
| 185 meters | Open |
| 300 meters | Closed |

| TERMINATOR ||
| --- | --- |
| Setting | JP3 |
| ⇨Unterminated | Open |
| Coaxial linear-bus terminated by card | Closed |

| WAIT STATE CONFIGURATION ||
| --- | --- |
| Bus Speed | JP4 |
| 6MHz - 8MHz | Open |
| ⇨ > 8MHz | Closed |

## ADDTRON TECHNOLOGY COMPANY, LTD.
## AE-220ST

| ROM SIZE | |
|---|---|
| Size | JP5 |
| 16KB | Pins 2 & 3 closed |
| 32KB | Pins 1 & 2 closed |

| INTERRUPT SETTINGS | | | | | | | |
|---|---|---|---|---|---|---|---|
| IRQ | JP6A | JP6B | JP6C | JP6D | JP6E | JP6F | JP6G |
| 2/9 | Closed | Open | Open | Open | Open | Open | Open |
| 3 | Open | Closed | Open | Open | Open | Open | Open |
| 4 | Open | Open | Closed | Open | Open | Open | Open |
| 5 | Open | Open | Open | Closed | Open | Open | Open |
| 10 | Open | Open | Open | Open | Closed | Open | Open |
| 11 | Open | Open | Open | Open | Open | Closed | Open |
| 12 | Open | Open | Open | Open | Open | Open | Closed |

| BASE I/O ADDRESS | | | | |
|---|---|---|---|---|
| Address | JP6H | JP6I | JP6J | JP6K |
| ⇨ 300h | Open | Open | Open | Closed |
| 320h | Open | Open | Closed | Open |
| 340h | Open | Closed | Open | Open |
| 360h | Closed | Open | Open | Open |

| BOOT ROM ADDRESS | | | | | |
|---|---|---|---|---|---|
| Address | JP6L | JP6M | JP6N | JP6O | JP6P |
| C0000h | Closed | Closed | Closed | Closed | Closed |
| C4000h | Closed | Closed | Closed | Closed | Open |
| ⇨ C8000h | Closed | Closed | Closed | Open | Closed |
| CC000h | Closed | Closed | Closed | Open | Open |
| D0000h | Closed | Closed | Open | Closed | Closed |
| D4000h | Closed | Closed | Open | Closed | Open |
| D8000h | Closed | Closed | Open | Open | Closed |
| DC000h | Closed | Closed | Open | Open | Open |
| E0000h | Closed | Open | Closed | Closed | Closed |

| NETWORK VERSION | |
|---|---|
| Version | JP6 |
| Ethernet Version 1 (uses DIX port only) | Open |
| ⇨ Ethernet Version 2 (can use all ports) | Closed |

## ADDTRON TECHNOLOGY COMPANY, LTD.
## AR-200

| | |
|---|---|
| **NIC Type** | Token Ring |
| **Transfer Rate** | 4/16Mbps |
| **Data Bus** | 16-bit ISA |
| **Topology** | Ring |
| **Wiring Type** | AUI transceiver via DB-15 |
| | Shielded/Unshielded twisted pair |
| **Boot ROM** | In firmware |

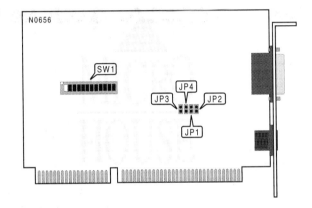

| CABLE TYPE | |
|---|---|
| **Type** | **JP1 - JP4** |
| AUI transceiver via DB-15 | Open |
| Unshielded twisted pair | Closed |

| MULTIPLE CARD CONFIGURATION | |
|---|---|
| **AR-200** | **SW1/9** |
| ⇨ Primary | On |
| Alternate | Off |

Note: If the AR-200 is the only card installed, use the "Primary setting".

| INTERRUPT REQUEST | | |
|---|---|---|
| **IRQ** | **SW1/7** | **SW1/8** |
| ⇨ 2 | Off | Off |
| 3 | Off | On |
| 7 | On | On |

## ADDTRON TECHNOLOGY COMPANY, LTD.
## AR-200

| NETWORK SPEED | |
|---|---|
| Speed | SW1/12 |
| ⇨16Mbps | On |
| 4Mbps | Off |
| Note: All cards on a segment must have this option set the same. | |

| ROM ADDRESS | | | | | | |
|---|---|---|---|---|---|---|
| Address | SW1/1 | SW1/2 | SW1/3 | SW1/4 | SW1/5 | SW1/6 |
| A0000h | Off | On | Off | Off | Off | Off |
| A4000h | Off | On | Off | Off | On | Off |
| A8000h | Off | On | Off | On | Off | Off |
| AC000h | Off | On | Off | On | On | Off |
| B0000h | Off | On | On | Off | Off | Off |
| B4000h | Off | On | On | Off | On | Off |
| B8000h | Off | On | On | On | Off | Off |
| BC000h | Off | On | On | On | On | Off |
| C0000h | On | Off | Off | Off | Off | Off |
| C4000h | On | Off | Off | Off | On | Off |
| C8000h | On | Off | Off | On | Off | Off |
| CC000h | On | Off | Off | On | On | Off |
| D0000h | On | Off | On | Off | Off | Off |
| D4000h | On | Off | On | Off | On | Off |
| D8000h | On | Off | On | On | Off | Off |
| DC000h | On | Off | On | On | On | Off |
| E0000h | On | On | Off | Off | Off | Off |
| E4000h | On | On | Off | Off | On | Off |
| E8000h | On | On | Off | On | Off | Off |
| EC000h | On | On | Off | On | On | Off |
| F0000h | On | On | On | Off | Off | Off |

| SHARED RAM CONFIGURATION | | |
|---|---|---|
| Size | SW1/10 | SW1/11 |
| ⇨ 16KB | On | Off |
| 8KB | Off | Off |
| 32KB | Off | On |
| 64KB | On | On |
| Note: This setting configures how much system memory the card will use. | | |

# ADDTRON TECHNOLOGY COMPANY, LTD.
# ET-200STS

| | |
|---|---|
| **NIC Type** | Ethernet |
| **Transfer Rate** | 10Mbps |
| **Data Bus** | 16-bit ISA |
| **Topology** | Linear Bus |
| **Wiring Type** | RG-58A/U 50ohm coaxial |
| | AUI transceiver via DB-15 port |
| **Boot ROM** | Available |

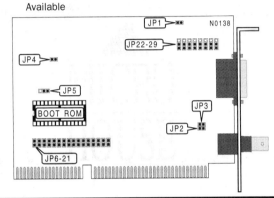

| ETHERNET VERSION ||
|---|---|
| **Version** | **JP1** |
| ⇨Version 2 (IEEE 802.3) | Closed |
| Version 1 | Open |

| SEGMENT LENGTH ||
|---|---|
| **Maximum Length** | **JP2** |
| ⇨185 meters | Closed |
| 300 meters | Open |

Note: Segment length is the total length of cable between the two farthest cards on the segment. Each card on the segment must have this option set the same.

| ONBOARD TERMINATOR ||
|---|---|
| **Setting** | **JP3** |
| ⇨Disabled | Open |
| Enabled | Closed |

Note: If the card is on either end of a linear bus network segment, the onboard terminator may be used instead of using an external terminator.

| DATA BUS SPEED ||
|---|---|
| **Speed** | **JP4** |
| ⇨6-8MHz | Open |
| 8-16MHz | Closed |

## ADDTRON TECHNOLOGY COMPANY, LTD.
## ET-200STS

### BOOT ROM SIZE

| Size | JP5 |
|---|---|
| ⇨8KB | Pins 2 & 3 Closed |
| 16KB | Pins 1 & 2 Closed |

### INTERRUPT REQUEST

| IRQ | JP6 | JP7 | JP8 | JP9 | JP10 | JP11 | JP12 |
|---|---|---|---|---|---|---|---|
| ⇨3 | Open | Closed | Open | Open | Open | Open | Open |
| 2/9 | Closed | Open | Open | Open | Open | Open | Open |
| 4 | Open | Open | Closed | Open | Open | Open | Open |
| 5 | Open | Open | Open | Closed | Open | Open | Open |
| 10 | Open | Open | Open | Open | Closed | Open | Open |
| 11 | Open | Open | Open | Open | Open | Closed | Open |
| 12 | Open | Open | Open | Open | Open | Open | Closed |

### I/O BASE ADDRESS

| Address | JP13 | JP14 | JP15 | JP16 |
|---|---|---|---|---|
| ⇨300h | Open | Open | Open | Closed |
| 320h | Open | Open | Closed | Open |
| 340h | Open | Closed | Open | Open |
| 360h | Closed | Open | Open | Open |

### BOOT ROM

| Setting | JP17 |
|---|---|
| ⇨Disabled | Open |
| Enabled | Closed |

### BASE MEMORY ADDRESS

| Address | JP18 | JP19 | JP20 | JP21 |
|---|---|---|---|---|
| ⇨D0000h | Closed | Open | Closed | Closed |
| C8000h | Closed | Closed | Open | Closed |
| CC000h | Closed | Closed | Open | Open |
| D4000h | Closed | Open | Closed | Open |
| D8000h | Closed | Open | Open | Closed |
| DC000h | Closed | Open | Open | Open |

### CABLE TYPE

| Mode | JP22 - JP29 |
|---|---|
| ⇨RG-58A/U 50ohm coaxial | Pins 2 & 3 Closed |
| AUI transceiver via DB-15 port | Pins 1 & 2 Closed |

## ADVANCED INTERLINK CORPORATION
## AICETHER-16

| | |
|---|---|
| **NIC Type** | Ethernet |
| **Transfer Rate** | 10Mbps |
| **Data Bus** | 16-bit ISA |
| **Topology** | Linear bus |
| **Wiring Type** | RG-58A/U 50ohm coaxial |
| | AUI transceiver via DB-15 port |
| **Boot ROM** | Available |

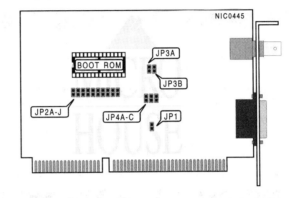

| COMPATIBILITY MODE ||
|---|---|
| **Setting** | **JP1** |
| ⇨Disabled | Open |
| Enabled | Closed |

Note: On some systems the data bus timing is not truly IBM PC/AT compatible. If the card is not initializing, enabling compatibility mode may allow the card to operate properly.

| INTERRUPT REQUEST |||||||||
|---|---|---|---|---|---|---|---|
| **IRQ** | **JP2A** | **JP2B** | **JP2C** | **JP2D** | **JP2E** | **JP2F** | **JP2G** |
| 2/9 | Open | Open | Open | Open | Open | Closed | Open |
| ⇨3 | Open | Open | Open | Open | Closed | Open | Open |
| 5 | Open | Open | Open | Closed | Open | Open | Open |
| 10 | Open | Open | Open | Open | Open | Open | Closed |
| 11 | Closed | Open | Open | Open | Open | Open | Open |
| 12 | Open | Closed | Open | Open | Open | Open | Open |
| 15 | Open | Open | Closed | Open | Open | Open | Open |

## ADVANCED INTERLINK CORPORATION
## AICETHER-16

### I/O BASE ADDRESS

| Address | JP2I | JP2J |
|---|---|---|
| 200h | Closed | Closed |
| 210h | Closed | Open |
| ▷300h | Open | Closed |
| 310h | Open | Open |

### CABLE TYPE

| Type | JP3A | JP3B |
|---|---|---|
| ▷RG-58A/U 50ohm coaxial | Closed | Closed |
| AUI transceiver via DB-15 port | Closed | Closed |

### BOOT ROM ADDRESS

| Address | JP4A | JP4B |
|---|---|---|
| C800h | Closed | Closed |
| CC00h | Closed | Open |
| D800h | Open | Closed |
| DC00h | Open | Open |

### BOOT ROM

| Setting | JP4C |
|---|---|
| ▷Disabled | Open |
| Enabled | Closed |

Note: Location of the boot ROM may vary on different revisions of this card.

### FACTORY CONFIGURED SETTINGS

| Jumper | Setting |
|---|---|
| JP2H | Open |

## ALTA RESEARCH CORPORATION
## ETHERCOMBO-16 T/C (VERSION 1.0)

| | |
|---|---|
| NIC Type | Ethernet |
| Transfer Rate | 10Mbps |
| Data Bus | 8/16-bit ISA |
| Topology | Star |
| | Linear bus |
| Wiring Type | Unshielded twisted pair |
| | RG58A/U 50ohm coaxial |
| Boot ROM | Available |

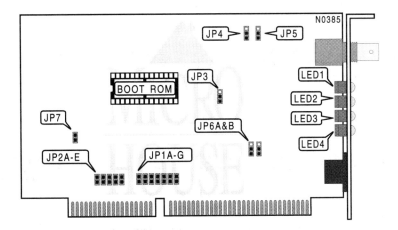

| I/O BASE ADDRESS AND BOOT ROM ADDRESS | | | |
|---|---|---|---|
| I/O Address | Boot ROM Address | JP1A | JP1B |
| ⇨300-31Fh | C8000h | Closed | Closed |
| 320-33Fh | CC000h | Open | Closed |
| 340-35Fh | D0000h | Closed | Open |
| 360-37Fh | D4000h | Open | Open |

| INTERRUPT REQUEST | | | | | | | | | | |
|---|---|---|---|---|---|---|---|---|---|---|
| IRQ | JP1C | JP1D | JP1E | JP1F | JP1G | JP2A | JP2B | JP2C | JP2D | JP2E |
| 2/9 | Closed | Open | Open | Open | Open | Open | Open | Open | Open | Open |
| ⇨3 | Open | Closed | Open | Open | Open | Open | Open | Open | Open | Open |
| 4 | Open | Open | Closed | Open | Open | Open | Open | Open | Open | Open |
| 5 | Open | Open | Open | Closed | Open | Open | Open | Open | Open | Open |
| 7 | Open | Open | Open | Open | Closed | Open | Open | Open | Open | Open |
| 10 | Open | Open | Open | Open | Open | Open | Open | Open | Open | Closed |
| 11 | Open | Open | Open | Open | Open | Open | Open | Open | Closed | Open |
| 12 | Open | Open | Open | Open | Open | Open | Open | Closed | Open | Open |
| 14 | Open | Open | Open | Open | Open | Closed | Open | Open | Open | Open |
| 15 | Open | Open | Open | Open | Open | Open | Closed | Open | Open | Open |

## ALTA RESEARCH CORPORATION
## ETHERCOMBO-16 T/C(VERSION 1.0)

| BOOT ROM | |
|---|---|
| **Setting** | **JP3** |
| ▷Disabled | Pins 2 & 3 closed |
| Enabled | Pins 1 & 2 closed |

| CABLE TYPE | |
|---|---|
| **Type** | **JP4** |
| ▷Unshielded twisted pair | Pins 2 & 3 closed |
| RG58A/U 50ohm coaxial | Pins 1 & 2 closed |

| NETWORK STANDARD | |
|---|---|
| **Standard** | **JP5** |
| ▷802.3 IEEE | Pins 1 & 2 closed |
| 802.2 IEEE | Pins 2 & 3 closed |

Note: The 802.2 IEEE setting allows the card to act as a bridge between an Ethernet PC network and a mainframe Token-ring network.

| POLARITY LED CONFIGURATION | | | |
|---|---|---|---|
| **Setting** | **LED4** | **Condition** | **JP6A & JP6B** |
| ▷Normal | On | Polarity reversed & corrected | Pins 1 & 2 closed |
| | Off | Polarity correct | |
| Reversed | On | Polarity correct | Pins 2 & 3 closed |
| | Off | Polarity reversed & corrected | |

Note: Changing this option to reversed does not reverse the polarity of the twisted pair wires, that is done automatically if a reversed polarity wire is detected. Changing this option to reversed simply changes the meaning of LED4. The indications for LED4 are valid only when the cable type is twisted pair.

| COMPATIBILITY MODE | |
|---|---|
| **Setting** | **JP7** |
| ▷Disabled | Closed |
| Enabled | Open |

Note: If the system in which the card is to be installed is an 80286 using an older C & T or VLSI chipset and the card is not loading the network shell properly, compatibility mode should be enabled (JP7 open).

| DIAGNOSTIC LED(S) | | | |
|---|---|---|---|
| **LED** | **Color** | **Status** | **Condition** |
| LED1 | Green | On | Twisted pair network connection is good |
| LED1 | Green | Off | Twisted pair network connection is broken |
| LED2 | Red | On | Data is being transmitted or received |
| LED2 | Red | Off | Data is not being transmitted or received |
| LED3 | Red | On | Collision detected on network |
| LED3 | Red | Off | No collisions detected on network |

Note: For the function of LED4, please see the Polarity LED Configuration table.

# AMERICAN RESEARCH CORPORATION
## ETHERFLEX PLUS/16B, PCN015

| | |
|---|---|
| **NIC Type** | Ethernet |
| **Transfer Rate** | 10Mbps |
| **Data Bus** | 16-bit ISA |
| **Topology** | Star |
| | Linear Bus |
| **Wiring Type** | Unshielded twisted pair |
| | RG58A/U 50ohm coaxial |
| | DIX transceiver via DB-15 port |
| **Boot ROM** | Available |

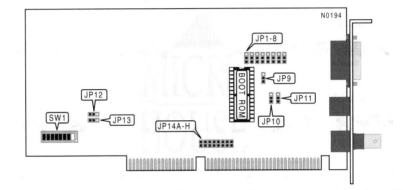

| I/O BASE ADDRESS | | | | |
|---|---|---|---|---|
| Address | SW1/1 | SW1/2 | SW1/3 | SW1/4 |
| 200h | On | On | On | On |
| 220h | On | On | On | Off |
| 240h | On | On | Off | On |
| 260h | On | On | Off | Off |
| ⇨280h | On | Off | On | On |
| 2A0h | On | Off | On | Off |
| 2C0h | On | Off | Off | On |
| 2E0h | On | Off | Off | Off |
| 300h | Off | On | On | On |
| 320h | Off | On | On | Off |
| 340h | Off | On | Off | On |
| 360h | Off | On | Off | Off |
| 380h | Off | Off | On | On |
| 3A0h | Off | Off | On | Off |
| 3C0h | Off | Off | Off | On |
| 3E0h | Off | Off | Off | Off |

# AMERICAN RESEARCH CORPORATION
## ETHERFLEX PLUS/16B, PCN015

| BOOT ROM ADDRESS | | | |
|---|---|---|---|
| Address | SW1/5 | SW1/6 | SW1/7 |
| C0000h | On | On | On |
| C4000h | On | On | Off |
| C8000h | On | Off | On |
| CC000h | On | Off | Off |
| D0000h | Off | On | On |
| D4000h | Off | On | Off |
| D8000h | Off | Off | On |
| ⇨DC000h | Off | Off | Off |

| BOOT ROM | |
|---|---|
| Setting | SW1/8 |
| ⇨Disabled | Off |
| Enabled | On |

| BOOT ROM SIZE | | |
|---|---|---|
| Size | JP1 | JP12 |
| ⇨16KB | Pins 1 & 2 Closed | Pins 1 & 2 Closed |
| 32KB | Pins 2 & 3 Closed | Pins 2 & 3 Closed |

| CABLE TYPE | | | |
|---|---|---|---|
| Type | JP3 - JP8 | JP10 | JP11 |
| Unshielded twisted pair | Pins 2 & 3 Closed | Pins 1 & 2 Closed | Pins 1 & 2 Closed |
| DIX transceiver via DB-15 port | Pins 1 & 2 Closed | Pins 2 & 3 Closed | Pins 2 & 3 Closed |
| RG-58A/U 50ohm coaxial | Pins 1 & 2 Closed | Pins 1 & 2 Closed | Pins 1 & 2 Closed |

| INTERRUPT REQUEST | | | | | | | | |
|---|---|---|---|---|---|---|---|---|
| IRQ | JP14A | JP14B | JP14C | JP14D | JP14E | JP14F | JP14G | JP14H |
| 2/9 | Open | Open | Open | Closed | Open | Open | Open | Open |
| ⇨3 | Open | Open | Open | Open | Open | Open | Open | Closed |
| 4 | Open | Open | Open | Open | Open | Open | Closed | Open |
| 5 | Open | Open | Open | Open | Open | Closed | Open | Open |
| 7 | Open | Open | Open | Open | Closed | Open | Open | Open |
| 10 | Open | Open | Closed | Open | Open | Open | Open | Open |
| 11 | Open | Closed | Open | Open | Open | Open | Open | Open |
| 15 | Closed | Open | Open | Open | Open | Open | Open | Open |

| FACTORY CONFIGURED SETTINGS | |
|---|---|
| Jumper | Setting |
| JP2 | Pins 2 & 3 Closed |
| JP9 | Open |
| JP13 | Pins 1 & 2 Closed |

## ANSEL COMMUNICATIONS, INC.
### NH2100-TA/NH2100-BA

| | |
|---|---|
| **NIC Type** | Ethernet |
| **Transfer Rate** | 10Mbps |
| **Data Bus** | 16-bit ISA |
| **Topology** | Linear bus |
| **Wiring Type** | Unshielded Twisted Pair (NH2100-TA) |
| | RG58A/U 50ohm coaxial (NH2100-BA) |
| | AUI transceiver via DB15 (NH2100-BA, NH2100-TA) |
| **Boot ROM** | Available |

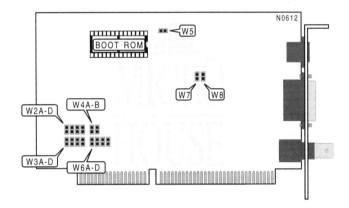

| DMA CHANNEL CONFIGURATION | | | | | | | | |
|---|---|---|---|---|---|---|---|---|
| Channel | W2A | W2B | W2C | W2D | W3A | W3B | W3C | W3D |
| DMA3 | Closed | Open | Open | Open | Closed | Open | Open | Open |
| DMA5 | Open | Closed | Open | Open | Open | Closed | Open | Open |
| DMA6 | Open | Open | Closed | Open | Open | Open | Closed | Open |
| DMA7 | Open | Open | Open | Closed | Open | Open | Open | Closed |

| BASE I/O ADDRESS | | | |
|---|---|---|---|
| Base I/O Address | Boot ROM Address | W4A | W4B |
| ⇨ 300h | C8000h | Closed | Closed |
| 320h | CC000h | Closed | Open |
| 340h | D0000h | Open | Closed |
| 360h | D4000h | Open | Open |

| BOOT ROM CONFIGURATION | |
|---|---|
| Mode | W5 |
| Enabled | Closed |
| Disabled | Open |

## ANSEL COMMUNICATIONS, INC.
### NH2100-TA/NH2100-BA

| CABLE SELECTION | |
|---|---|
| **Mode** | **W8** |
| Unshielded Twisted Pair/AUI transceiver via DB15 | Closed |
| RG58A/U 50ohm coaxial cable | Open |

| MISCELLANEOUS NOTES | |
|---|---|
| Note: | Jumper W7 is reserved for future use. |

# AT-LAN-TEC, INC.
## ALL-IN-ONE ETHERNET NE2000

| | |
|---|---|
| **NIC Type** | Ethernet |
| **Transfer Rate** | 10Mbps |
| **Data Bus** | 16-bit ISA |
| **Topology** | Linear bus |
| | Star |
| **Wiring Type** | RG58A/U 50ohm coaxial |
| | AUI transceiver via DB-15 port |
| | Unshielded twisted pair |
| **Boot ROM** | Available |

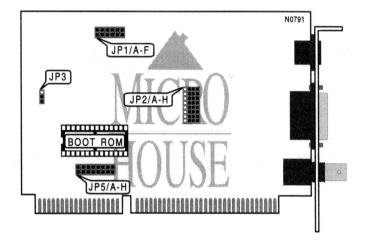

| REMOTE BOOT SELECT ||
|---|---|
| **Setting** | **JP1/A** |
| ⇨ Disabled | Open |
| Enabled | Closed |

# AT-LAN-TEC, INC.
## ALL-IN-ONE ETHERNET NE2000

| ROM ADDRESS SELECT | | | |
|---|---|---|---|
| Address | JP1/B | JP1/C | JP1/D |
| C800h | Open | Closed | Open |
| C000h | Open | Open | Open |
| C400h | Open | Open | Closed |
| CC00h | Open | Closed | Closed |
| D000h | Closed | Open | Open |
| D400h | Closed | Open | Closed |
| D800h | Closed | Closed | Open |
| DC00h | Closed | Closed | Closed |

| I/O ADDRESS SELECT | | |
|---|---|---|
| Address | JP1/E | JP1/F |
| 300h | Open | Open |
| 320h | Open | Closed |
| 340h | Closed | Open |
| 360h | Closed | Closed |

| CABLE TYPE SELECT | | | | | | | | |
|---|---|---|---|---|---|---|---|---|
| Cable Type | JP2/A | JP2/B | JP2/C | JP2/D | JP2/E | JP2/F | JP2/G | JP2/H |
| Twisted Pair | 1 & 2 | 1 & 2 | 1 & 2 | 1 & 2 | 1 & 2 | 1 & 2 | 1 & 2 | 1 & 2 |
| AUI | 2 & 3 | 2 & 3 | 2 & 3 | 2 & 3 | 2 & 3 | 2 & 3 | 2 & 3 | 2 & 3 |
| BNC | 3 & 4 | 3 & 4 | 3 & 4 | 3 & 4 | 3 & 4 | 3 & 4 | 3 & 4 | 3 & 4 |

Note: Pins designated are in the closed position.

| AT BUS SELECT | |
|---|---|
| Setting | JP3 |
| AT Bus Compatibility | Pins 1 & 2 closed |
| Not AT Bus Compatibility | Pins 2 & 3 closed |

| INTERRUPT SELECT | | | | | | | | |
|---|---|---|---|---|---|---|---|---|
| IRQ | JP5/A | JP5/B | JP5/C | JP5/D | JP5/E | JP5/F | JP5/G | JP5/H |
| IRQ2 | Closed | Open | Open | Open | Open | Open | Open | Open |
| IRQ3 | Open | Closed | Open | Open | Open | Open | Open | Open |
| IRQ4 | Open | Open | Closed | Open | Open | Open | Open | Open |
| IRQ5 | Open | Open | Open | Closed | Open | Open | Open | Open |
| IRQ10 | Open | Open | Open | Open | Closed | Open | Open | Open |
| IRQ11 | Open | Open | Open | Open | Open | Closed | Open | Open |
| IRQ12 | Open | Open | Open | Open | Open | Open | Closed | Open |
| IRQ15 | Open | Open | Open | Open | Open | Open | Open | Closed |

## CABLETRON SYSTEMS, INC.
## E2030/-X, E2040/-X (Fiber Optic)

| | |
|---|---|
| **NIC Type** | Ethernet |
| **Transfer Rate** | 10Mbps |
| **Data Bus** | 16-bit ISA |
| **Topology** | Star |
| | Linear Bus |
| **Wiring Type** | 200µ fiber optic cable |
| | AUI transceiver via DB-15 port |
| **Boot ROM** | Not Available |

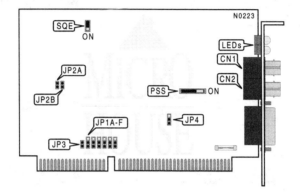

| CONNECTIONS ||
|---|---|
| **Purpose** | **Location** |
| SMA 905 Receive Connector (gray) | CN1 |
| SMA 905 Transmit Connector (black) | CN2 |

| INTERRUPT REQUEST ||||||||
|---|---|---|---|---|---|---|
| **IRQ** | **JP1A** | **JP1B** | **JP1C** | **JP1D** | **JP1E** | **JP1F** |
| 2/9 | Open | Open | Open | Open | Open | Pins 1 & 2 |
| ↻3 | Pins 1 & 2 | Open | Open | Open | Open | Open |
| 4 | Open | Pins 1 & 2 | Open | Open | Open | Open |
| 5 | Open | Open | Pins 1 & 2 | Open | Open | Open |
| 6 | Open | Open | Open | Pins 1 & 2 | Open | Open |
| 7 | Open | Open | Open | Open | Pins 1 & 2 | Open |
| 10 | Open | Pins 2 & 3 | Open | Open | Open | Open |
| 11 | Open | Open | Pins 2 & 3 | Open | Open | Open |
| 12 | Open | Open | Open | Pins 2 & 3 | Open | Open |
| 14 | Open | Open | Open | Open | Pins 2 & 3 | Open |
| 15 | Open | Open | Open | Open | Open | Pins 2 & 3 |
| Note: Pins designated should be in the closed position. |||||||

## CABLETRON SYSTEMS, INC.
## E2030/-X, E2040/-X (Fiber Optic)

| I/O BASE ADDRESS | | |
|---|---|---|
| Address | JP2A | JP2B |
| ⇨220h | Closed | Closed |
| 280h | Closed | Open |
| 300h | Open | Closed |
| 380h | Open | Open |

| COMPATIBILITY MODE | |
|---|---|
| Setting | JP3 |
| ⇨Disabled | Closed |
| Enabled | Open |

Note: On some systems the data bus timing is not truly IBM PC/AT compatible. If the card is not initializing, enabling compatibility mode may allow the card to operate properly.

| CARD SPEED | |
|---|---|
| Setting | JP4 |
| ⇨10MHz | Pins 1 & 2 Closed |
| 20MHz | Pins 2 & 3 Closed |

| CABLE TYPE | |
|---|---|
| Type | Switch PSS |
| ⇨200µ fiber optic cable | On |
| AUI transceiver via DB-15 port | Off |

| SIGNAL QUALITY ERROR (SQE) TEST | |
|---|---|
| Setting | Switch SQE |
| ⇨Disabled | Off |
| Enabled | On |

Note: Signal Quality Error (SQE) is a test of the collision circuitry and path.

| DIAGNOSTIC LED(S) | | | |
|---|---|---|---|
| LED | Color | Status | Condition |
| 1 | Green | On/Blinking | Data is being transmitted |
| 1 | Green | Off | Data is not being transmitted |
| 2 | Red | On/Blinking | Collision detected on network |
| 2 | Red | Off | Normal operation |
| 3 | Yellow | On/Blinking | Data is being received |
| 3 | Yellow | Off | Data is not being received |
| 4 | Green | On | Fiber optic network connection is good |
| 4 | Green | Off | Fiber optic network connection is broken |

## CABLETRON SYSTEMS, INC.
### T8025

| | |
|---|---|
| **NIC Type** | Token Ring |
| **Transfer Rate** | 4/16Mbps |
| **Data Bus** | 16-bit ISA |
| **Topology** | Ring |
| **Wiring Type** | AUI transceiver via DB-15 port |
| | Shielded twisted pair |
| | Unshielded twisted pair |
| **Boot ROM** | Available |

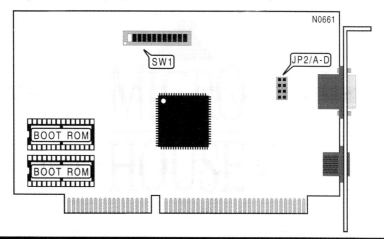

| BASE MEMORY ADDRESS SELECT | | | | | | |
|---|---|---|---|---|---|---|
| Address | SW1/1 | SW1/2 | SW1/3 | SW1/4 | SW1/5 | SW1/6 |
| C0000h - C1FFFh | On | Off | Off | Off | Off | Off |
| C2000h - C3FFFh | On | Off | Off | Off | Off | On |
| C4000h - C5FFFh | On | Off | Off | Off | On | Off |
| C6000h - C7FFFh | On | Off | Off | Off | On | On |
| C8000h - C9FFFh | On | Off | Off | On | Off | Off |
| CA000h - CBFFFh | On | Off | Off | On | Off | On |
| CC000h - CDFFFh | On | Off | Off | On | On | Off |
| CE000h - CFFFFh | On | Off | Off | On | On | On |
| D0000h - CDFFFh | On | Off | On | Off | Off | Off |
| D2000h - D3FFFh | On | Off | On | Off | Off | On |
| D4000h - D5FFFh | On | Off | On | Off | On | Off |
| D6000h - D7FFFh | On | Off | On | Off | On | On |
| D8000h - D9FFFh | On | Off | On | On | Off | Off |
| DA000h - DBFFFh | On | Off | On | On | Off | On |
| DC000h - DDFFFh | On | Off | On | On | On | Off |
| DE000h - DFFFFh | On | Off | On | On | On | On |

## CABLETRON SYSTEMS, INC.
## T8025

| DATA TRANSFER RATE ||
|---|---|
| **Rate** | **SW1/11** |
| 4Mbps | Off |
| 16Mbps | On |

| INTERRUPT SELECT |||
|---|---|---|
| **IRQ** | **SW1/7** | **SW1/8** |
| IRQ2 | Off | Off |
| IRQ3 | Off | On |
| IRQ5 | On | Off |
| IRQ7 | On | On |

| PRIMARY/ALTERNATE ADAPTER ||
|---|---|
| **Setting** | **SW1/12** |
| Primary | Off |
| Alternate | On |

| RAM PAGE SIZE |||
|---|---|---|
| **Size** | **SW1/9** | **SW1/10** |
| 8KB | Off | Off |
| 16KB | Off | On |
| 32KB | On | Off |
| 64KB | On | On |

| CABLE TYPE ||
|---|---|
| **Type** | **JP2/A - D** |
| Shielded twisted pair | Closed |
| Unshielded twisted pair | Open |

# CNET TECHNOLOGY, INC.
## CN120A

| | |
|---|---|
| **NIC Type** | ARCNET |
| **Transfer Rate** | 2.5Mbps |
| **Data Bus** | 8-bit ISA |
| **Topology** | Linear bus |
| **Wiring Type** | RG-62A/U 93ohm coaxial |
| **Boot ROM** | Available |

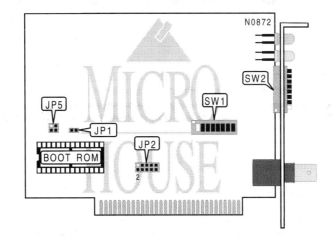

| NODE ADDRESS | | | | | | | | |
|---|---|---|---|---|---|---|---|---|
| Node | SW2/1 | SW2/2 | SW2/3 | SW2/4 | SW2/5 | SW2/6 | SW2/7 | SW2/8 |
| 0 | N/A | N/A | N/A | N/A | N/A | N/A | N/A | N/A |
| 1 | On | Off | Off | Off | Off | Off | Off | Off |
| 2 | Off | On | Off | Off | Off | Off | Off | Off |
| 3 | On | On | Off | Off | Off | Off | Off | Off |
| 4 | Off | Off | On | Off | Off | Off | Off | Off |
| 251 | On | On | Off | On | On | On | On | On |
| 252 | Off | Off | On | On | On | On | On | On |
| 253 | On | Off | On | On | On | On | On | On |
| 254 | Off | On | On | On | On | On | On | On |
| 255 | On | On | On | On | On | On | On | On |

Note: Node address 0 is used for messaging between nodes and must not be used.
A total of 255 node address settings are available. The switches are a binary representation of the decimal node addresses. Switch 1 is the Least Significant Bit and switch 8 is the Most Significant Bit. The switches have the following decimal values: switch 1=1, 2=2, 3=4, 4=8, 5=16, 6=32, 7=64, 8=128. Turn on the switches and add the values of the on switches to obtain the correct node address. (On=1, Off=0)

# CNET TECHNOLOGY, INC.
## CN120A

| BOOT ROM ||
|---|---|
| Setting | JP1 |
| ⇨Disabled | Open |
| Enabled | Closed |

| INTERRUPT SELECTION ||
|---|---|
| IRQ | JP2 |
| 2/9 | Pins 9 & 10 closed |
| 3 | Pins 7 & 8 closed |
| 4 | Pins 5 & 6 closed |
| 5 | Pins 3 & 4 closed |
| 7 | Pins 1 & 2 closed |

| BASE MEMORY ADDRESS & BOOT ROM ADDRESS ||||||||
|---|---|---|---|---|---|---|
| Base Address | Boot ROM Address | SW1/1 | SW1/2 | SW1/3 | SW1/4 | SW1/5 |
| C0000h | C2000h | On | On | On | On | On |
| C4000h | C6000h | On | On | Off | On | On |
| CC000h | CE000h | On | On | On | Off | On |
| ⇨D0000h | D2000h | On | On | Off | Off | On |
| D4000h | D6000h | On | On | On | On | Off |
| D8000h | DA000h | On | On | Off | On | Off |
| DC000h | DE000h | On | On | On | Off | On |
| E0000h | E2000h | On | On | Off | Off | Off |

| I/O BASE ADDRESS ||||
|---|---|---|---|
| Address | SW1/6 | SW1/7 | SW1/8 |
| 260h | On | On | On |
| 290h | Off | On | On |
| ⇨2E0h | On | Off | On |
| 2F0h | Off | Off | On |
| 300h | On | On | Off |
| 350h | Off | On | Off |
| 380h | On | Off | Off |
| 3E0h | Off | Off | Off |

| DIAGNOSTIC LED(S) ||||
|---|---|---|---|
| LED | Color | Status | Condition |
| LED1 | Red | Off | Power is not on |
| LED1 | Red | Blinking | Network error |
| LED1 | Red | On | Power is on |
| LED2 | Green | Off | Data is not being transmitted/received |
| LED2 | Green | On | Data is being transmitted/received |

# CNET TECHNOLOGY, INC.
## CN2000T REV. A

| | |
|---|---|
| **NIC Type** | Token-Ring |
| **Transfer Rate** | 4/16Mbps |
| **Data Bus** | 16-bit ISA |
| **Topology** | Ring |
| **Wiring Type** | Shielded twisted pair (AUI port) |
| | Unshielded twisted pair (RJ-45 jack) |
| **Boot ROM** | Available |

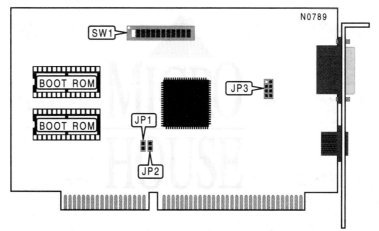

| BASE MEMORY ADDRESS SELECT ||||||| 
| ROM Address | SW1/1 | SW1/2 | SW1/3 | SW1/4 | SW1/5 | SW1/6 |
|---|---|---|---|---|---|---|
| ⇨ C0000H - C1FFFh | On | Off | Off | Off | Off | Off |
| C2000H - C3FFFh | On | Off | Off | Off | Off | On |
| C4000H - C5FFFh | On | Off | Off | Off | On | Off |
| C6000H - C7FFFh | On | Off | Off | Off | On | On |
| C8000H - C9FFFh | On | Off | Off | On | Off | Off |
| CA000H - CBFFFh | On | Off | Off | On | Off | On |
| CC000H - CDFFFh | On | Off | Off | On | On | Off |
| CE000H - CFFFFh | On | Off | Off | On | On | On |
| D0000H - D1FFFh | On | Off | On | Off | Off | Off |
| D2000H - D3FFFh | On | Off | On | Off | Off | On |
| D4000H - D5FFFh | On | Off | On | Off | On | Off |
| D6000H - D7FFFh | On | Off | On | Off | On | On |
| D8000H - D9FFFh | On | Off | On | On | Off | Off |
| DA000H - DBFFFh | On | Off | On | On | Off | On |
| DC000H - CDFFFh | On | Off | On | On | On | Off |
| DE000H - DFFFFh | On | Off | On | On | On | On |

# CNET TECHNOLOGY, INC.
## CN2000T REV. A

| PAGE RAM CONFIGURATION | | |
|---|---|---|
| Size | SW1/9 | SW1/10 |
| ⇨ 16KB | Off | On |
| 8KB | Off | Off |
| 32KB | On | Off |
| 64KB | On | On |

| CABLE TYPE | |
|---|---|
| Type | JP3 |
| ⇨ Shielded twisted pair (AUI port) | Open |
| Unshielded twisted pair (RJ-45 jack) | Closed |

| NETWORK SPEED | |
|---|---|
| Speed | SW1/11 |
| ⇨ 16Mbps | On |
| 4Mbps | Off |

| I/O ADDRESS SELECT | |
|---|---|
| Address | SW1/12 |
| A20H - A23H (RAM address - D8000) | On |
| A24H - A27H (RAM address - D4000) | Off |

| INTERRUPT SELECT | | |
|---|---|---|
| IRQ | SW1/7 | SW1/8 |
| ⇨ IRQ2/9 | Closed | Open |
| IRQ3 | Open | Closed |
| IRQ6 | Open | Open |
| IRQ7 | Open | Open |

| BIT SIZE SELECT | | |
|---|---|---|
| Size | JP1 | JP2 |
| ⇨ 8-bits | Open | Open |
| 16-bits | Closed | Closed |

# CNET TECHNOLOGY, INC.
## CN700E

| | |
|---|---|
| NIC Type | Ethernet |
| Transfer Rate | 10Mbps |
| Data Bus | 8-bit ISA |
| Topology | Star |
| Wiring Type | Unshielded twisted pair |
| | AUI transceiver via DB-15 port |
| Boot ROM | Available |

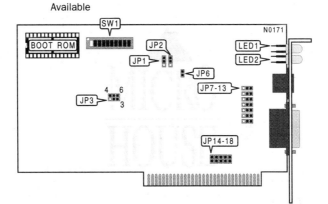

| FACTORY CONFIGURED SETTINGS ||
|---|---|
| Jumper | Setting |
| JP1 | Pins 1 & 2 Closed |
| JP2 | Pins 1 & 2 Closed |

| WAIT STATE ||
|---|---|
| Setting | JP3 |
| ▷Normal PC wait state (0 Wait States) | Pins 2 & 3 Closed |
| 0 wait states | Pins 1 & 4 closed |
| 1 wait states | Pins 2 & 5 closed |
| 2 wait states | Pins 3 & 6 closed |
| Note: Normal wait state is recommended in most PC installations. If performance is sluggish, test other wait state settings until the best selection is found for the system being used. ||

| LINK INTEGRITY TEST ||
|---|---|
| Setting | JP31 |
| ▷Enabled | Pins 1 & 2 Closed |
| Disabled | Pins 2 & 3 Closed |
| Note: Link Integrity is only used with 10Base-T wiring. When using an AUI transceiver, the test should be disabled. ||

| CABLE TYPE ||
|---|---|
| Type | JP7 - JP13 |
| ▷Unshielded twisted pair | Pins 2-3 Closed |
| AUI transceiver via DB-15 port | Pins 1-2 Closed |

## CNET TECHNOLOGY, INC.
## CN700E

### INTERRUPT REQUEST

| IRQ | JP14 | JP15 | JP16 | JP17 | JP18 |
|---|---|---|---|---|---|
| 2/9 | Open | Open | Open | Open | Closed |
| ⇨3 | Open | Open | Open | Closed | Open |
| 4 | Open | Open | Closed | Open | Open |
| 5 | Open | Closed | Open | Open | Open |
| 7 | Closed | Open | Open | Open | Open |

### BOOT ROM

| Setting | SW1/1 |
|---|---|
| ⇨Disabled | Off |
| Enabled | On |

### BOOT ROM ADDRESS

| Address | SW1/2 | SW1/3 | SW1/4 | SW1/5 | SW1/6 |
|---|---|---|---|---|---|
| C0000h | Off | On | On | On | On |
| C4000h | Off | On | On | On | Off |
| C8000h | Off | On | On | Off | On |
| CC000h | Off | On | On | Off | Off |
| D0000h | Off | On | Off | On | On |
| D4000h | Off | On | Off | On | Off |
| ⇨D8000h | Off | On | Off | Off | On |
| DC000h | Off | On | Off | Off | Off |

### I/O BASE ADDRESS

| Address | SW1/7 | SW1/8 | SW1/9 | SW1/10 |
|---|---|---|---|---|
| 200h | On | On | On | On |
| 220h | On | On | On | Off |
| 240h | On | On | Off | On |
| ⇨280h | On | Off | On | On |
| 2A0h | On | Off | On | Off |
| 2C0h | On | Off | Off | On |
| 300h | Off | On | On | On |
| 320h | Off | On | On | Off |
| 340h | Off | On | Off | On |

### DIAGNOSTIC LED(S)

| LED | Status | Condition |
|---|---|---|
| LED1 | On | Twisted pair network connection is good |
| LED1 | Off | Twisted pair network connection is broken |
| LED2 | On | Data is being transmitted or received |
| LED2 | Off | Data is not being transmitted or received |

## COGENT DATA TECHNOLOGIES, INC.
# EMASTER+ EM727 ATS

| | |
|---|---|
| **NIC Type** | Ethernet |
| **Transfer Rate** | 10Mbps |
| **Data Bus** | 16-bit ISA |
| **Topology** | Star |
| | Linear bus |
| **Wiring Type** | AUI transceiver via DB-15 |
| | RG-58A/U 50ohm coaxial |
| | Unshielded twisted pair |
| **Boot ROM** | Available |

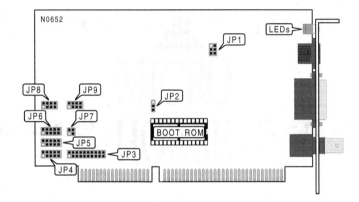

| BOOT ROM SIZE | | |
|---|---|---|
| Size | JP7/1 | JP7/2 |
| ⇨ 8KB | Closed | Open |
| 16KB | Open | Closed |

| CABLE TYPE | |
|---|---|
| Type | JP1 |
| ⇨ RG58A/U 50ohm coaxial | Pins 5 & 6 closed |
| AUI transceiver via DB-15 port | Pins 3 & 4 closed |
| Unshielded twisted pair | Pins 1 & 2 closed |

## COGENT DATA TECHNOLOGIES, INC.
### EMASTER+ EM727 ATS

| Channel | JP4/1 | JP4/2 | JP4/3 | JP4/4 | JP4/5 | JP5/1 | JP5/2 | JP5/3 | JP5/4 | JP5/5 |
|---|---|---|---|---|---|---|---|---|---|---|
| ▷ DMA5 | Open | Open | Closed | Open | Open | Open | Open | Closed | Open | Open |
| DMA0 | Closed | Open | Open | Open | Open | Closed | Open | Open | Open | Open |
| DMA3 | Open | Closed | Open | Open | Open | Open | Closed | Open | Open | Open |
| DMA6 | Open | Open | Open | Closed | Open | Open | Open | Open | Closed | Open |
| DMA7 | Open | Open | Open | Open | Closed | Open | Open | Open | Open | Closed |

Note: Each EM727 installed in the server must have a unique DMA setting.

### I/O BASE ADDRESS

| Address | JP9/1 | JP9/2 | JP9/3 | JP9/4 |
|---|---|---|---|---|
| ▷ 300h | Closed | Open | Open | Open |
| 320h | Open | Closed | Open | Open |
| 340h | Open | Open | Closed | Open |
| 360h | Open | Open | Open | Closed |

### INTERRUPT REQUEST

| IRQ | JP3/1 | JP3/2 | JP3/3 | JP3/4 | JP3/5 | JP3/6 | JP3/7 | JP3/8 | JP3/9 | JP3/10 |
|---|---|---|---|---|---|---|---|---|---|---|
| 3 | Closed | Open | Open | Open | Open | Open | Open | Open | Open | Closed |
| 4 | Open | Closed | Open | Open | Open | Open | Open | Open | Open | Open |
| 5 | Open | Open | Closed | Open | Open | Open | Open | Open | Open | Open |
| 7 | Open | Open | Open | Closed | Open | Open | Open | Open | Open | Open |
| 9 | Open | Open | Open | Open | Closed | Open | Open | Open | Open | Open |
| ▷ 10 | Open | Open | Open | Open | Open | Closed | Open | Open | Open | Open |
| 11 | Open | Open | Open | Open | Open | Open | Closed | Open | Open | Open |
| 12 | Open | Open | Open | Open | Open | Open | Open | Closed | Open | Open |
| 14 | Open | Open | Open | Open | Open | Open | Open | Open | Closed | Open |
| 15 | Open | Open | Open | Open | Open | Open | Open | Open | Open | Closed |

### LAN ID SELECT

| LAN | JP8/1 | JP8/2 | JP8/3 | JP8/4 |
|---|---|---|---|---|
| ▷A | Closed | Open | Open | Open |
| B | Open | Closed | Open | Open |
| C | Open | Open | Closed | Open |
| D | Open | Open | Open | Closed |

Note: Select a different card LAN for each EM727 installed. I/O base 360h with LAN D and I/O base 320h with LAN D are not supported.

## COGENT DATA TECHNOLOGIES, INC.
# EMASTER+ EM727 ATS

| Address | JP6/1 | JP6/2 | JP6/3 | JP6/4 | JP6/5 |
|---|---|---|---|---|---|
| **PROM BASE ADDRESS** | | | | | |
| ⇨ Disabled | Open | Open | Open | Open | Closed |
| C000h | Closed | Open | Open | Open | Open |
| D000h | Open | Closed | Open | Open | Open |
| D800h | Open | Open | Closed | Open | Open |
| DC00h | Open | Open | Open | Closed | Open |

| Setting | JP2 |
|---|---|
| **SA 19-17 LINES** | |
| ⇨ SA 19-17 lines enabled | Pins 1 & 2 closed |
| SA 19-17 lines disabled | Pins 2 & 3 closed |

| LED | Color | Status | Condition |
|---|---|---|---|
| **DIAGNOSTIC LED(s)** | | | |
| LED1 | Green | On | 10BaseT link integrity exists |
| LED1 | Green | Off | 10BaseT link integrity broken |
| LED2 | Yellow | Off | Data is not being transmitted/received |
| LED2 | Yellow | Blinking | Data is being transmitted/received |

## COMPEX, INC.
## ANET16-MC

| | |
|---|---|
| **NIC Type** | ARCNET |
| **Transfer Rate** | 2.5Mbps |
| **Data Bus** | 16-bit MCA |
| **Topology** | Linear Bus/Star |
| **Wiring Type** | Unshielded twisted pair |
| | RG-58A/U 50ohm coaxial |
| **Boot ROM** | Available |

| CABLE TYPE SELECTION | | |
|---|---|---|
| Cable Type | J3 | J4 |
| ⇨ RG-58A/U 50ohm coaxial | pins 2 & 3 closed | pins 2 & 3 closed |
| Unshielded twisted pair | pins 1 & 2 closed | pins 1 & 2 closed |

| MODE CONFIGURATION | |
|---|---|
| Type | J1 |
| ⇨ TRXnet | Closed |
| COMPEX | Open |

| BUS TYPE SELECTION | |
|---|---|
| Select | J2 |
| ⇨ Star | Closed |
| Linear Bus | Open |

| DIAGNOSTIC LED(S) | | | |
|---|---|---|---|
| LED | Color | Status | Condition |
| LED1 | Red | On | Network connection is good |
| LED1 | Red | Off | Network connection is broken |
| LED2 | Green | On | Data is being transmitted/received |
| LED2 | Green | Off | Data is not being transmitted/received |

## D-LINK
### DT-220

| | |
|---|---|
| **NIC Type** | Token-Ring |
| **Transfer Rate** | 4/16Mbps |
| **Data Bus** | 16-bit ISA |
| **Topology** | Ring |
| **Wiring Type** | Unshielded twisted pair |
| | Shielded twisted pair (DB-9 port) |
| **Boot ROM** | Available (Built-in Remote Program Load) |

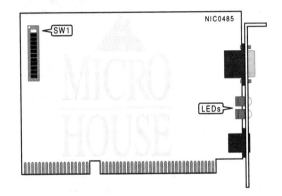

| BOOT ROM ADDRESS | | | | | | |
|---|---|---|---|---|---|---|
| **Address** | **SW1/1** | **SW1/2** | **SW1/3** | **SW1/4** | **SW1/5** | **SW1/6** |
| C0000h | Off | On | On | On | On | On |
| C2000h | Off | On | On | On | On | Off |
| C4000h | Off | On | On | On | Off | On |
| C6000h | Off | On | On | On | Off | Off |
| C8000h | Off | On | On | Off | On | On |
| CA000h | Off | On | On | Off | On | Off |
| CC000h | Off | On | On | Off | Off | On |
| CE000h | Off | On | On | Off | Off | Off |
| D0000h | Off | On | Off | On | On | On |
| D2000h | Off | On | Off | On | On | Off |
| D4000h | Off | On | Off | On | Off | On |
| D6000h | Off | On | Off | On | Off | Off |
| D8000h | Off | On | Off | Off | On | On |
| DA000h | Off | On | Off | Off | On | Off |
| DC000h | Off | On | Off | Off | Off | On |
| DE000h | Off | On | Off | Off | Off | Off |

## D-LINK
## DT-220

### INTERRUPT REQUEST

| IRQ | SW1/7 | SW1/8 |
|---|---|---|
| 2/9 | On | On |
| ⇨3 | On | Off |
| 7 | Off | Off |

### PRIMARY/SECONDARY CARD SELECT

| Setting | SW1/9 |
|---|---|
| ⇨Primary | Off |
| Secondary | On |

Note: If two NICs are installed in a system, one must be set to primary, and the other to secondary.

### SHARED MEMORY SIZE

| Size | SW1/10 | SW1/11 |
|---|---|---|
| 8KB | On | On |
| ⇨16KB | Off | On |
| 32KB | On | Off |
| 64KB | Off | Off |

Note: The 16KB option is the only setting that supports memory paging.

### NETWORK SEGMENT SPEED

| Speed | SW1/12 |
|---|---|
| 16Mbps | Off |
| 4Mbps | On |

Note: All NICs on the network segment must be set to the same speed.

### DIAGNOSTIC LED(S)

| Amber LED | Green LED | Condition |
|---|---|---|
| On | Off | Network speed is set to 4Mbps |
| Off | On | Network speed is set to 16Mbps |

Note: The amber LED will light momentarily when power is applied to the NIC. The appropriate LED will then light to indicate the network speed at which the card is set.

# D-LINK
## DE-500

| | |
|---|---|
| **NIC Type** | Ethernet |
| **Transfer Rate** | 10Mbps |
| **Data Bus** | VESA local bus |
| **Topology** | Linear Bus |
| | Star |
| **Wiring Type** | AUI Transceiver via DB-15 port |
| | RG-58A/U 50ohm coaxial |
| | Unshielded twisted pair |
| **Boot ROM** | Available |

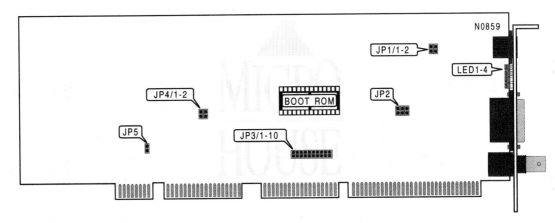

| CABLE TYPE SELECTION | | |
|---|---|---|
| **Cable Type** | **JP1/1** | **JP1/2** |
| ⇨ Unshielded twisted pair | Closed | Open |
| AUI Transceiver via DB-15 port | Open | Closed |
| RG-58A/U 50ohm coaxial | Closed | Closed |

| BASE I/O ADDRESS SELECTION | | |
|---|---|---|
| **Address** | **JP4/1** | **JP4/2** |
| ⇨ 300-31F | Closed | Closed |
| 320-33F | Closed | Open |
| 340-35F | Open | Closed |
| 360-37F | Open | Open |

# D-LINK
## DE-500

| INTERRUPT SELECTION |||||||||||
|---|---|---|---|---|---|---|---|---|---|---|
| IRQ | JP3/1 | JP3/2 | JP3/3 | JP3/4 | JP3/5 | JP3/6 | JP3/7 | JP3/8 | JP3/9 | JP3/10 |
| ⇨ 5 | Open | Open | Closed | Open | Open | Open | Open | Open | Open | Open |
| 3 | Closed | Open | Open | Open | Open | Open | Open | Open | Open | Open |
| 4 | Open | Closed | Open | Open | Open | Open | Open | Open | Open | Open |
| 6 | Open | Open | Open | Closed | Open | Open | Open | Open | Open | Open |
| 7 | Open | Open | Open | Open | Closed | Open | Open | Open | Open | Open |
| 9 | Open | Open | Open | Open | Open | Closed | Open | Open | Open | Open |
| 11 | Open | Open | Open | Open | Open | Open | Closed | Open | Open | Open |
| 12 | Open | Open | Open | Open | Open | Open | Open | Closed | Open | Open |
| 14 | Open | Open | Open | Open | Open | Open | Open | Open | Closed | Open |
| 15 | Open | Open | Open | Open | Open | Open | Open | Open | Open | Closed |

| BUS SPEED ||
|---|---|
| Size | JP5 |
| ⇨ 33Mhz | Closed |
| 50Mhz | Open |

| BOOT ROM CONFIGURATION ||
|---|---|
| Setting | JP2/1 |
| ⇨ Disable | Open |
| Enable | Closed |

| ROM ADDRESS |||
|---|---|---|
| Address | JP2/2 | JP2/3 |
| ⇨ D000:0-D000:1FFF | Closed | Open |
| C000:0-C000:1FFF | Open | Open |
| C800:0-C800:1FFF | Open | Closed |
| D800:0-D800:1FFF | Closed | Closed |

| DIAGNOSTIC LED(S) ||||
|---|---|---|---|
| LED | Color | Status | Condition |
| LED1 | Green | On | Power on |
| LED1 | Green | Blinking | Data is being transmitted |
| LED1 | Green | Off | Power off |
| LED2 | Green | On | Network connection is good (UTP only) |
| LED2 | Green | Blinking | Data is being received |
| LED2 | Green | Off | Network connection is broken (UTP only) |
| LED3 | Yellow | On | Network connection is broken (BNC only) |
| LED3 | Yellow | Blinking | Collision detected |
| LED3 | Yellow | Off | Normal operation |
| LED4 | Yellow | On | Card is transmitting for an extended time (jabber) |
| LED4 | Yellow | Off | Normal operation |

## DANPEX CORPORATION
## EN-2300B/EN-2300BT/EN-2300T

| | |
|---|---|
| NIC Type | Ethernet |
| Transfer Rate | 10Mbps |
| Data Bus | 16-bit ISA |
| Topology | Linear Bus, Star |
| Wiring Type | AUI transceiver via DB-15 port |
| | RG58A/U 50ohm coaxial |
| | Unshielded twisted pair |
| Boot ROM | Available |

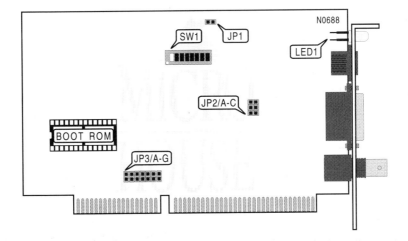

| CABLE TYPE | | | | |
|---|---|---|---|---|
| Card Type | Cable Type | JP2/A | JP2/B | JP2/C |
| EN-2300B | RG58A/U 50ohm coaxial | Closed | Open | N/A |
| EN-2300B | AUI transceiver via DB-15 port | Open | Closed | N/A |
| EN-2300BT | RG58A/U 50ohm coaxial | Closed | Open | Open |
| EN-2300BT | AUI transceiver via DB-15 port | Open | Closed | Open |
| EN-2300BT | Unshielded twisted pair | Open | Open | Closed |
| EN-2300T | Unshielded twisted pair | N/A | N/A | Closed |

| BASE I/O ADDRESS/INTERRUPT SELECT | | | | | | | | | | |
|---|---|---|---|---|---|---|---|---|---|---|
| Address | IRQ | SW1/5 | SW1/6 | JP3/A | JP3/B | JP3/C | JP3/D | JP3/E | JP3/F | JP3/G |
| 300h | IRQ3 | Off | Off | Open | Closed | Open | Open | Open | Open | Open |
| 320h | IRQ2 | On | Off | Closed | Open | Open | Open | Open | Open | Open |
| 340h | IRQ4 | Off | On | Open | Open | Closed | Open | Open | Open | Open |
| 360h | IRQ5 | On | On | Open | Open | Open | Closed | Open | Open | Open |

# DANPEX CORPORATION
# EN-2300B/2300BT/2300T

## BASE I/O ADDRESS/INTERRUPT SELECT (CONTINUED)

| Address | IRQ | SW1/5 | SW1/6 | JP3/A | JP3/B | JP3/C | JP3/D | JP3/E | JP3/F | JP3/G |
|---------|------|-------|-------|--------|--------|--------|--------|--------|--------|--------|
| 300h | IRQ2 | Off | Off | Closed | Open | Open | Open | Open | Open | Open |
| 320h | IRQ3 | On | Off | Open | Closed | Open | Open | Open | Open | Open |
| 340h | IRQ5 | Off | On | Open | Open | Open | Closed | Open | Open | Open |
| 360h | IRQ4 | On | On | Open | Open | Closed | Open | Open | Open | Open |
| 300h | IRQ4 | Off | Off | Open | Open | Closed | Open | Open | Open | Open |
| 320h | IRQ5 | On | Off | Open | Open | Open | Closed | Open | Open | Open |
| 340h | IRQ2 | Off | On | Closed | Open | Open | Open | Open | Open | Open |
| 360h | IRQ3 | On | On | Open | Closed | Open | Open | Open | Open | Open |
| 300h | IRQ10 | Off | Off | Open | Open | Open | Open | Closed | Open | Open |
| 320h | IRQ12 | On | Off | Open | Open | Open | Open | Open | Closed | Open |
| 340h | IRQ15 | Off | On | Open | Open | Open | Open | Open | Open | Closed |
| 360h | IRQ10 | On | On | Open | Open | Open | Open | Closed | Open | Open |
| 300h | IRQ12 | Off | Off | Open | Open | Open | Open | Open | Closed | Open |
| 320h | IRQ15 | On | Off | Open | Open | Open | Open | Open | Open | Closed |
| 340h | IRQ10 | Off | On | Open | Open | Open | Open | Closed | Open | Open |
| 360h | IRQ12 | On | On | Open | Open | Open | Open | Open | Closed | Open |
| 300h | IRQ15 | Off | Off | Open | Open | Open | Open | Open | Open | Closed |
| 320h | IRQ10 | On | Off | Open | Open | Open | Open | Closed | Open | Open |
| 340h | IRQ12 | Off | On | Open | Open | Open | Open | Open | Closed | Open |
| 360h | IRQ15 | On | On | Open | Open | Open | Open | Open | Open | Closed |

## ROM ADDRESS CONFIGURATION

| Memory Address | I/O Address | IRQ | SW1/2 | SW1/3 | SW1/4 |
|----------------|-------------|------|-------|-------|-------|
| ⇨ C800h | 300h | IRQ3 | Off | On | Off |
| CC00h | 320h | IRQ2 | On | On | Off |
| D000h | 340h | IRQ4 | Off | Off | On |
| D400h | 360h | IRQ5 | On | Off | On |

## ROM CONFIGURATION

| Setting | SW1/1 |
|---------|-------|
| Boot ROM enabled | On |
| Boot ROM disabled | Off |

## SYSTEM TYPE SELECT

| Type | SW1/7 |
|------|-------|
| ⇨ Other | Off |
| IBM PC Model 30-286, Compaq 286, C & T chipset systems | On |

# DANPEX CORPORATION
## EN-2300B/2300BT/2300T

### CONNECTIONS

| Purpose | Location |
|---|---|
| JP1 | LAN station LED connector |

### FACTORY CONFIGURED SETTINGS

| Switch | Setting |
|---|---|
| SW1/8 | N/A |

### DIAGNOSTIC LED

| LED | Color | Status | Condition |
|---|---|---|---|
| LED1 | Blue | Off | Data is not being transmitted/received |
| LED1 | Blue | Blinking | Data is being transmitted/received |

## DANPEX CORPORATION
## EN-9400P3

| | |
|---|---|
| **NIC Type** | Ethernet |
| **Transfer Rate** | 10Mbps |
| **Data Bus** | 32-bit PCI |
| **Topology** | Linear bus |
| | Star |
| **Wiring Type** | AUI Transceiver via DB-15 port |
| | RG-58A/U 50ohm coaxial |
| | Unshielded twisted pair |
| **Boot ROM** | Available |

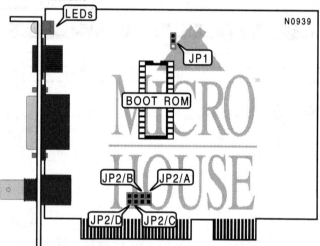

| CABLE TYPE SELECTION | |
|---|---|
| Type | JP1 |
| ⇨ RG-58A/U 50ohm coaxial | Pins 1 & 2 closed |
| AUI Transceiver via DB-15 port | Pins 2 & 3 closed |

| PCI INTERRUPT SELECTION | | | | |
|---|---|---|---|---|
| PCI INT | JP2/A | JP2/B | JP2/C | JP2/D |
| ⇨ INTA | Closed | Open | Open | Open |
| INTB | Open | Closed | Open | Open |
| INTC | Open | Open | Closed | Open |
| INTD | Open | Open | Open | Closed |

## DANPEX CORPORATION
## EN-9400P3

| DIAGNOSTIC LED(S) | | |
|---|---|---|
| LED | Status | Condition |
| 1 | On | Data is being transmitted/received |
| 1 | Off | Data is not being transmitted/received |
| 2 | On | Network connection is good |
| 2 | Off | Network connection is broken |

# DANPEX CORPORATION
## TR-4016

| | |
|---|---|
| **NIC Type** | Token-Ring |
| **Transfer Rate** | 4/16Mbps |
| **Data Bus** | 16-bit ISA |
| **Topology** | Ring |
| **Wiring Type** | Unshielded twisted pair |
| | Shielded twisted pair (DB-9 port) |
| **Boot ROM** | Available (Built-in Remote Program Load) |

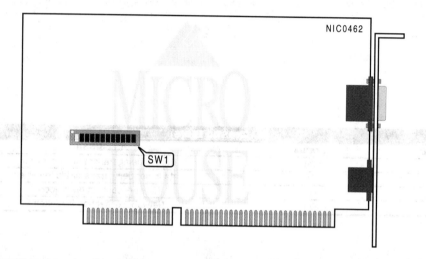

| CABLE TYPE |
|---|
| Note: The cable type is automatically detected when a cable is connected. Only one connector can be used at a time. |

| INTERRUPT REQUEST | | |
|---|---|---|
| IRQ | SW1/7 | SW1/8 |
| ⇨2 | Off | Off |
| 3 | Off | On |
| 7 | On | On |

| PRIMARY/ALTERNATE CARD | |
|---|---|
| Setting | SW1/9 |
| ⇨Primary | Off |
| Alternate | On |
| Note: If two cards are to be used in a single system, one of the cards must be set to primary and the other one must be set to alternate. | |

# DANPEX CORPORATION
## TR-4016

| BOOT ROM | |
|---|---|
| Setting | SW1/1 |
| ⇨Disabled | On |
| Enabled | Off |

| BOOT ROM ADDRESS | | | | | |
|---|---|---|---|---|---|
| Address | SW1/2 | SW1/3 | SW1/4 | SW1/5 | SW1/6 |
| A0000 - A1FFFh | Off | Off | Off | On | Off |
| A4000 - A5FFFh | On | Off | Off | On | Off |
| A8000 - A9FFFh | Off | On | Off | On | Off |
| AC000 - ADFFFh | On | On | Off | On | Off |
| B0000 - B1FFFh | Off | Off | Off | On | Off |
| B4000 - B5FFFh | On | Off | Off | On | Off |
| B8000 - B9FFFh | Off | On | Off | On | Off |
| BC000 - BDFFFh | On | On | Off | On | Off |
| C0000 - C1FFFh | Off | Off | On | Off | On |
| C4000 - C5FFFh | On | Off | On | Off | On |
| C8000 - C9FFFh | Off | On | On | Off | On |
| ⇨CC000 - CDFFFh | On | On | On | Off | On |
| D0000 - D1FFFh | Off | Off | On | Off | On |
| D4000 - D5FFFh | On | Off | On | Off | On |
| D8000 - D9FFFh | Off | On | On | Off | On |
| DC000 - DDFFFh | On | On | On | Off | On |
| E0000 - E1FFFh | Off | Off | On | On | On |
| E4000 - E5FFFh | On | Off | On | On | On |
| E8000 - E9FFFh | Off | On | On | On | On |
| EC000 - EDFFFh | On | On | On | On | On |
| F0000 - F1FFFh | Off | Off | On | On | On |

| SHARED RAM SIZE | | |
|---|---|---|
| Size | SW1/10 | SW1/11 |
| 8KB | On | On |
| ⇨16KB | Off | On |
| 32KB | On | Off |
| 64KB | Off | Off |

| NETWORK SEGMENT SPEED | |
|---|---|
| Speed | SW1/12 |
| ⇨16Mbps | On |
| 4Mbps | Off |

Note: All cards on the network segment must have this option set the same.

## DATAPOINT CORPORATION
## ARCNETPLUS HUB

**Card Type**  ARCNET hub
**Network Transfer Rate**  2.5Mbps/20Mbps
**Data Bus**  16-bit ISA
**Topology**  Star
**Wire Type**  RG-62A/U 93ohm coaxial
**Boot ROM**  Not available

| CONNECTIONS | | | |
|---|---|---|---|
| Function | Label | Function | Label |
| RG-62A/U 93ohm coaxial port 1 | CN1 | RG-62A/U 93ohm coaxial port 4 | CN4 |
| RG-62A/U 93ohm coaxial port 2 | CN2 | Internal Datapoint proprietary coaxial | CN5 |
| RG-62A/U 93ohm coaxial port 3 | CN3 | | |
| Note: CN5 is used to connect the hub card to an ARCNET, ARCNET PLUS, ARCNET HUB, or another ARCNETPLUS card installed in the same system. | | | |

## DATAPOINT CORPORATION
## ARCNETPLUS HUB

| \multicolumn{4}{c}{DIAGNOSTIC LED(S)} |||| 
|---|---|---|---|
| **LED** | **Color** | **Status** | **Condition** |
| LED1 | Green | On | Port 1 is transmitting/receiving at 2.5Mbps |
| LED1 | Green | Off | Port 1 is not transmitting/receiving at 2.5Mbps |
| LED2 | Yellow | On | Port 1 is transmitting/receiving at 20Mbps |
| LED2 | Yellow | Off | Port 1 is not transmitting/receiving at 20Mbps |
| LED3 | Green | On | Port 2 is transmitting/receiving at 2.5Mbps |
| LED3 | Green | Off | Port 2 is not transmitting/receiving at 2.5Mbps |
| LED4 | Yellow | On | Port 2 is transmitting/receiving at 20Mbps |
| LED4 | Yellow | Off | Port 2 is not transmitting/receiving at 20Mbps |
| LED5 | Green | On | Port 3 is transmitting/receiving at 2.5Mbps |
| LED5 | Green | Off | Port 3 is not transmitting/receiving at 2.5Mbps |
| LED6 | Yellow | On | Port 3 is transmitting/receiving at 20Mbps |
| LED6 | Yellow | Off | Port 3 is not transmitting/receiving at 20Mbps |
| LED7 | Green | On | Port 4 is transmitting/receiving at 2.5Mbps |
| LED7 | Green | Off | Port 4 is not transmitting/receiving at 2.5Mbps |
| LED8 | Yellow | On | Port 4 is transmitting/receiving at 20Mbps |
| LED8 | Yellow | Off | Port 4 is not transmitting/receiving at 20Mbps |
| LED9 | Green | On | Internal port is transmitting/receiving at 2.5Mbps |
| LED9 | Green | Off | Internal port is not transmitting/receiving at 2.5Mbps |
| LED10 | Yellow | On | Internal port is transmitting/receiving at 20Mbps |
| LED10 | Yellow | Off | Internal port is not transmitting/receiving at 20Mbps |

Note: The locations and colors of the individual LEDs within each pair may be reversed from the diagram.

## DIGITAL COMMUNICATIONS ASSOCIATES, INC.
### CLASSICBLUE ISA

| | |
|---|---|
| NIC Type | Token Ring |
| Transfer Rate | 4/16Mbps |
| Data Bus | 16-bit ISA |
| Topology | Ring |
| Wiring Type | Unshielded twisted pair |
| | Shielded twisted pair (DB-9) |
| Boot ROM | Not available |

| Address | SW1/1 | SW1/2 | SW1/3 | SW1/4 | SW1/5 | SW1/6 |
|---|---|---|---|---|---|---|
| ⇨ CC000h | Off | On | On | Off | Off | On |
| C0000h | Off | On | On | On | On | On |
| C2000h | Off | On | On | On | On | Off |
| C4000h | Off | On | On | On | Off | On |
| C6000h | Off | On | On | On | Off | Off |
| C8000h | Off | On | On | Off | On | On |
| CA000h | Off | On | On | Off | On | Off |
| CE000h | Off | On | On | Off | Off | Off |
| D0000h | Off | On | Off | On | On | On |
| D2000h | Off | On | Off | On | On | Off |
| D4000h | Off | On | Off | On | Off | On |
| D6000h | Off | On | Off | On | Off | Off |
| D8000h | Off | On | Off | Off | On | On |
| DA000h | Off | On | Off | Off | On | Off |
| DC000h | Off | On | Off | Off | Off | On |
| DE000h | Off | On | Off | Off | Off | Off |

Note: The default address for a ClassicBlue card as secondary LAN adapter is DC000h.

## DIGITAL COMMUNICATIONS ASSOCIATES, INC.
## CLASSICBLUE ISA

| DATA RATE SELECT | |
|---|---|
| Setting | SW1/12 |
| ⇨ 4Mbps | On |
| 16Mbps | Off |

| INTERRUPT SELECT | | |
|---|---|---|
| IRQ | SW1/7 | SW1/8 |
| ⇨ IRQ2 | On | On |
| IRQ3 | On | Off |
| IRQ6 | Off | On |
| IRQ7 | Off | Off |

| PRIMARY/SECONDARY LAN CARD SELECT | |
|---|---|
| Setting | SW1/9 |
| ⇨ ClassicBlue as primary LAN adapter | Off |
| ClassicBlue as secondary LAN adapter | On |

| RAM BUFFER SIZE SELECT | | |
|---|---|---|
| Size | SW1/10 | SW1/11 |
| ⇨ 16KB | Off | On |
| 8KB | On | On |
| 32KB | On | Off |
| 64KB | Off | Off |

## EAGLE TECHNOLOGY
## NE2000 (ASSY. #810-149-00X)

| | |
|---|---|
| **NIC Type** | Ethernet |
| **Transfer Rate** | 10Mbps |
| **Data Bus** | 16-bit ISA |
| **Topology** | Linear Bus |
| **Wiring Type** | RG–58A/U 50ohm coaxial |
| | AUI transceiver via DB-15 port |
| **Boot ROM** | Available |

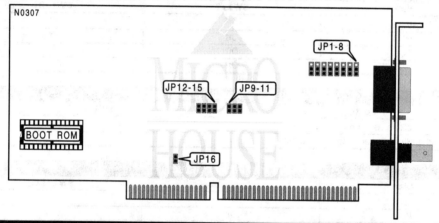

| CABLE TYPE | |
|---|---|
| **Type** | **JP1-JP8** |
| RG–58A/U 50ohm coaxial | Pins 2 & 3 closed |
| AUI transceiver via DB-15 port | Pins 1 & 2 closed |

| BOOT ROM | |
|---|---|
| **Setting** | **JP11** |
| ⇨Enabled | Closed |
| Disabled | Open |

| BASE I/O AND BOOT ROM ADDRESS | | | |
|---|---|---|---|
| **Address** | **Boot ROM** | **JP9** | **JP10** |
| ⇨300h | C8000h | Closed | Closed |
| 320h | CC000h | Open | Closed |
| 340h | D0000h | Closed | Open |
| 360h | N/A | Open | Open |
| Note: Address 360h cannot be used with a boot ROM. | | | |

# EAGLE TECHNOLOGY
## NE2000 (ASSY. #810-149-00X)

| INTERRUPT REQUEST | | | | |
|---|---|---|---|---|
| IRQ | JP12 | JP13 | JP14 | JP15 |
| 2/9 | Closed | Open | Open | Open |
| ⇨3 | Open | Closed | Open | Open |
| 4 | Open | Open | Closed | Open |
| 5 | Open | Open | Open | Closed |

| COMPATABILITY MODE | |
|---|---|
| Setting | JP16 |
| ⇨Disabled | Closed |
| Enabled | Open |

# EAGLE TECHNOLOGY
## NE2100

| | |
|---|---|
| **NIC Type** | Ethernet |
| **Transfer Rate** | 10Mbps |
| **Data Bus** | 16-bit ISA |
| **Topology** | Linear Bus |
| **Wiring Type** | RG-58A/U 50ohm coaxial |
| | AUI transceiver via DB-15 port |
| **Boot ROM** | Available |

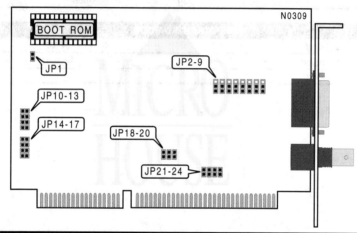

| WAIT STATE | |
|---|---|
| Setting | JP1 |
| ▷0 wait states | Open |
| 1 wait state | Closed |
| Note: Systems using 80ns or faster memory may need to enable one wait state. | |

| CABLE TYPE | |
|---|---|
| Type | JP2-JP9 |
| RG-58A/U 50ohm coaxial | Pins 2 & 3 closed |
| AUI transceiver via DB-15 port | Pins 1 & 2 closed |

| DMA CHANNEL | | | | | | | | |
|---|---|---|---|---|---|---|---|---|
| Channel | JP10 | JP11 | JP12 | JP13 | JP14 | JP15 | JP16 | JP17 |
| DMA3 | Closed | Open | Open | Open | Closed | Open | Open | Open |
| ▷DMA5 | Open | Closed | Open | Open | Open | Closed | Open | Open |
| DMA6 | Open | Open | Closed | Open | Open | Open | Closed | Open |
| DMA7 | Open | Open | Open | Closed | Open | Open | Open | Closed |

| BOOT ROM | |
|---|---|
| Setting | JP18 |
| ▷Disabled | Open |
| Enabled | Closed |

# EAGLE TECHNOLOGY
## NE2100

| BASE I/O AND BOOT ROM ADDRESS | | | |
|---|---|---|---|
| Address | Boot ROM | JP19 | JP20 |
| ⇨300h | C8000h | Closed | Closed |
| 320h | CC000h | Closed | Open |
| 340h | D0000h | Open | Closed |
| 360h | D4000h | Open | Open |

| INTERRUPT REQUEST | | | | |
|---|---|---|---|---|
| IRQ | JP21 | JP22 | JP23 | JP24 |
| 2/9 | Open | Open | Open | Closed |
| ⇨3 | Open | Open | Closed | Open |
| 4 | Open | Closed | Open | Open |
| 5 | Closed | Open | Open | Open |

## EDIMAX COMPUTER COMPANY
## EN-7016 (VERSION 1.0)

| | |
|---|---|
| **NIC Type** | Ethernet |
| **Transfer Rate** | 10Mbps |
| **Data Bus** | 16-bit ISA |
| **Topology** | Star |
| | Linear Bus |
| **Wiring Type** | RG-58A/U 50ohm coaxial |
| | Unshielded twisted pair |
| | AUI transceiver via DB-15 port |
| **Boot ROM** | Available |

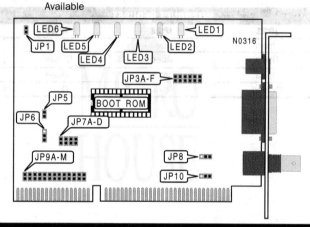

| CONNECTORS ||
|---|---|
| **Setting** | **Jumper** |
| External power LED | JP1 |

| CABLE TYPE ||||
|---|---|---|---|
| **Type** | **JP3A-JP3F** | **JP8** | **JP10** |
| ⇨RG-58A/U 50ohm coaxial | Open | Pins 1 & 2 closed | Pins 2 & 3 closed |
| Unshielded twisted pair | Closed | Pins 2 & 3 closed | Pins 2 & 3 closed |
| AUI transceiver via DB-15 port | Open | Pins 2 & 3 closed | Pins 1 & 2 closed |

| BOOT ROM ||
|---|---|
| **Setting** | **JP5** |
| ⇨Disabled | Open |
| Enabled | Closed |

| COMPATIBILITY MODE ENABLE ||
|---|---|
| **Mode** | **JP6** |
| ⇨Normal | Pins 1 & 2 closed |
| Compatibility mode enabled | Pins 2 & 3 closed |

Note: If card does not operate properly with your computer, try enabling the compatibility mode.

## EDIMAX COMPUTER COMPANY
## EN-7016 (VERSION 1.0)

| Address | JP7A | JP7B | JP7C | JP7D |
|---|---|---|---|---|
| C0000h | Closed | Closed | Closed | Closed |
| C4000h | Closed | Closed | Closed | Open |
| ⇨C8000h | Closed | Closed | Open | Closed |
| CC000h | Closed | Closed | Open | Open |
| D0000h | Closed | Open | Closed | Closed |
| D4000h | Closed | Open | Closed | Open |
| D8000h | Closed | Open | Open | Closed |
| DC000h | Closed | Open | Open | Open |

**BOOT ROM ADDRESS**

| IRQ | JP9A | JP9B | JP9C | JP9D | JP9E | JP9F | JP9G | JP9H | JP9I | JP9J | JP9K |
|---|---|---|---|---|---|---|---|---|---|---|---|
| ⇨3 | Open | Open | Open | Open | Open | Open | Open | Open | Open | Open | Closed |
| 4 | Open | Open | Open | Open | Open | Open | Open | Open | Open | Closed | Open |
| 5 | Open | Open | Open | Open | Open | Open | Open | Open | Closed | Open | Open |
| 6 | Open | Open | Open | Open | Open | Open | Open | Closed | Open | Open | Open |
| 7 | Open | Open | Open | Open | Open | Open | Closed | Open | Open | Open | Open |
| 9 | Open | Open | Open | Open | Open | Closed | Open | Open | Open | Open | Open |
| 10 | Open | Open | Open | Open | Closed | Open | Open | Open | Open | Open | Open |
| 11 | Open | Open | Open | Closed | Open | Open | Open | Open | Open | Open | Open |
| 12 | Open | Open | Closed | Open | Open | Open | Open | Open | Open | Open | Open |
| 14 | Open | Closed | Open | Open | Open | Open | Open | Open | Open | Open | Open |
| 15 | Closed | Open | Open | Open | Open | Open | Open | Open | Open | Open | Open |

**INTERRUPT SETTINGS**

| Address | JP9L | JP9M |
|---|---|---|
| ⇨300h | Closed | Closed |
| 320h | Closed | Open |
| 340h | Open | Closed |
| 360h | Open | Open |

**I/O BASE ADDRESS**

| LED | Color | Status | Condition |
|---|---|---|---|
| LED1 | Red | On | Card is transmitting for an extended amount of time |
| LED1 | Red | Off | Normal operation |
| LED2 | Green | On | Data is being transmitted |
| LED2 | Green | Off | Data is not being transmitted |
| LED3 | Green | On | Data is being received |
| LED3 | Green | Off | Data is not being received |
| LED4 | Red | On | A collision is detected on the network |
| LED4 | Red | Off | Normal operation |
| LED5 | Yellow | On | A reversed polarity twisted pair is detected |
| LED5 | Yellow | Off | Normal operation |
| LED6 | Green | On | Network connection is good |
| LED6 | Green | Off | Network connection is broken |

**DIAGNOSTIC LED(S)**

# GVC TECHNOLOGIES, INC.
## NIC-2500

| | |
|---|---|
| **NIC Type** | Token Ring |
| **Transfer Rate** | 4/16Mbps |
| **Data Bus** | 16-bit ISA |
| **Topology** | Token Ring |
| **Wiring Type** | Unshielded twisted pair |
| | Shielded twisted pair via DB-9 |
| **Boot ROM** | Available (in firmware) |

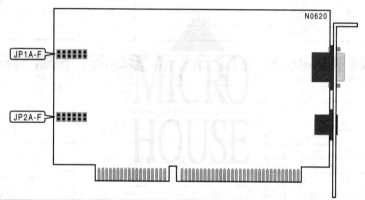

### BOOT ROM ADDRESS

| Address | JP1A | JP1B | JP1C | JP1D | JP1E | JP1F |
|---|---|---|---|---|---|---|
| C0000-C1FFFh | Closed | Open | Open | Open | Open | Open |
| C2000-C3FFFh | Closed | Open | Open | Open | Open | Closed |
| C4000-C5FFFh | Closed | Open | Open | Open | Closed | Open |
| C6000-C7FFFh | Closed | Open | Open | Open | Closed | Closed |
| C8000-C9FFFh | Closed | Open | Open | Closed | Open | Open |
| CA000-CBFFFh | Closed | Open | Open | Closed | Open | Closed |
| ⇨CC000-CDFFFh | Closed | Open | Open | Closed | Closed | Open |
| CE000-CFFFFh | Closed | Open | Open | Closed | Closed | Closed |
| D0000-D1FFFh | Closed | Open | Closed | Open | Open | Open |
| ⇨DC000-DDFFFh | Closed | Open | Closed | Closed | Closed | Open |

Note: CC000h or DC000h are the recommended default settings.

### INTERRUPT REQUEST

| IRQ | JP2A | JP2B |
|---|---|---|
| ⇨2/9 | Open | Open |
| 3 | Open | Closed |
| 6 | Closed | Open |
| 7 | Closed | Closed |

## GVC TECHNOLOGIES, INC.
### NIC-2500

| SHARED RAM CONFIGURATION | | |
|---|---|---|
| **Size** | **JP2C** | **JP2D** |
| 8KB | Open | Open |
| ⇨16KB | Open | Closed |

Note: This setting configures how much system memory the card will use.

| NETWORK SPEED | |
|---|---|
| **Speed** | **JP2E** |
| 4Mbps | Open |
| 16Mbps | Closed |

Note: All cards on a segment must have this option set the same.

| I/O BASE ADDRESS | |
|---|---|
| **Address** | **JP2F** |
| ⇨A20h | Open |
| A24h | Closed |

## HEWLETT-PACKARD COMPANY
## HP27236A

| | |
|---|---|
| NIC Type | Ethernet |
| Transfer Rate | 10Mbps |
| Data Bus | 16-bit ISA |
| Topology | Linear bus |
| | Star |
| Wiring Type | Unshielded twisted pair |
| Boot ROM | Available |

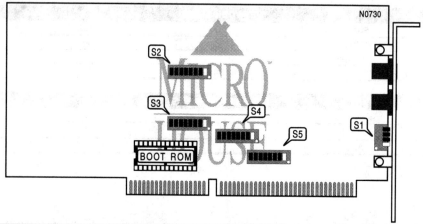

| BOOT ROM | |
|---|---|
| Setting | S2/7 |
| ⇨ Disabled | On |
| Enabled | Off |

| BASE MEMORY ADDRESS | | | | | | | | |
|---|---|---|---|---|---|---|---|---|
| Address | S3/1 | S3/2 | S3/S | S3/4 | S3/5 | S3/6 | S3/7 | S3/8 |
| ⇨ EC000h | On | On | On | Off | On | On | Off | Off |
| FF000h | On | On | On | On | On | On | On | On |
| EB000h | On | On | On | Off | On | Off | On | On |
| ED000h | On | On | On | Off | On | On | Off | On |
| EF000h | On | On | On | Off | On | On | On | On |
| Disabled | Off | Off | Off | Off | Off | Off | Off | Off |

| DMA CONFIGURATION | | | | | |
|---|---|---|---|---|---|
| DMA | S5/1 | S5/2 | S5/3 | S5/4 | S5/5 | S5/6 |
| ⇨ 1 | Off | Off | On | Off | Off | On |
| 2 | Off | On | Off | Off | On | Off |
| 3 | On | Off | Off | On | Off | Off |

## HEWLETT-PACKARD COMPANY
## HP27236A

| Address | S2/1 | S2/2 | S2/3 | S2/4 | S2/5 | S2/6 |
|---|---|---|---|---|---|---|
| ⇨ 300h | On | On | Off | Off | Off | Off |
| 200h | On | Off | Off | Off | Off | Off |
| 250h | On | Off | Off | On | Off | On |
| 260h | On | Off | Off | On | On | Off |
| 280h | On | Off | On | Off | Off | Off |
| 290h | On | Off | On | Off | Off | On |
| 2A0h | On | Off | On | Off | On | Off |
| 2B0h | On | Off | On | Off | On | On |
| 2C0h | On | Off | On | On | Off | Off |
| 2D0h | On | Off | On | On | Off | On |
| 330h | On | On | Off | Off | On | On |
| 340h | On | On | Off | On | Off | Off |
| 350h | On | On | Off | On | Off | On |
| 3F0h | On | On | On | On | On | On |

### INTERRUPT SETTINGS

| IRQ | S4/1 | S4/2 | S4/3 | S4/4 | S4/5 | S4/6 |
|---|---|---|---|---|---|---|
| ⇨ IRQ3 | Off | Off | Off | Off | On | Off |
| IRQ4 | Off | Off | Off | On | Off | Off |
| IRQ5 | Off | Off | On | Off | Off | Off |
| IRQ7 | On | Off | Off | Off | Off | Off |

### LOOPBACK

| Mode | S1/1 |
|---|---|
| ⇨ Normal | To the right position |
| Test | To the left position |

### FACTORY CONFIGURED SETTINGS

| Switch | Setting |
|---|---|
| S1/1 | N/A |
| S1/2 | N/A |
| S1/3 | N/A |
| S2/8 | N/A |
| S4/7 | N/A |
| S4/8 | N/A |
| S5/7 | N/A |
| S5/8 | N/A |

# ICL
## TOKENTEAM AT 16/4

| | |
|---|---|
| **NIC Type** | Token Ring |
| **Transfer Rate** | 4/16Mbps |
| **Data Bus** | 16-bit ISA |
| **Topology** | Ring |
| **Wiring Type** | Shielded twisted pair |
| | Unshielded twisted pair |
| **Boot ROM** | Available |

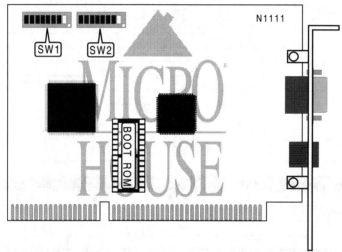

| SPEED CONFIGURATION ||
|---|---|
| **Speed** | **SW1/1** |
| ⇨ 4Mbps | Off |
| 16Mbps | On |

| BASE I/O ADDRESS SELECTION |||
|---|---|---|
| **Address** | **SW1/2** | **SW1/3** |
| ⇨ A20h | Off | Off |
| A24h | On | Off |
| A50h | Off | On |
| A54h | On | On |

# ICL
# TOKENTEAM AT 16/4

| Select | SW1/4 | SW1/5 | SW1/6 | SW1/7 | SW1/8 |
|---|---|---|---|---|---|
| ⇨ CC000h | Off | On | On | Off | Off |
| C0000h | Off | Off | Off | Off | Off |
| C2000h | On | Off | Off | Off | Off |
| C4000h | Off | On | Off | Off | Off |
| C6000h | On | On | Off | Off | Off |
| C8000h | Off | Off | On | Off | Off |
| CA000h | On | Off | On | Off | Off |
| CE000h | On | On | On | Off | Off |
| D0000h | Off | Off | Off | On | Off |
| D2000h | On | Off | Off | On | Off |
| D4000h | Off | On | Off | On | Off |
| D6000h | On | On | Off | On | Off |
| D8000h | Off | Off | On | On | Off |
| DA000h | On | Off | On | On | Off |
| DC000h | Off | On | On | On | Off |
| DE000h | On | On | On | On | Off |

## INTERRUPT SELECTION

| IRQ | SW2/1 | SW2/2 | SW2/3 |
|---|---|---|---|
| 2/9 | Off | Off | Off |
| 3 | On | Off | On |
| ⇨ 5 | Off | On | On |
| 7 | On | On | Off |
| 10 | On | Off | Off |
| 11 | Off | On | Off |
| 12 | On | On | Off |
| 15 | Off | Off | On |

## DMA CHANNEL SELECTION

| Channel | SW2/4 | SW2/5 | SW2/6 |
|---|---|---|---|
| 0 | On | On | Off |
| 1 | Off | Off | On |
| 3 | On | Off | On |
| 5 | Off | Off | On |
| 6 | On | Off | Off |
| 7 | Off | On | Off |
| Disabled | On | On | On |

## TECHNICAL NOTE

Note: SW2/7 and SW2/8 are not used.

# IMC NETWORK CORPORATION
## PCNIC FO

**NIC Type**           Ethernet
**Transfer Rate**      10Mbps
**Data Bus**           8/16-bit ISA
**Topology**           Linear bus
**Wiring Type**        50/125μ, 62.5/125μ, 85/125μ, 100/140μ Fiber optic cable
**Boot ROM**           Not available

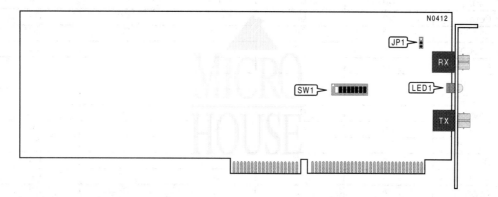

| DIAGNOSTIC LED(S) | | | |
|---|---|---|---|
| **LED** | **Color** | **Status** | **Condition** |
| LED1 | Green | On | Network connection is good |
| LED1 | Green | Off | Network connection is broken |

| FACTORY CONFIGURED SETTING | |
|---|---|
| **Switch** | **Setting** |
| SW1/1 | Unused |
| SW1/2 | Unused |

| COMPATIBILITY MODE | |
|---|---|
| **Setting** | **JP1** |
| ⇨Disabled | Pins 1 & 2 closed |
| Enabled | Pins 2 & 3 closed |

Note: On some systems the data bus timing is not truly IBM PC/AT compatible. If the card will not initialize, enabling compatibility mode may allow the card to operate properly.

# IMC NETWORK CORPORATION
## PCNIC FO

| I/O BASE ADDRESS (LAST TWO DIGITS) | | | | | |
|---|---|---|---|---|---|
| **Address** | **SW1/3** | **SW3/4** | **SW3/5** | **SW3/6** | **SW3/7** |
| x00 - x07 | On | On | On | On | On |
| x08 - x0Fh | Off | On | On | On | On |
| x10 - x07 | On | Off | On | On | On |
| x18 - x1Fh | Off | Off | On | On | On |
| x20 - x07 | On | On | Off | On | On |
| x28 - x2Fh | Off | On | Off | On | On |
| x30 - x07 | On | Off | Off | On | On |
| x38 - x3Fh | Off | Off | Off | On | On |
| x40 - x47h | On | On | On | Off | On |
| x40 - x4Fh | Off | On | On | Off | On |
| x50 - x57h | On | Off | On | Off | On |
| x50 - x5Fh | Off | Off | On | Off | On |
| x60 - x67h | On | On | Off | Off | On |
| x60 - x6Fh | Off | On | Off | Off | On |
| x70 - x77h | On | Off | Off | Off | On |
| x70 - x7Fh | Off | Off | Off | Off | On |
| x80 - x87h | On | On | On | On | Off |
| x88 - x8Fh | Off | On | On | On | Off |
| x90 - x97h | On | Off | On | On | Off |
| x90 - x9Fh | Off | Off | On | On | Off |
| xA0 - xA7h | On | On | Off | On | Off |
| xA0 - xAFh | Off | On | Off | On | Off |
| xB0 - xB7h | On | Off | Off | On | Off |
| xB0 - xBFh | Off | Off | Off | On | Off |
| xC0 - xC7h | On | On | On | Off | Off |
| xC0 - xCFh | Off | On | On | Off | Off |
| xD0 - xD7h | On | Off | On | Off | Off |
| xD0 - xDFh | Off | Off | On | Off | Off |
| xE0 - xE7h | On | On | Off | Off | Off |
| xE0 - xEFh | Off | On | Off | Off | Off |
| xF0 - xF7h | On | Off | Off | Off | Off |
| xF0 - xFFh | Off | Off | Off | Off | Off |

Note: The two digits selected here must be added to the digit selected in the following table.

| I/O BASE ADDRESS (FIRST DIGIT) | |
|---|---|
| **Address** | **SW1/8** |
| 2xx - 2xxh | On |
| 3xx - 3xxh | Off |

# INTEL CORPORATION
## TOKEN EXPRESS ISA 16/4 LAN

| | |
|---|---|
| **NIC Type** | Token-Ring |
| **Transfer Rate** | 4/16Mbps |
| **Data Bus** | 8/16-bit ISA |
| **Topology** | Ring |
| **Wiring Type** | Unshielded twisted pair (RJ-45 port) |
| | Shielded twisted pair (DB-9 port) |
| **Boot ROM** | Available |

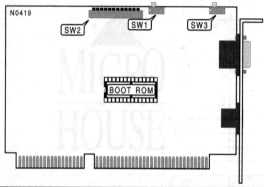

| NETWORK SEGMENT SPEED ||
|---|---|
| **Setting** | **SW1** |
| ⇨4Mbps | Off |
| 16Mbps | On |

Note: All cards on the network segment must have this option set the same.

| BOOT ROM ||
|---|---|
| **Setting** | **SW2/1** |
| ⇨Disabled | Off |
| Enabled | On |

| I/O BASE ADDRESS |||
|---|---|---|
| **Address** | **SW2/9** | **SW2/10** |
| ⇨A20-A23h , A30-A3Fh | On | On |
| A24-A27h , A40-A4Fh | Off | On |
| A50-A53h , A60-A6Fh | On | Off |
| A54-A57h , A70-A7Fh | Off | Off |

| DMA CHANNEL |||
|---|---|---|
| **Channel** | **SW2/11** | **SW2/12** |
| ⇨DMA5 | Off | Off |
| DMA6 | Off | On |
| DMA7 | On | Off |
| Programmed I/O | On | On |

Note: If the card is installed into an 8-bit slot, programmed I/O is selected regardless of the switch settings.

# INTEL CORPORATION
## TOKEN EXPRESS ISA 16/4 LAN

| BOOT ROM ADDRESS ||||||
|---|---|---|---|---|---|
| Address | SW2/2 | SW2/3 | SW2/4 | SW2/5 | SW2/6 |
| C0000 - C1FFFh | Off | Off | Off | Off | Off |
| C2000 - C3FFFh | Off | Off | Off | Off | On |
| C4000 - C5FFFh | Off | Off | Off | On | Off |
| C6000 - C7FFFh | Off | Off | Off | On | On |
| C8000 - C9FFFh | Off | Off | On | Off | Off |
| CA000 - CBFFFh | Off | Off | On | Off | On |
| CC000 - CDFFFh | Off | Off | On | On | Off |
| CE000 - CFFFFh | Off | Off | On | On | On |
| D0000 - D1FFFh | Off | On | Off | Off | Off |
| D2000 - D3FFFh | Off | On | Off | Off | On |
| D4000 - D5FFFh | Off | On | Off | On | Off |
| D6000 - D7FFFh | Off | On | Off | On | On |
| D8000 - D9FFFh | Off | On | On | Off | Off |
| DA000 - DBFFFh | Off | On | On | Off | On |
| DC000 - DDFFFh | Off | On | On | On | Off |
| DE000 - DFFFFh | Off | On | On | On | On |
| E0000 - E1FFFh | On | Off | Off | Off | Off |
| E2000 - E3FFFh | On | Off | Off | Off | On |
| E4000 - E5FFFh | On | Off | Off | On | Off |
| E6000 - E7FFFh | On | Off | Off | On | On |
| E8000 - E9FFFh | On | Off | On | Off | Off |
| EA000 - EBFFFh | On | Off | On | Off | On |
| EC000 - EDFFFh | On | Off | On | On | Off |
| EE000 - EFFFFh | On | Off | On | On | On |
| F0000 - F1FFFh | On | On | Off | Off | Off |
| F2000 - F3FFFh | On | On | Off | Off | On |
| F4000 - F5FFFh | On | On | Off | On | Off |
| F6000 - F7FFFh | On | On | Off | On | On |
| F8000 - F9FFFh | On | On | On | Off | Off |
| FA000 - FBFFFh | On | On | On | Off | On |
| FC000 - FDFFFh | On | On | On | On | Off |
| FE000 - FFFFFh | On | On | On | On | On |

| INTERRUPT REQUEST |||
|---|---|---|
| IRQ | SW2/7 | SW2/8 |
| ➪2/9 | Off | Off |
| 3 | Off | On |
| 10 | On | Off |
| 11 | On | On |

| CABLE TYPE ||
|---|---|
| Type | SW3 |
| ➪Shielded twisted pair (DB-9 port) | On |
| Unshielded twisted pair (RJ-45 port) | Off |

## KATRON TECHNOLOGIES, INC.
## 16/4 TOKEN RING ADAPTER

**NIC Type**         Token ring
**Transfer Rate**    4/16Mbps
**Data Bus**         16-bit ISA
**Topology**         Linear bus
**Wiring Type**      AUI transceiver via DB-15 port
                     Unshielded twisted pair
**Boot ROM**         Not available

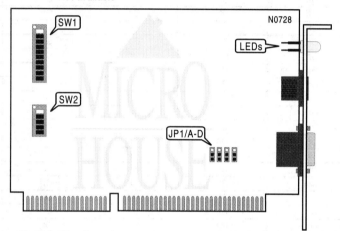

| Address | SW2/1 | SW2/2 | SW2/3 | SW2/4 | SW2/5 |
|---|---|---|---|---|---|
| ⇨ CC000h | Off | Off | On | On | Off |
| C0000h | Off | Off | Off | Off | Off |
| C2000h | Off | Off | Off | Off | On |
| C4000h | Off | Off | Off | On | Off |
| C6000h | Off | Off | Off | On | On |
| C8000h | Off | Off | On | Off | Off |
| CA000h | Off | Off | On | Off | On |
| CE000h | Off | Off | On | On | On |
| D0000h | Off | On | Off | Off | Off |
| D2000h | Off | On | Off | Off | On |
| D4000h | Off | On | Off | On | Off |
| D6000h | Off | On | Off | On | On |
| D8000h | Off | On | On | Off | Off |
| DA000h | Off | On | On | Off | On |
| DC000h | Off | On | On | On | Off |
| DE000h | Off | On | On | On | On |

BASE ADDRESS SELECT

## KATRON COMPUTERS, INC.
## 16/4 TOKEN RING ADAPTER

| INTERRUPT SELECT | | | | |
|---|---|---|---|---|
| IRQ | SW1/7 | SW1/8 | SW1/9 | SW1/10 |
| ⇨ IRQ2 | On | Off | Off | Off |
| IRQ3 | Off | On | Off | Off |
| IRQ6 | Off | Off | On | Off |
| IRQ7 | Off | Off | Off | On |

| PRIMARY/ALTERNATE CARD | |
|---|---|
| Setting | SW1/1 |
| ⇨ Primary | Off |
| Alternate | On |

| SHARED RAM SIZE | | | | |
|---|---|---|---|---|
| Size | SW1/3 | SW1/4 | SW1/5 | SW1/6 |
| ⇨ 16KB | Off | On | Off | Off |
| 8KB | On | Off | Off | Off |
| 32KB | Off | Off | On | Off |
| 64KB | Off | Off | Off | On |

| DATA RATE | |
|---|---|
| Setting | SW1/2 |
| ⇨ 4Mbps | Off |
| 16Mbps | On |

| CABLE TYPE | | | | |
|---|---|---|---|---|
| Type | JP1/A | JP1/B | JP1/C | JP1/D |
| AUI transceiver via DB-15 port | Pins 2 & 3 | Pins 2 & 3 | Pins 2 & 3 | Pins 2 & 3 |
| Unshielded twisted pair | Pins 1 & 2 | Pins 1 & 2 | Pins 1 & 2 | Pins 1 & 2 |

Note: All Pins designated should be placed in the closed position.

| DIAGNOSTIC LEDs | | |
|---|---|---|
| LED1 (Green) | LED2 (Yellow) | Condition |
| Off | Off | Network connection is broken, or software initialization is unsuccessful. |
| Off | Blinking | Data is being transmitted, but the network connection is broken. |
| On | Off | Adapter failure |
| On | On | Network connection is good. |

# KINGSTON TECHNOLOGY CORPORATION
## ETHERRX KNE2121

| | |
|---|---|
| NIC Type | Ethernet |
| Transfer Rate | 10Mbps |
| Data Bus | 16-bit ISA |
| Topology | Linear bus |
| Wiring Type | Unshielded twisted pair |
| | RG58A/U 50ohm coaxial |
| Boot ROM | Available |

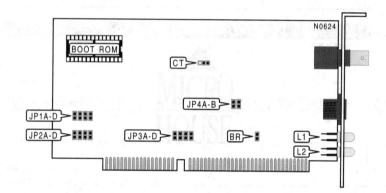

| BOOT ROM | |
|---|---|
| Mode | BR |
| Disabled | Open |
| Enabled | Closed |

| CABLING TYPE | |
|---|---|
| Cable type | CT |
| Unshielded twisted pair | Pins 1 & 2 closed |
| RG58A/U 50ohm coaxial | Pins 2 & 3 closed |

| DMA CHANNEL | | | | | | | | |
|---|---|---|---|---|---|---|---|---|
| Channel | JP1A | JP1B | JP1C | JP1D | JP2A | JP2B | JP2C | JP2D |
| DMA3 | Open | Open | Open | Closed | Open | Open | Open | Closed |
| DMA5 | Open | Open | Closed | Open | Open | Open | Closed | Open |
| DMA6 | Open | Closed | Open | Open | Open | Closed | Open | Open |
| DMA7 | Closed | Open | Open | Open | Closed | Open | Open | Open |

## KINGSTON TECHNOLOGY CORPORATION
## ETHERRX KNE2121

### INTERRUPT SETTINGS

| IRQ | JP3A | JP3B | JP3C | JP3D |
|---|---|---|---|---|
| 3 | Closed | Open | Open | Open |
| 4 | Open | Closed | Open | Open |
| 5 | Open | Open | Closed | Open |
| 9 | Open | Open | Open | Closed |

### BASE I/O ADDRESS

| Address | JP4A | JP4B |
|---|---|---|
| ⇨ 300h | Closed | Closed |
| 320h | Open | Closed |
| 340h | Closed | Open |
| 360h | Open | Open |

### DIAGNOSTIC LED(S)

| LED | Color | Status | Condition |
|---|---|---|---|
| L1 | Green | Off | 10BaseT link integrity exists |
| L1 | Green | On | 10BaseT link integrity broken |
| L2 | Amber | On | Normal polarity |
| L2 | Amber | Off | Reversed polarity detected and corrected. |

# KLEVER COMPUTERS, INC.
## K1604

| | |
|---|---|
| **NIC Type** | Token-Ring |
| **Transfer Rate** | 4/16Mbps |
| **Data Bus** | 8/16-bit ISA |
| **Topology** | Ring |
| **Wiring Type** | Unshielded twisted pair (RJ-45) |
| | Shielded twisted pair (DB-9 port) |
| **Boot ROM** | Not available |

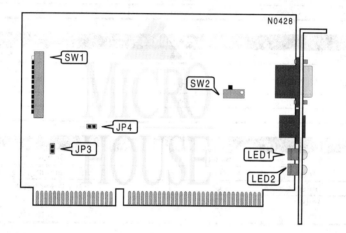

| DATA BUS SIZE | | |
|---|---|---|
| Size | JP3 | JP4 |
| ⇨16-bit | On | On |
| 8-bit | Off | Off |

| INTERRUPT REQUEST | | |
|---|---|---|
| IRQ | SW1/7 | SW1/8 |
| ⇨2/9 | Off | Off |
| 3 | Off | On |
| 6 | On | Off |
| 7 | On | On |

| SHARED RAM SIZE | | |
|---|---|---|
| Size | SW1/9 | SW1/10 |
| 8KB | Off | Off |
| ⇨16KB | Off | On |
| 32KB | On | Off |
| 64KB | On | On |
| Note: RAM paging is only available in the 16KB mode. | | |

## KLEVER COMPUTERS, INC.
## K1604

| Address | SW1/1 | SW1/2 | SW1/3 | SW1/4 | SW1/5 | SW1/6 |
|---|---|---|---|---|---|---|
| C0000 - C1FFFh | On | Off | Off | Off | Off | Off |
| C2000 - C3FFFh | On | Off | Off | Off | Off | On |
| C4000 - C5FFFh | On | Off | Off | Off | On | Off |
| C6000 - C7FFFh | On | Off | Off | Off | On | On |
| C8000 - C9FFFh | On | Off | Off | On | Off | Off |
| CA000 - CBFFFh | On | Off | Off | On | Off | On |
| CC000 - CDFFFh | On | Off | Off | On | On | Off |
| CE000 - CFFFFh | On | Off | Off | On | On | On |
| D0000 - D1FFFh | On | Off | On | Off | Off | Off |
| D2000 - D3FFFh | On | Off | On | Off | Off | On |
| D4000 - D5FFFh | On | Off | On | Off | On | Off |
| D6000 - D7FFFh | On | Off | On | Off | On | On |
| D8000 - D9FFFh | On | Off | On | On | Off | Off |
| DA000 - DBFFFh | On | Off | On | On | Off | On |
| DC000 - DDFFFh | On | Off | On | On | On | Off |
| DE000 - DFFFFh | On | Off | On | On | On | On |
| E0000 - E1FFFh | On | On | Off | Off | Off | Off |
| E2000 - E3FFFh | On | On | Off | Off | Off | On |
| E4000 - E5FFFh | On | On | Off | Off | On | Off |
| E6000 - E7FFFh | On | On | Off | Off | On | On |
| E8000 - E9FFFh | On | On | Off | On | Off | Off |
| EA000 - EBFFFh | On | On | Off | On | Off | On |
| EC000 - EDFFFh | On | On | Off | On | On | Off |
| EE000 - EFFFFh | On | On | Off | On | On | On |
| F0000 - F1FFFh | On | On | On | Off | Off | Off |
| F2000 - F3FFFh | On | On | On | Off | Off | On |
| F4000 - F5FFFh | On | On | On | Off | On | Off |
| F6000 - F7FFFh | On | On | On | Off | On | On |
| F8000 - F9FFFh | On | On | On | On | Off | Off |
| FA000 - FBFFFh | On | On | On | On | Off | On |
| FC000 - FFFFFh | On | On | On | On | On | Off |
| FE000 - FFFFFh | On | On | On | On | On | On |

Note: If a second card is to be installed in the same system, the second card should use address DC000 as its default. A maximum of two cards can be installed in a single system.

## KLEVER COMPUTERS, INC.
## K1604

| NETWORK SEGMENT SPEED ||
|---|---|
| **Speed** | **SW1/11** |
| ⇨16Mbps | On |
| 4Mbps | Off |
| Note: All cards on the network segment must have this option set identically. ||

| PRIMARY/ALTERNATE CARD ||
|---|---|
| **Setting** | **SW1/12** |
| ⇨Primary | Off |
| Alternate | On |
| Note: If two cards are to be used in a single system, one of the cards must be set to primary and the other one must be set to alternate. ||

| CABLE TYPE ||
|---|---|
| **Type** | **SW2** |
| ⇨Unshielded twisted pair (RJ-45 port) | On |
| Shielded twisted pair (DB-9 port) | Off |

| DIAGNOSTIC LED(S) ||||
|---|---|---|---|
| **LED** | **Color** | **Status** | **Condition** |
| LED1 | Green | On | 16Mbps network speed |
| LED1 | Red | On | 4Mbps network speed |
| LED2 | Red | Blinking | Data is being transmitted or received |
| LED2 | Off | Off | Data is not being transmitted or received |
| Note: LED1 is a two color LED. ||||

## LONGSHINE MICROSYSTEM, INC.
## LCS-8636

| | |
|---|---|
| **NIC Type** | Token Ring |
| **Transfer Rate** | 4/16Mbps |
| **Data Bus** | 16-bit ISA |
| **Topology** | Ring |
| **Wiring Type** | AUI transceiver via DB-15 port |
| | Unshielded twisted pair |
| **Boot ROM** | Not available |

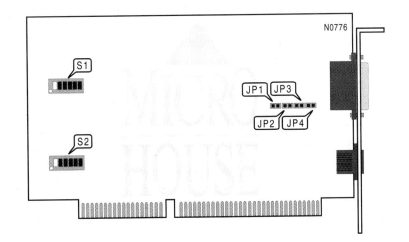

| BASE MEMORY ADDRESS SELECT | | | | | | |
|---|---|---|---|---|---|---|
| **Address** | **S1/1** | **S1/2** | **S1/3** | **S1/4** | **S1/5** | **S1/6** |
| ⇨CC00h | On | Off | Off | On | On | Off |
| B000h | Off | On | On | Off | Off | Off |
| B200h | Off | On | On | Off | Off | On |
| B400h | Off | On | On | Off | On | Off |
| B600h | Off | On | On | Off | On | On |
| B800h | Off | On | On | On | Off | Off |
| BA00h | Off | On | On | On | Off | On |
| BC00h | Off | On | On | On | On | Off |
| BE00h | Off | On | On | On | On | On |
| C000h | On | Off | Off | Off | Off | Off |
| C200h | On | Off | Off | Off | Off | On |
| C400h | On | Off | Off | Off | On | Off |
| C600h | On | Off | Off | Off | On | On |
| C800h | On | Off | Off | On | Off | Off |
| CA00h | On | Off | Off | On | Off | On |
| CE00h | Off | On | Off | On | On | On |
| D000h | On | Off | On | Off | Off | Off |
| D200h | On | Off | On | Off | Off | On |

## LONGSHINE MICROSYSTEM, INC.
## LCS-8636

| Address | BASE MEMORY ADDRESS SELECT(CONTINUED) | | | | | |
|---|---|---|---|---|---|---|
| | S1/1 | S1/2 | S1/3 | S1/4 | S1/5 | S1/6 |
| D400h | On | Off | On | Off | On | Off |
| D600h | On | Off | On | Off | On | On |
| D800h | On | Off | On | On | Off | Off |
| DA00h | On | Off | On | On | Off | On |
| DC00h | On | Off | On | On | On | Off |
| DE00h | On | Off | On | On | On | On |
| E000h | On | On | Off | Off | Off | Off |
| E200h | On | On | Off | Off | Off | On |
| E400h | On | On | Off | Off | On | Off |
| E600h | On | On | Off | Off | On | On |
| E800h | On | On | Off | On | Off | Off |
| EA00h | On | On | Off | On | Off | On |
| EC00h | On | On | Off | On | On | Off |
| EE00h | On | On | Off | On | On | On |

| INTERRUPT SELECT | | |
|---|---|---|
| IRQ | S2/1 | S2/2 |
| ⇨ IRQ2 | Off | Off |
| IRQ3 | Off | On |
| IRQ6 | On | Off |
| IRQ7 | On | On |

| SHARED RAM | | |
|---|---|---|
| Size | S2/3 | S2/4 |
| ⇨ 16KB | Off | On |
| 8KB | Off | Off |
| 32KB | On | Off |
| 64KB | On | On |

| DATA TRANSFER RATE | |
|---|---|
| Rate | S2/5 |
| ⇨ 16Mbps | On |
| 4Mbps | Off |

| I/O ADDRESS SELECT | |
|---|---|
| Address | S2/6 |
| ⇨ A20 - A23h | Off |
| A24 - A27h | On |

| CABLE TYPE | | | | |
|---|---|---|---|---|
| Type | JP1 | JP2 | JP3 | JP4 |
| Thick Ethernet | Open | Open | Open | Open |
| Unshielded twisted pair | Closed | Closed | Closed | Closed |

## MADGE NETWORKS, LTD.
## SMART 16/4 AT BRIDGENODE

| | |
|---|---|
| **NIC Type** | Token Ring |
| **Transfer Rate** | 4/16 Mbps |
| **Data Bus** | 16-bit ISA |
| **Topology** | Ring |
| **Wiring Type** | AUI transceiver via DB-15 port |
| | Shielded twisted pair |
| | Unshielded twisted pair |
| **Boot ROM** | Available |

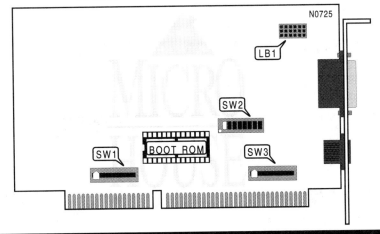

| INTERRUPT SELECT | | | | | | | | |
|---|---|---|---|---|---|---|---|---|
| IRQ | SW1/1 | SW1/2 | SW1/3 | SW1/4 | SW1/5 | SW1/6 | SW1/7 | SW1/8 |
| IRQ3 | Off | Off | Off | Off | Off | Off | On | Off |
| IRQ2/9 | Off | Off | Off | Off | Off | Off | Off | On |
| IRQ5 | Off | Off | Off | Off | Off | On | Off | Off |
| IRQ7 | Off | Off | Off | Off | On | Off | Off | Off |
| IRQ10 | Off | Off | Off | On | Off | Off | Off | Off |
| IRQ11 | Off | Off | On | Off | Off | Off | Off | Off |
| IRQ12 | Off | On | Off | Off | Off | Off | Off | Off |
| IRQ15 | On | Off | Off | Off | Off | Off | Off | Off |

Note: Switch 1 is a slide switch, which allows only one position to be 'on' at a given time.

| ADDRESS SELECT | | | |
|---|---|---|---|
| I/O Address | Base Memory Address | SW2/1 | SW2/2 |
| ⇨0A20 - 02AFh | CC000 - CDFFFh | On | On |
| 1A20 - 1A2Fh | DC000 - DDFFFh | Off | On |
| 2A20 - 2A2Fh | CE000 - CFFFFh | On | Off |
| 3A20 - 3A2Fh | DE000 - DFFFFh | Off | Off |

| BOOT ROM | |
|---|---|
| Setting | SW2/3 |
| ⇨Disabled | Off |
| Enabled | On |

# MADGE NETWORKS, LTD.
## SMART 16/4 AT BRIDGENODE

| BUSMASTER DMA ||
|---|---|
| Setting | SW2/4 |
| ⇨Enabled | On |
| Disabled | Off |

| BUS AUTODETECT ||
|---|---|
| Setting | SW2/5 |
| ⇨Enabled | On |
| Disabled(Force 8-bit) | Off |

| BUS COMPATABILITY ||
|---|---|
| Setting | SW2/6 |
| ⇨Normal AT bus timing | On |
| Alternative bus timing | Off |

| CLOCK SOURCE ||
|---|---|
| Setting | SW2/7 |
| ⇨Asynchronous bus | Off |
| Synchronous bus | On |

| DATA TRANSFER RATE ||
|---|---|
| Setting | SW2/8 |
| ⇨4Mbps | On |
| 16Mbps | Off |

| DMA CHANNEL SELECT |||||
|---|---|---|---|---|
| DMA | SW3/1 | SW3/2 | SW3/3 | SW3/4 |
| ⇨5 | Off | Off | On | Off |
| 1 | On | Off | Off | Off |
| 3 | Off | On | Off | Off |
| 6 | Off | Off | Off | On |

Note: Switch 1 is a slide switch, which allows only one position to be 'on' at a given time.

| TWISTED PAIR CABLE TYPE SELECT ||
|---|---|
| Setting | LB1 |
| Shielded twisted pair | ⬇ |
| Unshielded twisted pair | ⬆ |

Note: Jumper LB1 is used with a link block pictured above. Remove the block and replace with arrow in up or down position, depending on cable type.

# MADGE NETWORKS, LTD.
## SMART 16/4 CLIENT RINGNODE

| | |
|---|---|
| NIC Type | Token Ring |
| Transfer Rate | 4/16Mbps |
| Data Bus | 16-bit ISA |
| Topology | Ring |
| Wiring Type | Shielded/Unshielded twisted pair |
| Boot ROM | Available |

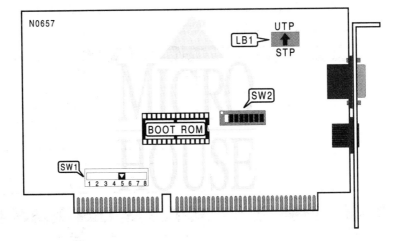

| BOOT ROM | |
|---|---|
| Setting | SW2/3 |
| ⇨ Disabled | Off |
| Enabled | On |

| COMPATIBILITY MODE | |
|---|---|
| Setting | SW2/6 |
| ⇨ Normal ISA bus timing | On |
| Alternate ISA bus timing | Off |

Note: Alternate ISA bus timing is used if the host computer is not 100% ISA bus compatible (such as the IBM PS/2 model 30-286) or if the Madge Ringnode diagnostics program fails to locate the NIC.

| DMA CHANNEL BUS | |
|---|---|
| Setting | SW2/7 |
| ⇨ Asynchronous bus (uses host system's clock) | Off |
| Synchronous bus (uses NIC on-board clock) | On |

Note: The synchronous bus speed is used for systems that have a clock speed less than 10MHz.

# MADGE NETWORKS, LTD.
## SMART 16/4 CLIENT RINGNODE

| EXPANSION SLOT SIZE ||
|---|---|
| Slot Size | SW2/5 |
| ⇨ Automatic select | On |
| Forced 8-bit mode | Off |
| Note: If an 8-bit transfer mode is required and the NIC is installed in a 16-bit slot, use forced 8-bit mode. If the NIC is to be used in a Novell Netware 386 file server, automatic select must be used. ||

| INTERRUPT REQUEST ||
|---|---|
| IRQ | SW1 |
| ⇨ 3 | Position 7 |
| 2/9 | Position 8 |
| 5 | Position 6 |
| 7 | Position 5 |
| 10 | Position 4 |
| 11 | Position 3 |
| 12 | Position 2 |
| 15 | Position 1 |

| NETWORK SPEED ||
|---|---|
| Speed | SW2/7 |
| ⇨ 16Mbps | Off |
| 4Mbps | On |
| Note: All cards on a segment must have this option set the same. ||

| BASE I/O AND BOOT ROM ADDRESS ||||
|---|---|---|---|
| Address | Boot ROM Address | SW2/1 | SW2/2 |
| ⇨ 0A20-0A2Fh | CC000-CDFFFh | On | On |
| 1A20-1A2Fh | DC000-DDFFFh | Off | On |
| 2A20-2A2Fh | CE000-CFFFFh | On | Off |
| 3A20-3A2Fh | DE000-DFFFFh | Off | Off |

| CABLE TYPE ||
|---|---|
| Type | LB1 |
| Shielded twisted pair (DB-9 connector) | STP |
| Unshielded twisted pair (RJ-45 jack) | UTP |

## MADGE NETWORKS, LTD.
# SMART 16/4 FIBER AT RINGNODE

**NIC Type**         Token Ring
**Transfer Rate**    4/16Mbps
**Data Bus**         16-bit ISA
**Topology**         Ring
**Wiring Type**      62.5/125µm multimode graded index optical fiber
**Boot ROM**         Available

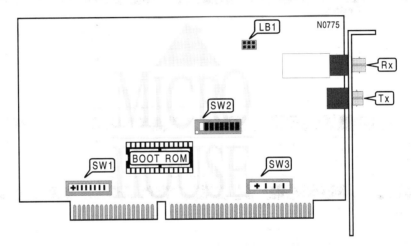

| ADDRESS SELECT | | | |
|---|---|---|---|
| I/O Address | Base Memory Address | SW2/1 | SW2/2 |
| ⇨ 0A20 - 0A2Fh | CC000 - CDFFFh | On | On |
| 1A20 - 1A2Fh | DC000 - DFFFFh | Off | On |
| 2A20 - 2A2Fh | CE000 - CFFFFh | On | Off |
| 3A20 - 3A2Fh | DE000 - DFFFFh | Off | Off |

| AT BUS TIMING COMPATIBILITY | |
|---|---|
| Setting | SW2/6 |
| ⇨ Normal | On |
| Alternate | Off |

| BOOT ROM | |
|---|---|
| Setting | SW2/3 |
| ⇨ Disabled | Off |
| Enabled | On |

| BUS MASTER DMA | |
|---|---|
| Setting | SW2/4 |
| ⇨ Enabled | On |
| Disabled | Off |

## MADGE NETWORKS, LTD.
## SMART 16/4 FIBER AT RINGNODE

| CLOCK SOURCE | |
|---|---|
| Setting | SW2/7 |
| ⇨ Asynchronous | Off |
| Synchronous | On |

| CONCENTRATOR TYPE | |
|---|---|
| Setting | LB1 |
| ⇨ 802.5J | 802.5J ← LOCAL WRAP |
| Local wrap | 802.5J → LOCAL WRAP |

Note: Concentrator type is determined by placing a link block over the LB1 jumper block. The block has an arrow printed on the front and should resemble the diagram shown above.

| DATA TRANSFER RATE | |
|---|---|
| Speed | SW2/8 |
| ⇨ 16Mbps | Off |
| 4Mbps | On |

| DMA CONFIGURATION | |
|---|---|
| DMA | SW3 |
| ⇨ 5 | Position 3 |
| 1 | Position 1 |
| 3 | Position 2 |
| 6 | Position 4 |

Note: Switch 3 is a four position slide switch. Each position should be silk-screened on the switch itself.

| 8/16 BIT MODE | |
|---|---|
| Setting | SW2/5 |
| ⇨ 8/16-bit mode | On |
| Force 8-bit mode | Off |

## MADGE NETWORKS, LTD.
# SMART 16/4 FIBER AT RINGNODE

| INTERRUPT SELECT ||
|---|---|
| **IRQ** | **SW1** |
| ⇨ IRQ3 | Position 7 |
| IRQ2/9 | Position 8 |
| IRQ5 | Position 6 |
| IRQ7 | Position 5 |
| IRQ10 | Position 4 |
| IRQ11 | Position 3 |
| IRQ12 | Position 2 |
| IRQ15 | Position 1 |

Note: Switch 1 is an eight position slide switch. Each position should be silk-screened on the switch itself.

| CONNECTIONS ||
|---|---|
| **Purpose** | **Location** |
| Fiber-optic transmitter | Tx |
| Fiber-optic receiver | Rx |

# MADGE NETWORKS, LTD.
## STRAIGHT BLUE 16/4 ISA PLUS

| | |
|---|---|
| **NIC Type** | Token Ring |
| **Transfer Rate** | 4/16Mbps |
| **Data Bus** | 8-bit ISA/16-bit ISA |
| **Topology** | Token ring |
| **Wiring Type** | Unshielded twisted pair |
| | Shielded twisted pair via DB9 |
| **Boot ROM** | Available (in firmware) |

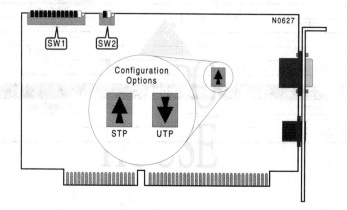

| ADDRESS CONFIGURATION | | | | | | | |
|---|---|---|---|---|---|---|---|
| Setting | SW1/1 | SW1/2 | SW1/3 | SW1/4 | SW1/5 | SW1/6 | SW1/9 |
| ⇨Primary | Off | On | On | Off | Off | On | Off |
| Secondary | Off | On | Off | Off | Off | On | On |

| INTERRUPT REQUEST | | |
|---|---|---|
| IRQ | SW1/7 | SW1/8 |
| ⇨2 | On | On |
| 3 | On | Off |
| 6 | Off | On |
| 7 | Off | Off |

| SHARED RAM CONFIGURATION | | |
|---|---|---|
| Size | SW1/10 | SW1/11 |
| 8KB | On | On |
| 16KB | Off | On |
| 32KB | On | Off |
| 64KB | Off | Off |

# MADGE NETWORKS, LTD.
## STRAIGHT BLUE 16/4 ISA PLUS

| NETWORK SPEED ||
|---|---|
| **Speed** | **SW1/12** |
| 4Mbps | On |
| ⇨16Mbps | Off |

Note: All cards on a segment must have this option set the same.

| COMPATIBILITY MODE ||
|---|---|
| **Mode** | **SW1/1** |
| ⇨Normal operation | Off |
| Compatibility mode | On |

| DATA BUS ||
|---|---|
| **Type** | **SW2/2** |
| 8-bit ISA | Off |
| 16-bit ISA | On |

# MAXTECH CORPORATION
## NIC-2004 SERIES

| | |
|---|---|
| **NIC Type** | Ethernet |
| **Transfer Rate** | 10Mbps |
| **Data Bus** | 16-bit ISA |
| **Topology** | Linear bus |
| | Star |
| **Wiring Type** | AUI transceiver via DB-15 port |
| | RG-58A/U 50ohm coaxial |
| | RG-58C/U 50ohm coaxial |
| | Unshielded Twisted Pair |
| **Boot ROM** | Available |

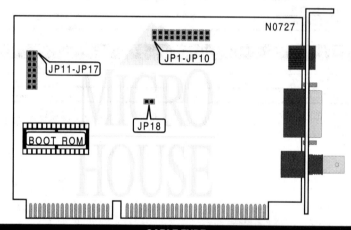

| CABLE TYPE | | |
|---|---|---|
| **Type** | **JP5** | **JP6** |
| ⇨ Thin Ethernet | Open | Closed |
| Thick Ethernet | Closed | Open |
| TPI (10Base-T compatible squelch) | Open | Open |
| TPI (reduced squelch) | Closed | Closed |

| BASE I/O ADDRESS | | | |
|---|---|---|---|
| **Address** | **JP15** | **JP16** | **JP17** |
| ⇨ 300h | Open | Open | Open |
| Software | Open | Open | Closed |
| 240h | Open | Closed | Open |
| 280h | Open | Closed | Closed |
| 2C0h | Closed | Open | Open |
| 320h | Closed | Open | Closed |
| 340h | Closed | Closed | Open |
| 360h | Closed | Closed | Closed |

# MAXTECH CORPORATION
## NIC-2004 SERIES

| BOOT ROM ADDRESS | | | | | |
|---|---|---|---|---|---|
| Address | Size | JP1 | JP2 | JP3 | JP4 |
| ⇨ Disabled | No Boot PROM | Open | Open | Open | Open |
| C000h | 8K/16K | Open | Open | Closed | Open |
| C400h | 8K/16K | Open | Open | Closed | Closed |
| C800h | 8K/16K | Open | Closed | Open | Open |
| CC00h | 8K/16K | Open | Closed | Open | Closed |
| D000h | 8K/16K | Open | Closed | Closed | Open |
| D400h | 8K/16K | Open | Closed | Closed | Closed |
| D800h | 8K/16K | Closed | Open | Open | Open |
| DC00h | 8K/16K | Closed | Open | Open | Closed |
| C000h | 32K/32K | Closed | Open | Closed | Open |
| C800h | 32K/32K | Closed | Open | Closed | Closed |
| D000h | 32K/32K | Closed | Closed | Open | Open |
| D800h | 32K/32K | Closed | Closed | Open | Closed |
| C000h | 64K/64K | Closed | Closed | Closed | Open |
| D000h | 64K/64K | Closed | Closed | Closed | Closed |

| INTERRUPT REQUEST | | | | |
|---|---|---|---|---|
| IRQ | JP7 | JP8 | JP9 | JP10 |
| ⇨ IRQ3 | Closed | Open | Open | Open |
| IRQ4 | Closed | Open | Open | Closed |
| IRQ5 | Closed | Open | Closed | Open |
| IRQ9 | Closed | Open | Closed | Closed |
| IRQ10 | Closed | Closed | Open | Open |
| IRQ11 | Closed | Closed | Open | Closed |
| IRQ12 | Closed | Closed | Closed | Open |
| IRQ15 | Closed | Closed | Closed | Closed |

| LINK INTEGRITY | |
|---|---|
| Setting | JP11 |
| ⇨ Enabled | Open |
| Disabled | Closed |

| I/O CHRDY MODE | |
|---|---|
| Mode | JP12 |
| ⇨ CHRDY generated after command strobe | Open |
| CHRDY generated after BALE goes high | Closed |

## MAXTECH CORPORATION
## NIC-2004 SERIES

| I/O 16 CONTROL ||
|---|---|
| Mode | JP13 |
| ⇨ I/O 16 generated only on address decode | Open |
| I/O 16 generated after IORD or IOWR go active | Closed |

| MEMORY / I/O MODE ||
|---|---|
| Mode | JP14 |
| ⇨ I/O mode | Open |
| Shared memory mode | Closed |

| CARD CONFIGURATION ||
|---|---|
| Select | JP18 |
| ⇨ Software configuration | Closed |
| Hardware (jumpers) configuration | Open |

## MICRODYNE CORPORATION
## NTR2000

| | |
|---|---|
| **NIC Type** | Token Ring |
| **Transfer Rate** | 4/16Mbps |
| **Data Bus** | 16-bit ISA |
| **Topology** | Ring |
| **Wiring Type** | AUI Transceiver via DB-15 port |
| **Boot ROM** | Available |

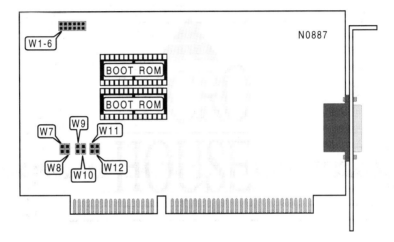

| INTERRUPT SELECTION | | |
|---|---|---|
| IRQ | W7 | W8 |
| ⇨ 2 | Open | Open |
| 3 | Open | Closed |
| 6 | Closed | Open |
| 7 | Closed | Closed |

| NETWORK SPEED SELECTION | |
|---|---|
| Speed | W11 |
| ⇨ 4Mbps | Open |
| 16Mbps | Closed |

| SHARED RAM SIZE | | |
|---|---|---|
| Size | W9 | W10 |
| 8Kb | Open | Open |
| 16Kb | Open | Closed |
| 32Kb | Closed | Open |
| 64Kb | Closed | Closed |

# MICRODYNE CORPORATION
## NTR2000

| ROM ADDRESS | | | | | | |
|---|---|---|---|---|---|---|
| Address | W1 | W2 | W3 | W4 | W5 | W6 |
| ⇨ CC00h | Closed | Open | Open | Closed | Closed | Open |
| C000h | Closed | Open | Open | Open | Open | Open |
| C200h | Closed | Open | Open | Open | Open | Closed |
| C400h | Closed | Open | Open | Open | Closed | Open |
| C600h | Closed | Open | Open | Open | Closed | Closed |
| C800h | Closed | Open | Open | Closed | Open | Open |
| CA00h | Closed | Open | Open | Closed | Open | Closed |
| CE00h | Closed | Open | Open | Closed | Closed | Closed |
| D000h | Closed | Open | Closed | Open | Open | Open |
| DC00h | Closed | Open | Closed | Closed | Closed | Open |

Note: Address DC00h is manufacturer's suggested alternate default setting

| ADAPTER | | |
|---|---|---|
| Selection | Port Address/Shared Memory | W12 |
| Primary | 00A20h/D8000h | Open |
| Alternate | 00A24h/D4000h | Closed |

## MULTI-TECH SYSTEMS, INC.
# EN301TP8/EN301TP16

| | |
|---|---|
| **NIC Type** | Ethernet |
| **Transfer Rate** | 10Mbps |
| **Data Bus** | 8/16-bit ISA |
| **Topology** | Star |
| **Wiring Type** | Unshielded twisted pair |
| **Boot ROM** | Available |

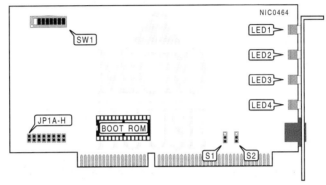

### MISCELLANEOUS NOTES
Although the 16-bit card is shown here, the 8-bit version's jumper settings and locations are identical.

### INTERRUPT REQUEST

| IRQ | JP1A | JP1B | JP1C | JP1D | JP1E | JP1F | JP1G | JP1H |
|---|---|---|---|---|---|---|---|---|
| 2/9 | Open | Open | Open | Open | Open | Open | Open | Closed |
| 5 | Open | Open | Open | Open | Open | Open | Closed | Open |
| 6 | Open | Open | Open | Open | Open | Closed | Open | Open |
| 7 | Open | Open | Open | Open | Closed | Open | Open | Open |
| 10 | Open | Open | Open | Closed | Open | Open | Open | Open |
| 11 | Open | Open | Closed | Open | Open | Open | Open | Open |
| 12 | Open | Closed | Open | Open | Open | Open | Open | Open |
| 15 | Closed | Open | Open | Open | Open | Open | Open | Open |

### SIGNAL QUALITY ERROR (SQE) TEST

| Setting | S1 |
|---|---|
| ⇨Disabled | Pins 2 & 3 closed |
| Enabled | Pins 1 & 2 closed |

Note: Signal Quality Error (SQE) is a test of the collision circuitry and path.

### LINK INTEGRITY TEST

| Type | S2 |
|---|---|
| ⇨Enabled | Pins 1 & 2 closed |
| Disabled | Pins 2 & 3 closed |

## MULTI-TECH SYSTEMS, INC.
### EN301TP8 / EN301TP16

### I/O BASE ADDRESS

| Address | SW1/1 | SW1/2 | SW1/3 | SW1/4 | SW1/5 |
|---|---|---|---|---|---|
| 200 - 21Ah | On | On | On | On | Off |
| 220 - 23Ah | Off | On | On | On | Off |
| 240 - 25Ah | On | Off | On | On | Off |
| 260 - 27Ah | Off | Off | On | On | Off |
| 280 - 29Ah | On | On | Off | On | Off |
| 2A0 - 2BAh | Off | On | Off | On | Off |
| 2C0 - 2DAh | On | Off | Off | On | Off |
| 2E0 - 2FAh | Off | Off | Off | On | Off |
| ➪300 - 31Ah | On | On | On | Off | Off |
| 320 - 33Ah | Off | On | On | Off | Off |
| 340 - 35Ah | On | Off | On | Off | Off |
| 360 - 37Ah | Off | Off | On | Off | Off |
| 380 - 39Ah | On | On | Off | Off | Off |
| 3A0 - 3BAh | Off | On | Off | Off | Off |
| 3C0 - 3DAh | On | Off | Off | Off | Off |
| 3E0 - 3FAh | Off | Off | Off | Off | Off |

### BOOT ROM ADDRESS

| Address | SW1/6 | SW1/7 | SW1/8 |
|---|---|---|---|
| C0000 - C2FFFh | On | On | On |
| C8000 - CAFFFh | Off | On | On |
| ➪D0000 - D2FFFh | On | Off | On |
| D8000 - DAFFFh | Off | Off | On |
| E0000 - E2FFFh | On | On | Off |
| E8000 - EAFFFh | Off | On | Off |
| F0000 - F2FFFh | On | Off | Off |
| F8000 - FAFFFh | Off | Off | Off |

### DIAGNOSTIC LED(S)

| LED | Color | Status | Condition |
|---|---|---|---|
| LED1 | Green | On | Data is being transmitted |
| LED1 | Green | Off | Data is not being transmitted |
| LED2 | Green | On | Data is being received |
| LED2 | Green | Off | Data is not being received |
| LED3 | Yellow | On | Collision detected on network |
| LED3 | Yellow | Off | No collisions detected on network |
| LED4 | Red | On | Card was transmitting continuously (Jabber) (Twisted pair drivers are now disabled) |
| LED4 | Red | Off | Normal operation |

# NDC COMMUNICATIONS, INC.
## ND1120

| | |
|---|---|
| **NIC Type** | Token-Ring |
| **Transfer Rate** | 4/16Mbps |
| **Data Bus** | 16-bit ISA |
| **Topology** | Ring |
| **Wiring Type** | Unshielded twisted pair (IBM Type 3) |
| | Shielded twisted pair via DB-9 port (IBM Type 1) |
| **Boot ROM** | Available |

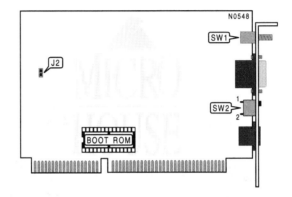

| PRIMARY/ALTERNATE CARD ||
|---|---|
| **Setting** | **SW1** |
| ⇨Primary | Open (Switch not pushed in) |
| Alternate | Closed (Switch pushed in) |
| Note: If two of these cards are to be used in a single system, one of the cards must be set to primary and the other card must be set to alternate. ||

| CABLE TYPE ||
|---|---|
| **Type** | **SW2** |
| ⇨Shielded twisted pair via DB-9 port (IBM Type 1) | Position 1 |
| Unshielded twisted pair (IBM Type 3) | Position 2 |

| COMPATIBILITY MODE ||
|---|---|
| **Setting** | **J2** |
| ⇨128KB address block may be shared with other adapters installed in the system provided there are no conflicts in that memory location. | Closed |
| 128KB address block can only be shared with other 16-bit adapters provided there are no conflicts in that memory location. | Open |

| BOOT ROM CONFIGURATION NOTES |
|---|
| If a Boot ROM is installed, the card must be configured to operate in 8-bit mode (using software setup). |

# NETWORTH, INC.
## UBUTPTRA TOKEN RING

**NIC Type:** Token-Ring
**Transfer Rate:** 4/16Mbps
**Data Bus:** 32-bit EISA, 16-bit ISA, or 8-bit ISA
**Topology:** Ring
**Wiring Type:** Shielded twisted pair
Unshielded twisted pair
**Boot ROM:** Available

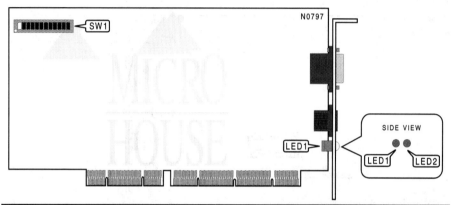

| I/O ADDRESS ||
|---|---|
| **Address** | **SW1/1** |
| A20h | Off |
| A24h | On |

| DATA TRANSFER RATE ||
|---|---|
| **Rate** | **SW1/2** |
| 16Mbps | On |
| 4Mbps | Off |

| SHARED RAM SIZE |||
|---|---|---|
| **Setting** | **SW1/3** | **SW1/4** |
| 16KB | On | Off |
| 8KB | Off | Off |
| 32KB | Off | On |
| 64KB | On | On |

# NETWORTH, INC.
## UBUTPTRA TOKEN RING

| INTERRUPT SELECT |||
|---|---|---|
| IRQ | SW1/5 | SW1/6 |
| IRQ9 (IRQ2 for 8-bit ISA) | Off | Off |
| IRQ3 | On | Off |
| IRQ6 | Off | On |
| IRQ7 | On | On |

| BOOT ROM ADDRESS ||||||| 
|---|---|---|---|---|---|---|
| Address | SW1/7 | SW1/8 | SW1/9 | SW1/10 | SW1/11 | SW1/12 |
| CC000h | Off | On | On | Off | Off | On |
| C0000h | Off | Off | Off | Off | Off | On |
| C2000h | On | Off | Off | Off | Off | On |
| C4000h | Off | On | Off | Off | Off | On |
| C6000h | On | On | Off | Off | Off | On |
| C8000h | Off | Off | On | Off | Off | On |
| CA000h | On | Off | On | Off | Off | On |
| CE000h | On | On | On | Off | Off | On |
| D0000h | Off | Off | Off | On | Off | On |
| D2000h | On | Off | Off | On | Off | On |
| D4000h | Off | On | Off | On | Off | On |
| D6000h | On | On | Off | On | Off | On |
| D8000h | Off | Off | On | On | Off | On |
| DA000h | On | Off | On | On | Off | On |
| DC000h | Off | On | On | On | Off | On |
| DE000h | On | On | On | On | Off | On |

| DIAGNOSTIC LEDS ||||
|---|---|---|---|
| LED | Color | Status | Condition |
| LED1 | Green | On | 16Mbps data rate enabled |
| LED1 | Green | Off | 4Mbps data rate enabled |
| LED2 | Yellow | On | Data is being transmitted or received |
| LED2 | Yellow | Off | Data is not being transmitted or received |

## NETWORTH, INC.
## UTP16C

| | |
|---|---|
| **NIC Type** | Ethernet |
| **Transfer Rate** | 10Mbps |
| **Data Bus** | 16-bit ISA |
| **Topology** | Linear bus |
| | Star |
| **Wiring Type** | Unshielded twisted pair |
| **Boot ROM** | Available |

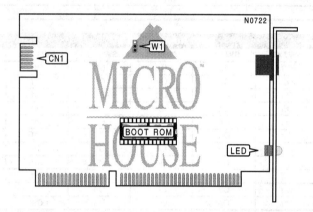

| BOOT ROM ||
|---|---|
| **Setting** | **W1** |
| ▷ Disabled | Pins 2 & 3 closed |
| Enabled | Pins 1 & 2 closed |

| DIAGNOSTIC LED |||
|---|---|---|
| **Color** | **Status** | **Condition** |
| Green | On | Network connection is good |
| Green | Off | Network connection is broken |

| MISCELLANEOUS TECHNICAL NOTE ||
|---|---|
| Note: | CN1 is used by the manufacturer for testing purposes and should not be used. |

## OLICOM INTERNATIONAL
## ISA 16/4 ADAPTER

| | |
|---|---|
| **NIC Type** | Token-Ring |
| **Transfer Rate** | 4/16Mbps |
| **Data Bus** | 8/16-bit ISA |
| **Topology** | Ring |
| **Wiring Type** | Unshielded twisted pair (RJ-45 port) |
| | Shielded twisted pair (DB-9 port) |
| **Boot ROM** | Available |

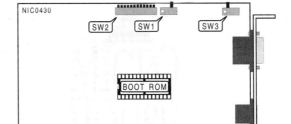

| NETWORK SEGMENT SPEED ||
|---|---|
| **Setting** | **SW1** |
| ⇨4Mbps | Off |
| 16Mbps | On |

Note: All cards on the network segment must have this option set the same.

| BOOT ROM ||
|---|---|
| **Setting** | **SW2/1** |
| ⇨Disabled | On |
| Enabled | Off |

| I/O BASE ADDRESS |||
|---|---|---|
| **Address** | **SW2/9** | **SW2/10** |
| ⇨A20-A23h , A30-A3Fh | Off | Off |
| A24-A27h , A40-A4Fh | On | Off |
| A50-A53h , A60-A6Fh | Off | On |
| A54-A57h , A70-A7Fh | On | On |

## OLICOM INTERNATIONAL
## ISA 16/4 ADAPTER

| DMA CHANNEL | | |
|---|---|---|
| Channel | SW2/11 | SW2/12 |
| ⇨DMA5 | On | On |
| DMA6 | On | Off |
| DMA7 | Off | On |
| Programmed I/O | Off | Off |

Note: If the card is installed into an 8-bit slot, programmed I/O is automatically selected regardless of the switch settings.

| BOOT ROM ADDRESS | | | | | |
|---|---|---|---|---|---|
| Address | SW2/2 | SW2/3 | SW2/4 | SW2/5 | SW2/6 |
| C0000 - C1FFFh | On | On | On | On | On |
| C2000 - C3FFFh | On | On | On | On | Off |
| C4000 - C5FFFh | On | On | On | Off | On |
| C6000 - C7FFFh | On | On | On | Off | Off |
| C8000 - C9FFFh | On | On | Off | On | On |
| CA000 - CBFFFh | On | On | Off | On | Off |
| CC000 - CDFFFh | On | On | Off | Off | On |
| CE000 - CFFFFh | On | On | Off | Off | Off |
| D0000 - D1FFFh | On | Off | On | On | On |
| D2000 - D3FFFh | On | Off | On | On | Off |
| D4000 - D5FFFh | On | Off | On | Off | On |
| D6000 - D7FFFh | On | Off | On | Off | Off |
| D8000 - D9FFFh | On | Off | Off | On | On |
| DA000 - DBFFFh | On | Off | Off | On | Off |
| DC000 - DDFFFh | On | Off | Off | Off | On |
| DE000 - DFFFFh | On | Off | Off | Off | Off |
| E0000 - E1FFFh | Off | On | On | On | On |
| E2000 - E3FFFh | Off | On | On | On | Off |
| E4000 - E5FFFh | Off | On | On | Off | On |
| E6000 - E7FFFh | Off | On | On | Off | Off |
| E8000 - E9FFFh | Off | On | Off | On | On |
| EA000 - EBFFFh | Off | On | Off | On | Off |
| EC000 - EDFFFh | Off | On | Off | Off | On |
| EE000 - EFFFFh | Off | On | Off | Off | Off |
| F0000 - F1FFFh | Off | Off | On | On | On |
| F2000 - F3FFFh | Off | Off | On | On | Off |
| F4000 - F5FFFh | Off | Off | On | Off | On |
| F6000 - F7FFFh | Off | Off | On | Off | Off |
| F8000 - F9FFFh | Off | Off | Off | On | On |
| FA000 - FBFFFh | Off | Off | Off | On | Off |
| FC000 - FDFFFh | Off | Off | Off | Off | On |
| FE000 - FFFFFh | Off | Off | Off | Off | Off |

## OLICOM INTERNATIONAL
ISA 16/4 ADAPTER

| INTERRUPT REQUEST | | |
|---|---|---|
| IRQ | SW2/7 | SW2/8 |
| ⇨2/9 | Off | Off |
| 3 | On | Off |
| 10 | Off | On |
| 11 | On | On |

| CABLE TYPE | |
|---|---|
| Type | SW3 |
| ⇨Shielded twisted pair (DB-9 port) | On |
| Unshielded twisted pair (RJ-45 port) | Off |

# OLICOM INTERNATIONAL
## OC-3109

| | |
|---|---|
| **NIC Type** | Token-Ring |
| **Transfer Rate** | 4Mbps |
| **Data Bus** | 8-bit ISA |
| **Topology** | Ring |
| **Wiring Type** | Shielded twisted pair (DB-9 port) |
| **Boot ROM** | Available |

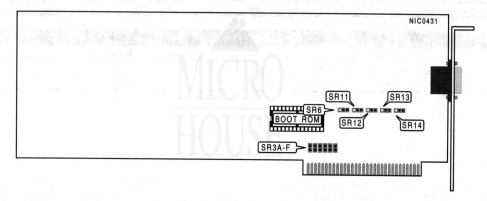

### INTERRUPT REQUEST

| IRQ | SR3A | SR3B | SR3C | SR3D | SR3E | SR3F |
|---|---|---|---|---|---|---|
| ⇨2 | Closed | Open | Open | Open | Open | Open |
| 3 | Open | Closed | Open | Open | Open | Open |
| 4 | Open | Open | Closed | Open | Open | Open |
| 5 | Open | Open | Open | Closed | Open | Open |
| 6 | Open | Open | Open | Open | Closed | Open |
| 7 | Open | Open | Open | Open | Open | Closed |

### CLOCK SELECT

| Setting | SR6 |
|---|---|
| ⇨Onboard clock | Pins 1 & 2 closed |
| System bus clock | Pins 2 & 3 closed |

Note: If the system's bus clock speed is greater than 8MHz the onboard clock must be enabled.

### BOOT ROM SIZE

| Size | SR11 | SR12 |
|---|---|---|
| ⇨8/16KB | Pins 1 & 2 closed | Pins 1 & 2 closed |
| 32KB | Pins 2 & 3 closed | Pins 1 & 2 closed |
| 64KB | Pins 2 & 3 closed | Pins 2 & 3 closed |

## OLICOM INTERNATIONAL
## OC-3109

| BOOT ROM ||
|---|---|
| Setting | SR13 |
| ⇨Disabled | Pins 1 & 2 closed |
| Enabled | Pins 2 & 3 closed |

| PRIMARY/ALTERNATE SELECT, I/O BASE ADDRESS, AND BOOT ROM ADDRESS ||||
|---|---|---|---|
| Setting | I/O Address | Boot ROM Address | SR14 |
| ⇨Primary | 0A20 - 0A2Fh | D0000 - D7FFFh | Pins 1 & 2 closed |
| Alternate | 1A20 - 1A2Fh | D8000 - DFFFFh | Pins 2 & 3 closed |

Note: If two cards are to be used in a single system, one of the cards must be set to primary and the other must be set to alternate.

## PROTEON, INC.
# PRONET-10 P1308

| | |
|---|---|
| **NIC Type** | Token Ring |
| **Transfer Rate** | 4/16Mbps |
| **Data Bus** | 16-bit ISA |
| **Topology** | Star |
| **Wiring Type** | Unshielded twisted pair |
| | AUI transceiver via DB-15 port |
| **Boot ROM** | Available |

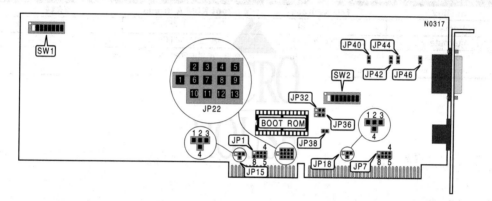

| Node | SW1/1 | SW1/2 | SW1/3 | SW1/4 | SW1/5 | SW1/6 | SW1/7 | SW1/8 |
|---|---|---|---|---|---|---|---|---|
| 0 | - | - | - | - | - | - | - | - |
| 1 | Off | On | On | On | On | On | On | On |
| 2 | On | Off | On | On | On | On | On | On |
| 3 | Off | Off | On | On | On | On | On | On |
| 4 | On | On | Off | On | On | On | On | On |
| 251 | Off | Off | On | Off | Off | Off | Off | Off |
| 252 | On | On | Off | Off | Off | Off | Off | Off |
| 253 | Off | On | Off | Off | Off | Off | Off | Off |
| 254 | On | Off | Off | Off | Off | Off | Off | Off |
| 255 | Off | Off | Off | Off | Off | Off | Off | Off |

Note: Node address 0 is used for messaging between nodes and must not be used.
A total of 255 node address settings are available. The switches are a binary representation of the decimal node addresses. Switch 1 is the Least Significant Bit and switch 8 is the Most Significant Bit. The switches have the following decimal values: switch 1=1, 2=2, 3=4, 4=8, 5=16, 6=32, 7=64, 8=128. Turn off the switches and add the values of the off switches to obtain the correct node address. (On=0, off=1)

## PROTEON, INC.
## PRONET-10 P1308

| 16-BIT DMA CHANNEL ||| 
|---|---|---|
| **Channel** | **JP1** | **JP15** |
| ⇨DMA5 | Pins 3 & 4 and 5 & 6 closed | Pins 2 & 3 closed |
| DMA6 | Pins 3 & 4 and 2 & 7 closed | Pins 2 & 4 closed |
| DMA7 | Pins 3 & 4 and 1 & 8 closed | Pins 1 & 2 closed |
| Disabled | Open | Open |

Note: For use in an AT-class machine only.

| 8-BIT DMA CHANNEL |||
|---|---|---|
| **Channel** | **JP7** | **JP18** |
| ⇨DMA1 | Pins 3 & 4 and 2 & 7 closed | Pins 1 & 2 closed |
| DMA2 | Pins 3 & 4 and 1 & 8 closed | Pins 2 & 3 closed |
| DMA3 | Pins 3 & 4 and 5 & 6 closed | Pins 2 & 4 closed |
| Disabled | Open | Open |

Note: Can be used in either a PC/XT-class machine or an AT-class machine.

| INTERRUPT REQUEST ||
|---|---|
| **IRQ** | **JP22** |
| ⇨2 | Pins 2 & 6 closed |
| 3 | Pins 5 & 9 closed |
| 4 | Pins 4 & 8 closed |
| 5 | Pins 3 & 7 closed |
| 10 | Pins 9 & 13 closed |
| 11 | Pins 8 & 12 closed |
| 12 | Pins 7 & 11 closed |
| 14 | Pins 1 & 6 closed |
| 15 | Pins 6 & 10 closed |

| COMPATIBILITY MODE |||
|---|---|---|
| **Setting** | **JP32** | **JP36** |
| ⇨Compatibility mode enabled | Pins 1 & 2 closed | Pins 1 & 2 closed |
| Extended mode enabled | Pins 2 & 3 closed | Pins 2 & 3 closed |

Note: Extended mode provides 16-bit data transfers and hardware checksums in an AT-class machine when the card is installed in a 16-bit slot. When the card is installed in an 8-bit slot, extended mode only provides hardware checksums.

| BOOT ROM ||
|---|---|
| **Setting** | **JP38** |
| ⇨Disabled | Open |
| Enabled | Closed |

## PROTEON, INC.
## PRONET-10 P1308

| CABLE TYPE | | | | | |
|---|---|---|---|---|---|
| Type | | JP40 | JP42 | JP44 | JP46 |
| Unshielded twisted pair | | Closed | Closed | Closed | Closed |
| AUI transceiver via DB-15 port | | Open | Open | Open | Open |

| BOOT ROM ADDRESS | | | | |
|---|---|---|---|---|
| Address | SW2/1 | SW2/2 | SW2/3 | SW2/4 |
| C0000h | On | On | On | On |
| C4000h | On | On | On | Off |
| C8000h | On | On | Off | On |
| CC000h | On | On | Off | Off |
| D0000h | On | Off | On | On |
| D4000h | On | Off | On | Off |
| ⇨D8000h | On | Off | Off | On |
| DC000h | On | Off | Off | Off |
| E0000h | Off | On | On | On |
| E4000h | Off | On | On | Off |
| E8000h | Off | On | Off | On |
| EC000h | Off | On | Off | Off |
| F0000h | Off | Off | On | On |
| F4000h | Off | Off | On | Off |
| F8000h | Off | Off | Off | On |
| FC000h | Off | Off | Off | Off |

| I/O BASE ADDRESS | | | | |
|---|---|---|---|---|
| Address | SW2/5 | SW2/6 | SW2/7 | SW2/8 |
| 300h[1] | On | On | On | On |
| 308h | On | On | On | Off |
| 310h[1] | On | On | Off | On |
| 318h | On | On | Off | Off |
| 320h[1] | On | Off | On | On |
| 328h | On | Off | On | Off |
| 330h[1] | On | Off | Off | On |
| 338h | On | Off | Off | Off |
| 340h[1] | Off | On | On | On |
| 348h | Off | On | On | Off |
| 350h[1] | Off | On | Off | On |
| 358h | Off | On | Off | Off |
| 360h[1] | Off | Off | On | On |
| 368h | Off | Off | On | Off |
| 370h[1] | Off | Off | Off | On |
| 378h | Off | Off | Off | Off |

Note: Use any of the above settings when compatibility mode (JP32 & JP36) is enabled.
Note [1]: Use only these settings when extended mode (JP32 & JP36) is enabled.

## PURE DATA, LTD.
### PDI7023-16CAT

| | |
|---|---|
| **NIC Type** | Ethernet |
| **Transfer Rate** | 10Mbps |
| **Data Bus** | 16-bit ISA |
| **Topology** | Star |
| | Linear bus |
| **Wiring Type** | Unshielded twisted pair |
| | RG58A/U 50ohm coaxial |
| | AUI transceiver via DB-15 port |
| **Boot ROM** | Not available |

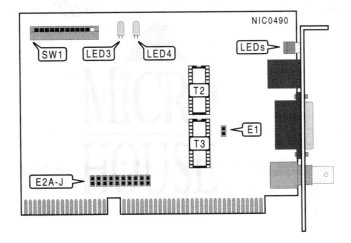

| COAXIAL SEGMENT LENGTH | |
|---|---|
| **Maximum Length** | **E1** |
| ▷200 meters | Closed |
| 300 meters | Open |

| INTERRUPT REQUEST | | | | | | | | | | |
|---|---|---|---|---|---|---|---|---|---|---|
| **IRQ** | **E2A** | **E2B** | **E2C** | **E2D** | **E2E** | **E2F** | **E2G** | **E2H** | **E2I** | **E2J** |
| 2/9 | Open | Open | Open | Open | Closed | Open | Open | Open | Open | Open |
| 3 | Open | Open | Open | Open | Open | Open | Open | Open | Open | Closed |
| ▷4 | Open | Open | Open | Open | Open | Open | Open | Open | Closed | Open |
| 5 | Open | Open | Open | Open | Open | Open | Open | Closed | Open | Open |
| 6 | Open | Open | Open | Open | Open | Open | Closed | Open | Open | Open |
| 7 | Open | Open | Open | Open | Open | Closed | Open | Open | Open | Open |
| 10 | Open | Open | Open | Closed | Open | Open | Open | Open | Open | Open |
| 11 | Open | Open | Closed | Open | Open | Open | Open | Open | Open | Open |
| 12 | Open | Closed | Open | Open | Open | Open | Open | Open | Open | Open |
| 15 | Closed | Open | Open | Open | Open | Open | Open | Open | Open | Open |

## PURE DATA, LTD.
### PDI7023-16CAT

| CABLE TYPE | |
|---|---|
| Type | Isolation Transformer Location |
| ⇨Unshielded twisted pair | N/A |
| RG58A/U 50ohm coaxial | T2 |
| AUI transceiver via DB-15 port | T3 |

Note: The cable type is automatically selected, but the isolation transformer must be moved to the proper socket to enable the BNC connector or DB-15 port.

| I/O BASE ADDRESS | | | |
|---|---|---|---|
| Address | SW1/1 | SW1/2 | SW1/3 |
| 260h | On | On | On |
| 290h | Off | On | On |
| 2E0h | On | Off | On |
| 2F0h | Off | Off | On |
| ⇨300h | On | On | Off |
| 350h | Off | On | Off |
| 380h | On | Off | Off |
| 3E0h | Off | Off | Off |

| EXTENDED TRANSMIT TIMEOUT | |
|---|---|
| Setting | SW1/5 |
| ⇨Enabled | On |
| Disabled | Off |

Note: With this option enabled, if the card is transmitting for an extended amount of time the transmission will be interrupted and continued after a preset time. This keeps one card from dominating the network.

| LINK INTEGRITY TEST | |
|---|---|
| Setting | SW1/7 |
| ⇨Enabled | On |
| Disabled | Off |

Note: The link integrity test is valid only when the cable type is unshielded twisted pair.

## PURE DATA, LTD.
## PDI7023-16CAT

| FACTORY CONFIGURED SETTINGS ||
|---|---|
| Switch | Setting |
| SW1/4 | Off |
| SW1/6 | On |
| SW1/8 | Off |

| DIAGNOSTIC LED(S) ||||
|---|---|---|---|
| LED | Color | Status | Condition |
| LED1 | Green | On | Data is being transmitted or received |
| LED1 | Green | Off | Data is not being transmitted or received |
| LED2 | Red | On | Card is being accessed by host computer |
| LED2 | Red | Off | Card is not being accessed by host computer |
| LED3 | Red | On | Collision detected on network |
| LED3 | Red | Off | Normal operation |
| LED4 | Green | On | Cable type is AUI via DB-15 or RG-58A/U coaxial |
| LED4 | Green | Off | Cable type is Twisted pair |

# PURE DATA, LTD.
## PDI8023-16CAT

| | |
|---|---|
| **NIC Type** | Ethernet |
| **Transfer Rate** | 10Mbps |
| **Data Bus** | 16-bit ISA |
| **Topology** | Star |
| | Linear bus |
| **Wiring Type** | Unshielded twisted pair |
| | RG58A/U 50ohm coaxial |
| | AUI transceiver via DB-15 port |
| **Boot ROM** | Available |

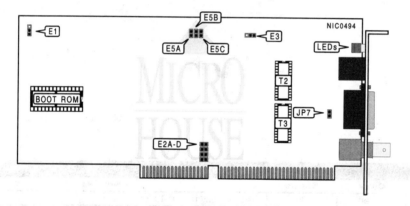

| SOFTWARE CONFIGURATION RESET | |
|---|---|
| **Setting** | **E1** |
| ⇨Normal operation | Pins 2 & 3 closed |
| Reset | Pins 1 & 2 closed |

Note: Many of the options are configured using the setup utility provided with the card. To reset those configurations to the factory defaults, move E1 to the Reset position (pins 1 & 2 closed) and apply power for a few seconds. Always make sure that you turn off the power to the computer before you change any jumpers.

| CARD I/D NUMBER | | | | |
|---|---|---|---|---|
| **Number** | **E2A** | **E2B** | **E2C** | **E2D** |
| ⇨1 | Open | Open | Open | Closed |
| 2 | Open | Open | Closed | Open |
| 3 | Open | Closed | Open | Open |
| 4 | Closed | Open | Open | Open |

Note: Up to four cards can be installed into a system at the same time. If more than one card is installed each one must have a separate card I/D number.

## PURE DATA, LTD.
## PDI8023-16CAT

| LINK INTEGRITY TEST ||
|---|---|
| **Setting** | **E3** |
| ⇨Enabled | Pins 1 & 2 closed |
| Disabled | Pins 2 & 3 closed |

Note: The link integrity test is only valid when the cable type is unshielded twisted pair.

| CABLE TYPE |||||
|---|---|---|---|---|
| **Type** | **E5A** | **E5B** | **E5C** | **Isolation Transformer Location** |
| ⇨RG58A/U 50ohm coaxial | Open | Closed | Open | T3 |
| Unshielded twisted pair | Open | Open | Closed | N/A |
| AUI transceiver via DB-15 port | Closed | Open | Open | T2 |

Note: If you leave E5B closed you can select the cable type through the software setup utility included with the card. However, you must still move the Isolation transformer to the appropriate socket.

| SEGMENT LENGTH ||
|---|---|
| **Maximum Length** | **JP7** |
| ⇨185 meters | Pins 2 & 3 closed |
| 300 meters | Pins 1 & 2 closed |

Note: Segment length is the total length of cable between the two farthest cards on the segment. Each card on the segment must have this option set the same.

| DIAGNOSTIC LED(S) ||||
|---|---|---|---|
| **LED** | **Color** | **Status** | **Condition** |
| LED1 | Green | On | Data is being transmitted or received |
| LED1 | Green | Off | Data is not being transmitted or received |
| LED2 | Red | On | Card is being accessed by host computer |
| LED2 | Red | Off | Card is not being accessed by host computer |

# PURE DATA LTD.
## PDI8025

| | |
|---|---|
| Transfer Rate | 4/16Mbps |
| Data Bus | 16-bit ISA |
| Topology | Ring |
| Wiring Type | Shielded/Unshielded twisted pair |
| Boot ROM | Available |

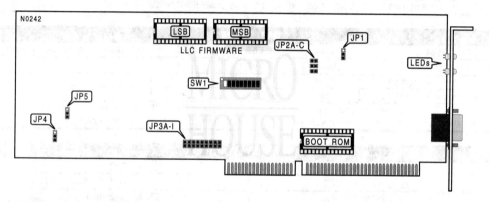

| WATCHDOG TIMER ||
|---|---|
| Setting | JP1 |
| ⇨Watchdog timer enabled | Pins 2 & 3 Closed |
| Watchdog timer disabled | Pins 1 & 2 Closed |

Note: The watchdog timer acts as a checksum for each data packet.

| DMA CHANNEL ||||
|---|---|---|---|
| Channel | JP2A | JP2B | JP2C |
| ⇨DMA5 | Closed | Open | Open |
| DMA6 | Open | Closed | Open |
| DMA7 | Open | Open | Closed |

| INTERRUPT REQUEST ||||||||||
|---|---|---|---|---|---|---|---|---|---|
| IRQ | JP3A | JP3B | JP3C | JP3D | JP3E | JP3F | JP3G | JP3H | JP3I |
| ⇨3 | Open | Open | Open | Open | Open | Open | Open | Open | Closed |
| 4 | Open | Open | Open | Open | Open | Open | Open | Closed | Open |
| 5 | Open | Open | Open | Open | Open | Open | Closed | Open | Open |
| 7 | Open | Open | Open | Open | Open | Closed | Open | Open | Open |
| 9 | Open | Open | Open | Open | Closed | Open | Open | Open | Open |
| 10 | Open | Open | Open | Closed | Open | Open | Open | Open | Open |
| 11 | Open | Open | Closed | Open | Open | Open | Open | Open | Open |
| 12 | Open | Closed | Open | Open | Open | Open | Open | Open | Open |
| 15 | Closed | Open | Open | Open | Open | Open | Open | Open | Open |

# PURE DATA LTD.
## PDI8025

| LLC EPROM ENABLE | |
|---|---|
| **Setting** | **JP4** |
| ⇨External EPROMs enabled | Pins 2 & 3 Closed |
| External EPROMs disabled | Pins 1 & 2 Closed |
| Note: The LLC EPROMs contain the Logic Link Control program. These chips are installed on the card at the factory. The card will function properly with the LLC chips disabled. ||

| CARD TIMING | |
|---|---|
| **Setting** | **JP5** |
| ⇨Oscillator select internal (6MHz) | Pins 2 & 3 Closed |
| Oscillator select external (Host bus speed of 8 or 10MHz) | Pins 1 & 2 Closed |

| DIAGNOSTIC LED(S) | | |
|---|---|---|
| **Color** | **Status** | **Condition** |
| Red | On/Blinking | I/O Bus activity detected |
| Red | Off | No I/O Bus activity detected |
| Green | On | Card is attached to network |
| Green | Off | Card is not initializing |

## PURE DATA LTD.
## PDI8025

| I/O | Memory | SW1/1 | SW1/2 | SW1/3 | SW1/4 | SW1/5 | SW1/6 | SW1/7 | SW1/8 |
|---|---|---|---|---|---|---|---|---|---|
| 0A20h | Disabled | On | On | On | On | On | On | On | On |
| 0A20h | C0000 | On | On | On | On | Off | On | On | On |
| 0A20h | C4000 | On | On | On | Off | On | On | On | On |
| 0A20h | CC000 | On | On | On | Off | Off | On | On | On |
| 0A20h | D0000 | On | On | Off | On | On | On | On | On |
| 0A20h | D4000 | On | On | Off | On | Off | On | On | On |
| 0A20h | D8000 | On | On | Off | Off | On | On | On | On |
| 0A20h | DC000 | On | On | Off | Off | Off | On | On | On |
| 1A20h | Disabled | On | Off | On | On | On | On | On | On |
| 1A20h | C0000 | On | Off | On | On | Off | On | On | On |
| 1A20h | C4000 | On | Off | On | Off | On | On | On | On |
| 1A20h | CC000 | On | Off | On | Off | Off | On | On | On |
| 1A20h | D0000 | On | Off | Off | On | On | On | On | On |
| 1A20h | D4000 | On | Off | On | On | Off | On | On | On |
| 1A20h | D8000 | On | Off | Off | Off | On | On | On | On |
| 1A20h | DC000 | On | Off | Off | Off | Off | On | On | On |
| 2A20h | Disabled | Off | On | On | On | On | On | On | On |
| 2A20h | C0000 | Off | On | On | On | Off | On | On | On |
| 2A20h | C4000 | Off | On | On | Off | On | On | On | On |
| 2A20h | CC000 | Off | On | On | Off | Off | On | On | On |
| 2A20h | D0000 | Off | On | Off | On | On | On | On | On |
| 2A20h | D4000 | Off | On | Off | On | Off | On | On | On |
| 2A20h | D8000 | Off | On | Off | Off | On | On | On | On |
| 2A20h | DC000 | Off | On | Off | Off | Off | On | On | On |
| 3A20h | Disabled | Off | Off | On | On | On | On | On | On |
| 3A20h | C0000 | Off | Off | On | On | Off | On | On | On |
| 3A20h | C4000 | Off | Off | On | Off | On | On | On | On |
| 3A20h | CC000 | Off | Off | On | Off | Off | On | On | On |
| 3A20h | D0000 | Off | Off | Off | On | On | On | On | On |
| 3A20h | D4000 | Off | Off | Off | On | Off | On | On | On |
| 3A20h | D8000 | Off | Off | Off | Off | On | On | On | On |
| 3A20h | DC000 | Off | Off | Off | Off | Off | On | On | On |

# PURE DATA LTD.
## PDI9025-16

| | |
|---|---|
| NIC Type | Token Ring |
| Transfer Rate | 4/16Mbps |
| Data Bus | 16-bit ISA |
| Topology | Ring |
| Wiring Type | Shielded/Unshielded twisted pair |
| Boot ROM | Available |

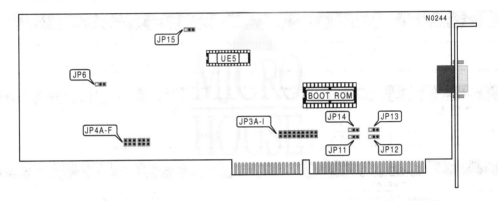

### INTERRUPT REQUEST

| IRQ | JP3A | JP3B | JP3C | JP3D | JP3E | JP3F | JP3G | JP3H | JP3I |
|---|---|---|---|---|---|---|---|---|---|
| 3 | Closed | Open | Open | Open | Open | Open | Open | Open | Open |
| 4 | Open | Closed | Open | Open | Open | Open | Open | Open | Open |
| 5 | Open | Open | Closed | Open | Open | Open | Open | Open | Open |
| 6 | Open | Open | Open | Closed | Open | Open | Open | Open | Open |
| 7 | Open | Open | Open | Open | Closed | Open | Open | Open | Open |
| ⇨2/9 | Open | Open | Open | Open | Open | Closed | Open | Open | Open |
| 10 | Open | Open | Open | Open | Open | Open | Closed | Open | Open |
| 11 | Open | Open | Open | Open | Open | Open | Open | Closed | Open |
| 12 | Open | Open | Open | Open | Open | Open | Open | Open | Closed |

### DMA CHANNEL

| Channel | JP4A | JP4B | JP4C | JP4D | JP4E | JP4F |
|---|---|---|---|---|---|---|
| ⇨DMA5 | Open | Open | Closed | Open | Open | Closed |
| DMA6 | Open | Closed | Open | Open | Closed | Open |
| DMA7 | Closed | Open | Open | Closed | Open | Open |

### CARD TIMING

| Setting | JP6 |
|---|---|
| ⇨Oscillator select internal (6MHz) | Pins 2 & 3 Closed |
| Oscillator select external (Host bus speed of 8 or 10MHz) | Pins 1 & 2 Closed |

## PURE DATA LTD.
## PDI9025-16

| BOOT ROM SIZE | | |
|---|---|---|
| ROM size | JP11 | JP12 |
| ⇨8/16KB | Pins 2 & 3 Closed | Pins 2 & 3 Closed |
| 32KB | Pins 1 & 2 Closed | Pins 2 & 3 Closed |
| 64KB | Pins 1 & 2 Closed | Pins 1 & 2 Closed |

| BOOT ROM | |
|---|---|
| Setting | JP13 |
| Disabled | Pins 2 & 3 Closed |
| Enabled | Pins 1 & 2 Closed |

| ALTERNATE/PRIMARY CARD SELECT | | | |
|---|---|---|---|
| Setting | Address | Boot ROM Address | JP14 |
| ⇨Primary | 0A20-0A2Fh | D0000-D7FFFh | Pins 2 & 3 Closed |
| Alternate | 1A20-1A2Fh | D8000-DFFFFh | Pins 1 & 2 Closed |

| BRIDGE PAL ENABLED | |
|---|---|
| Setting | JP15 |
| ⇨Bridge PAL disabled | Pins 2 & 3 Closed |
| Bridge PAL enabled | Pins 1 & 2 Closed |

Note: If installing two cards for a bridge configuration, both cards must be set to enable the bridge PALs and bridge PAL chips must be installed at UE5 of each card.

## PURE DATA, LTD.
### PDN508

| | |
|---|---|
| NIC Type | ARCNET |
| Transfer Rate | 2.5Mbps |
| Data Bus | NEC MultiSpeed Interface |
| Topology | Star |
| Wiring Type | RG-62A/U 93ohm coaxial |
| Boot ROM | Not available |

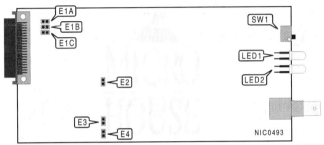

### NODE ADDRESS
Note: The node address is selected using the software supplied with the card.

### INTERRUPT REQUEST

| IRQ | E1A | E1B | E1C |
|---|---|---|---|
| ⇨2 | Open | Open | Closed |
| 3 | Open | Closed | Open |
| 5 | Closed | Open | Open |

### I/O BASE ADDRESS

| Address | E2 |
|---|---|
| ⇨2E0h | Closed |
| 300h | Open |

### CABLE LENGTH AND RESPONSE/RECONFIGURATION TIMEOUTS

| Cable Length | Response Time | Reconfiguration Time | E3 | E4 |
|---|---|---|---|---|
| ⇨< 20,000 feet | 74.7µs | 840ms | Open | Open |
| > 20,000 feet | 283.4µs | 1680ms | Open | Closed |
| > 20,000 feet | 561.8µs | 1680ms | Closed | Open |
| > 20,000 feet | 1118.6µs | 1680ms | Closed | Closed |

Note: All NICs on the network segment must have this option set the same.

### NIC POWER

| Setting | SW1 |
|---|---|
| ⇨On | On |
| Off | Off |

### DIAGNOSTIC LED(S)

| LED | Color | Status | Condition |
|---|---|---|---|
| LED1 | Red | On | Data is being transmitted or received |
| LED1 | Red | Off | Data is not being transmitted or received |
| LED2 | Green | On | Card is reconfiguring |
| LED2 | Green | Off | Normal operation |

## RACAL-INTERLAN, INC.
## 16-BIT ETHERNET CONTROLLER

| | |
|---|---|
| **NIC Type:** | Ethernet |
| **Transfer Rate:** | 10Mbps |
| **Data Bus:** | 16-bit ISA |
| **Topology:** | Linear Bus |
| **Wiring Type** | AUI transceiver via DB-15 port |
| | RG58A/U 50ohm coaxial |
| **Boot ROM** | Available |

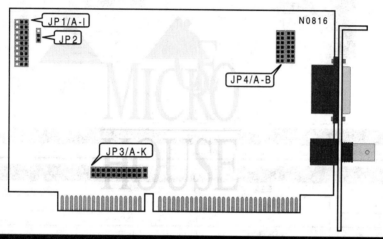

| MEMORY ADDRESS - JP1 | | | | |
|---|---|---|---|---|
| **Address** | **Jumper A** | **Jumper B** | **Jumper C** | **Jumper D** |
| ⇨ D0000h | Pins 2 & 3 closed | Pins 1 & 2 closed | Pins 1 & 2 closed | Pins 1 & 2 closed |
| C0000h | Pins 1 & 2 closed | Pins 1 & 2 closed | Pins 1 & 2 closed | Pins 1 & 2 closed |
| C2000h | Pins 1 & 2 closed | Pins 1 & 2 closed | Pins 1 & 2 closed | Pins 2 & 3 closed |
| C4000h | Pins 1 & 2 closed | Pins 1 & 2 closed | Pins 2 & 3 closed | Pins 1 & 2 closed |
| C6000h | Pins 1 & 2 closed | Pins 1 & 2 closed | Pins 2 & 3 closed | Pins 2 & 3 closed |
| C8000h | Pins 1 & 2 closed | Pins 2 & 3 closed | Pins 1 & 2 closed | Pins 1 & 2 closed |
| CA000h | Pins 1 & 2 closed | Pins 2 & 3 closed | Pins 1 & 2 closed | Pins 2 & 3 closed |
| CC000h | Pins 1 & 2 closed | Pins 2 & 3 closed | Pins 2 & 3 closed | Pins 1 & 2 closed |
| CE000h | Pins 1 & 2 closed | Pins 2 & 3 closed | Pins 2 & 3 closed | Pins 2 & 3 closed |
| D2000h | Pins 2 & 3 closed | Pins 1 & 2 closed | Pins 1 & 2 closed | Pins 2 & 3 closed |
| D4000h | Pins 2 & 3 closed | Pins 1 & 2 closed | Pins 2 & 3 closed | Pins 1 & 2 closed |
| D6000h | Pins 2 & 3 closed | Pins 1 & 2 closed | Pins 2 & 3 closed | Pins 2 & 3 closed |
| D8000h | Pins 2 & 3 closed | Pins 2 & 3 closed | Pins 1 & 2 closed | Pins 1 & 2 closed |
| DA000h | Pins 2 & 3 closed | Pins 2 & 3 closed | Pins 1 & 2 closed | Pins 2 & 3 closed |
| DC000h | Pins 2 & 3 closed | Pins 2 & 3 closed | Pins 2 & 3 closed | Pins 1 & 2 closed |
| DE000h | Pins 2 & 3 closed | Pins 2 & 3 closed | Pins 2 & 3 closed | Pins 2 & 3 closed |

# RACAL-INTERLAN, INC.
## 16-BIT ETHERNET CONTROLLER

| I/O ADDRESS - JP1 | | | | | |
|---|---|---|---|---|---|
| Address | Jumper E | Jumper F | Jumper G | Jumper H | Jumper I |
| ⇨ 360h | 2 & 3 closed | 1 & 2 closed | 2 & 3 closed | 2 & 3 closed | 1 & 2 closed |
| 200h | 1 & 2 closed | 1 & 2 closed | 1 & 2 closed | 1 & 2 closed | 1 & 2 closed |
| 210h | 1 & 2 closed | 1 & 2 closed | 1 & 2 closed | 1 & 2 closed | 2 & 3 closed |
| 220h | 1 & 2 closed | 1 & 2 closed | 1 & 2 closed | 2 & 3 closed | 1 & 2 closed |
| 230h | 1 & 2 closed | 1 & 2 closed | 1 & 2 closed | 2 & 3 closed | 2 & 3 closed |
| 240h | 1 & 2 closed | 1 & 2 closed | 2 & 3 closed | 1 & 2 closed | 1 & 2 closed |
| 250h | 1 & 2 closed | 1 & 2 closed | 2 & 3 closed | 1 & 2 closed | 2 & 3 closed |
| 260h | 1 & 2 closed | 1 & 2 closed | 2 & 3 closed | 2 & 3 closed | 1 & 2 closed |
| 270h | 1 & 2 closed | 1 & 2 closed | 2 & 3 closed | 2 & 3 closed | 2 & 3 closed |
| 280h | 1 & 2 closed | 2 & 3 closed | 1 & 2 closed | 1 & 2 closed | 1 & 2 closed |
| 290h | 1 & 2 closed | 2 & 3 closed | 1 & 2 closed | 1 & 2 closed | 2 & 3 closed |
| 2A0h | 1 & 2 closed | 2 & 3 closed | 1 & 2 closed | 2 & 3 closed | 1 & 2 closed |
| 2B0h | 1 & 2 closed | 2 & 3 closed | 1 & 2 closed | 2 & 3 closed | 2 & 3 closed |
| 2C0h | 1 & 2 closed | 2 & 3 closed | 2 & 3 closed | 1 & 2 closed | 1 & 2 closed |
| 2D0h | 1 & 2 closed | 2 & 3 closed | 2 & 3 closed | 1 & 2 closed | 2 & 3 closed |
| 2E0h | 1 & 2 closed | 2 & 3 closed | 2 & 3 closed | 2 & 3 closed | 1 & 2 closed |
| 2F0h | 1 & 2 closed | 2 & 3 closed | 2 & 3 closed | 2 & 3 closed | 2 & 3 closed |
| 300h | 2 & 3 closed | 1 & 2 closed | 1 & 2 closed | 1 & 2 closed | 1 & 2 closed |
| 310h | 2 & 3 closed | 1 & 2 closed | 1 & 2 closed | 1 & 2 closed | 2 & 3 closed |
| 320h | 2 & 3 closed | 1 & 2 closed | 1 & 2 closed | 2 & 3 closed | 1 & 2 closed |
| 330h | 2 & 3 closed | 1 & 2 closed | 1 & 2 closed | 2 & 3 closed | 2 & 3 closed |
| 340h | 2 & 3 closed | 1 & 2 closed | 2 & 3 closed | 1 & 2 closed | 1 & 2 closed |
| 350h | 2 & 3 closed | 1 & 2 closed | 2 & 3 closed | 1 & 2 closed | 2 & 3 closed |
| 370h | 2 & 3 closed | 1 & 2 closed | 2 & 3 closed | 2 & 3 closed | 2 & 3 closed |
| 380h | 2 & 3 closed | 2 & 3 closed | 1 & 2 closed | 1 & 2 closed | 1 & 2 closed |
| 390h | 2 & 3 closed | 2 & 3 closed | 1 & 2 closed | 1 & 2 closed | 2 & 3 closed |
| 3A0h | 2 & 3 closed | 2 & 3 closed | 1 & 2 closed | 2 & 3 closed | 1 & 2 closed |
| 3B0h | 2 & 3 closed | 2 & 3 closed | 1 & 2 closed | 2 & 3 closed | 2 & 3 closed |
| 3C0h | 2 & 3 closed | 2 & 3 closed | 2 & 3 closed | 1 & 2 closed | 1 & 2 closed |
| 3D0h | 2 & 3 closed | 2 & 3 closed | 2 & 3 closed | 1 & 2 closed | 2 & 3 closed |
| 3E0h | 2 & 3 closed | 2 & 3 closed | 2 & 3 closed | 2 & 3 closed | 1 & 2 closed |
| 3F0h | 2 & 3 closed | 2 & 3 closed | 2 & 3 closed | 2 & 3 closed | 2 & 3 closed |

| BOOT ROM | |
|---|---|
| Setting | JP2 |
| ⇨ Disabled | Pins 1 & 2 closed |
| Enabled | Pins 2 & 3 closed |

## RACAL-INTERLAN, INC.
## 16-BIT ETHERNET CONTROLLER

| INTERRUPT SELECT - JP3 | | | | | | |
|---|---|---|---|---|---|---|
| IRQ | Jumper F | Jumper G | Jumper H | Jumper I | Jumper J | Jumper K |
| ⇨ IRQ7 | Open | Closed | Open | Open | Open | Open |
| IRQ3 | Open | Open | Open | Open | Open | Closed |
| IRQ4 | Open | Open | Open | Open | Closed | Open |
| IRQ5 | Open | Open | Open | Closed | Open | Open |
| IRQ6 | Open | Open | Closed | Open | Open | Open |
| IRQ9 | Closed | Open | Open | Open | Open | Open |

| INTERRUPT SELECT - JP3 | | | | | | |
|---|---|---|---|---|---|---|
| IRQ | Jumper A | Jumper B | Jumper C | Jumper D | Jumper E | |
| IRQ10 | Open | Open | Open | Open | Closed | |
| IRQ11 | Open | Open | Open | Closed | Open | |
| IRQ12 | Open | Open | Closed | Open | Open | |
| IRQ14 | Open | Closed | Open | Open | Open | |
| IRQ15 | Closed | Open | Open | Open | Open | |

| CABLE TYPE | |
|---|---|
| Type | JP4 |
| ⇨ RG-58A/U 50ohm coaxial | See Configuration "A" |
| AUI transceiver via DB-15 port | See Configuration "B" |

CONFIGURATION "A"     CONFIGURATION "B"

# RACAL INTERLAN, INC.
## INTERLAN AT SERIES

| | |
|---|---|
| NIC Type | Ethernet |
| Transfer Rate | 10Mbps |
| Data Bus | 16-bit ISA |
| Topology | Linear bus |
| Wiring Type | Unshielded twisted pair |
| | RG58A/U 50ohm coaxial |
| | AUI transceiver via DB-15 |
| Boot ROM | Available |

| CABLE TYPE ||
|---|---|
| Type | NTS |
| Unshielded twisted pair | Configuration 2 |
| ⇨RG58A/U 50ohm coaxial | Configuration 1 |
| AUI transceiver via DB-15 port | Configuration 3 |

| BASE I/O ADDRESS |||||
|---|---|---|---|---|
| Address | JP1A | JP1B | JP1C | JP1D |
| 300h | Closed | Open | Open | Open |
| 320h | Open | Closed | Open | Open |
| 340h | Open | Open | Closed | Open |
| 360h | Open | Open | Open | Closed |

| INTERRUPT SETTINGS |||||
|---|---|---|---|---|
| IRQ | JP2A | JP2B | JP2C | JP2D |
| 2/9 | Open | Closed | Open | Open |
| 5 | Closed | Open | Open | Open |
| 12 | Open | Open | Closed | Open |
| 15 | Open | Open | Open | Closed |

## RACAL INTERLAN, INC.
### INTERLAN AT SERIES

| \multicolumn{6}{c}{BOOT ROM ADDRESS} ||||||
|---|---|---|---|---|---|
| Address | JP3A | JP3B | JP3C | JP3D | JP3E |
| No Boot Rom | Closed | Open | Open | Open | Open |
| C0000h | Open | Closed | Open | Open | Open |
| C4000h | Open | Open | Closed | Open | Open |
| ⇨ C8000h | Open | Open | Open | Closed | Open |
| CC000h | Open | Open | Open | Open | Closed |

| BASE MEMORY ADDRESS | | | | |
|---|---|---|---|---|
| Address | JP4A | JP4B | JP4C | JP4D |
| D0000 | Closed | Open | Open | Open |
| D4000 | Open | Closed | Open | Open |
| D8000 | Open | Open | Closed | Open |
| DC000 | Open | Open | Open | Closed |

| LINK INTEGRITY ||
|---|---|
| Function | LDJ |
| ⇨ Link integrity enabled | Pins 1 & 2 closed |
| Link integrity disabled | Pins 2 & 3 closed |

Note: If card is used in a pre-10BaseT cabling network, disable the link integrity function.

| MAIN BOARD COMPATIBILITY | | |
|---|---|---|
| Setting | W14 | W21 |
| NIC installed in a Compaq EISA system | Pins 1 & 2 closed | Pins 2 & 3 closed |
| NIC installed in a true IBM XT system | Pins 2 & 3 closed | Pins 1 & 2 closed |
| NIC installed in any other 100% IBM compatible system | Pins 1 & 2 closed | Pins 1 & 2 closed |

# RACAL INTERLAN, INC.
## NI5210/8-UTP, NI5210/16-UTP

| | |
|---|---|
| NIC Type | Ethernet |
| Transfer Rate | 10Mbps |
| Data Bus | 8-bit ISA |
| Topology | Star |
| Wiring Type | Unshielded twisted pair |
| | AUI transceiver via DB-15 |
| Boot ROM | Available |

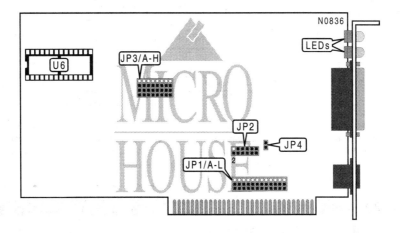

| CABLE TYPE ||
|---|---|
| Type | JP3/A-H |
| Unshielded twisted pair | Pins 1, 2, & 3 closed |
| AUI transceiver via DB-15 port | Pins 2, 3, & 4 closed |

| BOOT ROM ADDRESS |||||
|---|---|---|---|---|
| Address | JP1/G | JP1/H | JP1/I | JP1/J |
| C0000h | Pins 1 & 2 closed | Pins 1 & 2 closed | Pins 1 & 2 closed | Pins 1 & 2 closed |
| C4000h | Pins 1 & 2 closed | Pins 1 & 2 closed | Pins 1 & 2 closed | Pins 2 & 3 closed |
| ⇨ C8000h | Pins 1 & 2 closed | Pins 1 & 2 closed | Pins 2 & 3 closed | Pins 1 & 2 closed |
| CC000h | Pins 1 & 2 closed | Pins 1 & 2 closed | Pins 2 & 3 closed | Pins 2 & 3 closed |
| D0000h | Pins 1 & 2 closed | Pins 2 & 3 closed | Pins 1 & 2 closed | Pins 1 & 2 closed |
| D4000h | Pins 1 & 2 closed | Pins 2 & 3 closed | Pins 1 & 2 closed | Pins 2 & 3 closed |
| D8000h | Pins 1 & 2 closed | Pins 2 & 3 closed | Pins 2 & 3 closed | Pins 1 & 2 closed |
| DC000h | Pins 1 & 2 closed | Pins 2 & 3 closed | Pins 2 & 3 closed | Pins 2 & 3 closed |
| E0000h | Pins 2 & 3 closed | Pins 1 & 2 closed | Pins 1 & 2 closed | Pins 1 & 2 closed |
| E4000h | Pins 2 & 3 closed | Pins 1 & 2 closed | Pins 1 & 2 closed | Pins 2 & 3 closed |
| E8000h | Pins 2 & 3 closed | Pins 1 & 2 closed | Pins 2 & 3 closed | Pins 1 & 2 closed |
| EC000h | Pins 2 & 3 closed | Pins 1 & 2 closed | Pins 2 & 3 closed | Pins 2 & 3 closed |

## RACAL INTERLAN, INC.
## NI5210/8-UTP, NI5210/16-UTP

| Address | JP1/A | JP1/B | JP1/C | JP1/D | JP1/E | JP1/F |
|---|---|---|---|---|---|---|
| \multicolumn{7}{c}{BASE I/O ADDRESS} |
| 200h | Pins 1 & 2 | Pins 1 & 2 | Pins 1 & 2 | Pins 1 & 2 | Pins 1 & 2 | Pins 1 & 2 |
| 208h | Pins 1 & 2 | Pins 1 & 2 | Pins 1 & 2 | Pins 1 & 2 | Pins 1 & 2 | Pins 2 & 3 |
| 210h | Pins 1 & 2 | Pins 1 & 2 | Pins 1 & 2 | Pins 1 & 2 | Pins 2 & 3 | Pins 1 & 2 |
| 218h | Pins 1 & 2 | Pins 1 & 2 | Pins 1 & 2 | Pins 1 & 2 | Pins 2 & 3 | Pins 2 & 3 |
| 220h | Pins 1 & 2 | Pins 1 & 2 | Pins 1 & 2 | Pins 2 & 3 | Pins 1 & 2 | Pins 1 & 2 |
| 3D8h | Pins 2 & 3 | Pins 2 & 3 | Pins 2 & 3 | Pins 1 & 2 | Pins 2 & 3 | Pins 2 & 3 |
| 3E0h | Pins 2 & 3 | Pins 2 & 3 | Pins 2 & 3 | Pins 2 & 3 | Pins 1 & 2 | Pins 1 & 2 |
| 3E8h | Pins 2 & 3 | Pins 2 & 3 | Pins 2 & 3 | Pins 2 & 3 | Pins 1 & 2 | Pins 2 & 3 |
| 3F0h | Pins 2 & 3 | Pins 2 & 3 | Pins 2 & 3 | Pins 2 & 3 | Pins 2 & 3 | Pins 1 & 2 |
| 3F8h | Pins 2 & 3 | Pins 2 & 3 | Pins 2 & 3 | Pins 2 & 3 | Pins 2 & 3 | Pins 2 & 3 |

Notes: Pins designated should be in the closed position.
A total of 64 base I/O addresses are available. The jumpers are a binary representation of the decimal memory addresses. Jumper JP1/F is the Least Significant Bit and JP1/A is the Most Significant Bit. The jumpers have the following decimal values: JP1/F=8, JP1/E=16, JP1/D=32, JP1/C=64, JP1/B=128, JP1/A=256. Configure the jumpers and add 512 and the values of the jumpers closing pins 2 & 3 to obtain the correct memory base I/O address.

### INTERRUPT SETTINGS

| IRQ | JP2 |
|---|---|
| 2 | Pins 1 & 2 closed |
| 3 | Pins 3 & 4 closed |
| 4 | Pins 5 & 6 closed |
| 5 | Pins 7 & 8 closed |
| 6 | Pins 9 & 10 closed |
| 7 | Pins 11 & 12 closed |

### IBM PC/XT COMPATIBILITY

| Setting | JP4 |
|---|---|
| NIC installed in main board slot 8 | Closed |
| NIC installed in main board slot 1-7 | Open |

Note: This jumper only applies when the NIC is installed in an IBM PC/XT.

### ONBOARD RAM/BOOT ROM CONFIGURATION

| Function | JP1/K | JP1/L |
|---|---|---|
| 8KB of RAM installed (NI5120-UTP-8) | Pins 2 & 3 closed | N/A |
| 16KB of RAM installed (NI5120-UTP-16) | Pins 1 & 2 closed | Pins 1 & 2 closed |
| 8KB of RAM and Boot ROM installed | Pins 1 & 2 closed | Pins 2 & 3 closed |

Note: Socket U6 can either be an 8K RAM chip or a Boot ROM.

## RACAL INTERLAN, INC.
# NI5210/8-UTP, NI5210/16-UTP

| DIAGNOSTIC LEDS ||||
|---|---|---|---|
| **LED** | **Color** | **Status** | **Condition** |
| LED1 | Red | Off | Data is not being received |
| LED1 | Red | On | Data is being received |
| LED2 | Yellow | Off | Data is not being transmitted |
| LED2 | Yellow | On | Data is being transmitted |
| LED3 | Green | Off | UTP link status OK |
| LED3 | Green | On | UTP link broken |
| LED4 | Red | Off | AUI port selected |
| LED4 | Red | On | UTP port selected |
| Note: LED3 is only active when the UTP port is selected. ||||

## RACAL INTERLAN INC.
## NI6510-10BT

| | |
|---|---|
| NIC Type | Ethernet |
| Transfer Rate | 10Mbps |
| Data Bus | 16-bit ISA |
| Topology | Linear bus |
| | Star |
| Wiring Type | RG58A/U 50ohm coaxial |
| | AUI Transceiver via DB-15 port |
| Boot ROM | Available |

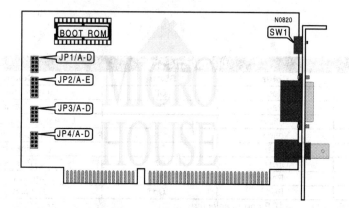

| I/O ADDRESS SELECT | | | | |
|---|---|---|---|---|
| Address | JP1/A | JP1/B | JP1/C | JP1/D |
| ⇨ 300h | Closed | Open | Open | Open |
| 320h | Open | Closed | Open | Open |
| 340h | Open | Open | Closed | Open |
| 360h | Open | Open | Open | Closed |

| BASE MEMORY ADDRESS SELECT | | | | | |
|---|---|---|---|---|---|
| Address | JP2/A | JP2/B | JP2/C | JP2/D | JP2/E |
| ⇨ Boot disabled | Open | Open | Open | Open | Closed |
| D0000 | Closed | Open | Open | Open | Open |
| D4000 | Open | Closed | Open | Open | Open |
| D8000 | Open | Open | Closed | Open | Open |
| DC000 | Open | Open | Open | Closed | Open |

## RACAL INTERLAN INC.
## NI6510

| INTERRUPT SELECT | | | | |
|---|---|---|---|---|
| IRQ | JP3/A | JP3/B | JP3/C | JP3/D |
| ▷ IRQ9 | Closed | Open | Open | Open |
| IRQ12 | Open | Closed | Open | Open |
| IRQ15 | Open | Open | Closed | Open |
| IRQ5 | Open | Open | Open | Closed |

| DMA CHANNEL SELECT | | | | |
|---|---|---|---|---|
| Channel | JP4/A | JP4/B | JP4/C | JP4/D |
| ▷ DMA3 | Open | Closed | Open | Open |
| DMA0 | Closed | Open | Open | Open |
| DMA5 | Open | Open | Closed | Open |
| DMA6 | Open | Open | Open | Closed |

| CABLE SELECT | |
|---|---|
| Setting | SW1 |
| ▷ Standard Ethernet | Up |
| Thin Ethernet | Down |

## RACAL INTERLAN, INC.
## NP600/XL, NP600/14

| | |
|---|---|
| NIC Type | Ethernet |
| Transfer Rate | 10Mbps |
| Data Bus | 16-bit ISA |
| Topology | Linear bus |
| | Star |
| Wiring Type | RG58A/U 50ohm coaxial |
| | AUI transceiver via DB-15 port |
| Boot ROM | N/A |

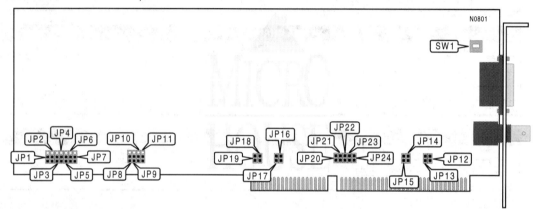

| I/O ADDRESS | | | | | | | |
|---|---|---|---|---|---|---|---|
| Address | JP1 | JP2 | JP3 | JP4 | JP5 | JP6 | JP7 |
| 200 | Pins 1 & 2 | Pins 2 & 3 | Pins 2 & 3 | Pins 2 & 3 | Pins 2 & 3 | Pins 2 & 3 | Pins 2 & 3 |
| 208 | Pins 1 & 2 | Pins 2 & 3 | Pins 2 & 3 | Pins 2 & 3 | Pins 2 & 3 | Pins 2 & 3 | Pins 1 & 2 |
| 210 | Pins 1 & 2 | Pins 2 & 3 | Pins 2 & 3 | Pins 2 & 3 | Pins 2 & 3 | Pins 1 & 2 | Pins 2 & 3 |
| 218 | Pins 1 & 2 | Pins 2 & 3 | Pins 2 & 3 | Pins 2 & 3 | Pins 2 & 3 | Pins 1 & 2 | Pins 1 & 2 |
| 220 | Pins 1 & 2 | Pins 2 & 3 | Pins 2 & 3 | Pins 2 & 3 | Pins 1 & 2 | Pins 2 & 3 | Pins 2 & 3 |
| 228 | Pins 1 & 2 | Pins 2 & 3 | Pins 2 & 3 | Pins 2 & 3 | Pins 1 & 2 | Pins 2 & 3 | Pins 1 & 2 |
| 230 | Pins 1 & 2 | Pins 2 & 3 | Pins 2 & 3 | Pins 2 & 3 | Pins 1 & 2 | Pins 1 & 2 | Pins 2 & 3 |
| 238 | Pins 1 & 2 | Pins 2 & 3 | Pins 2 & 3 | Pins 2 & 3 | Pins 1 & 2 | Pins 1 & 2 | Pins 1 & 2 |
| 240 | Pins 1 & 2 | Pins 2 & 3 | Pins 2 & 3 | Pins 1 & 2 | Pins 2 & 3 | Pins 2 & 3 | Pins 2 & 3 |
| 248 | Pins 1 & 2 | Pins 2 & 3 | Pins 2 & 3 | Pins 1 & 2 | Pins 2 & 3 | Pins 2 & 3 | Pins 1 & 2 |
| 250 | Pins 1 & 2 | Pins 2 & 3 | Pins 2 & 3 | Pins 1 & 2 | Pins 2 & 3 | Pins 1 & 2 | Pins 2 & 3 |
| 258 | Pins 1 & 2 | Pins 2 & 3 | Pins 2 & 3 | Pins 1 & 2 | Pins 2 & 3 | Pins 1 & 2 | Pins 1 & 2 |
| 260 | Pins 1 & 2 | Pins 2 & 3 | Pins 2 & 3 | Pins 1 & 2 | Pins 1 & 2 | Pins 2 & 3 | Pins 2 & 3 |
| 268 | Pins 1 & 2 | Pins 2 & 3 | Pins 2 & 3 | Pins 1 & 2 | Pins 1 & 2 | Pins 2 & 3 | Pins 1 & 2 |
| 270 | Pins 1 & 2 | Pins 2 & 3 | Pins 2 & 3 | Pins 1 & 2 | Pins 1 & 2 | Pins 1 & 2 | Pins 2 & 3 |
| 278 | Pins 1 & 2 | Pins 2 & 3 | Pins 2 & 3 | Pins 1 & 2 | Pins 1 & 2 | Pins 1 & 2 | Pins 1 & 2 |
| 280 | Pins 1 & 2 | Pins 2 & 3 | Pins 1 & 2 | Pins 2 & 3 | Pins 2 & 3 | Pins 2 & 3 | Pins 2 & 3 |
| 288 | Pins 1 & 2 | Pins 2 & 3 | Pins 1 & 2 | Pins 2 & 3 | Pins 2 & 3 | Pins 2 & 3 | Pins 1 & 2 |

Note: Pins designated are in the closed position.

## RACAL INTERLAN, INC.
## NP600/XL (NP600/14)

| I/O ADDRESS ||||||||
|---|---|---|---|---|---|---|---|
| Address | JP1 | JP2 | JP3 | JP4 | JP5 | JP6 | JP7 |
| 290 | Pins 1 & 2 | Pins 2 & 3 | Pins 1 & 2 | Pins 2 & 3 | Pins 2 & 3 | Pins 1 & 2 | Pins 2 & 3 |
| 298 | Pins 1 & 2 | Pins 2 & 3 | Pins 1 & 2 | Pins 2 & 3 | Pins 2 & 3 | Pins 1 & 2 | Pins 1 & 2 |
| 2A0 | Pins 1 & 2 | Pins 2 & 3 | Pins 1 & 2 | Pins 2 & 3 | Pins 1 & 2 | Pins 2 & 3 | Pins 2 & 3 |
| 2A8 | Pins 1 & 2 | Pins 2 & 3 | Pins 1 & 2 | Pins 2 & 3 | Pins 1 & 2 | Pins 2 & 3 | Pins 1 & 2 |
| 2B0 | Pins 1 & 2 | Pins 2 & 3 | Pins 1 & 2 | Pins 2 & 3 | Pins 1 & 2 | Pins 1 & 2 | Pins 2 & 3 |
| 2B8 | Pins 1 & 2 | Pins 2 & 3 | Pins 1 & 2 | Pins 2 & 3 | Pins 1 & 2 | Pins 1 & 2 | Pins 1 & 2 |
| 2C0 | Pins 1 & 2 | Pins 2 & 3 | Pins 1 & 2 | Pins 1 & 2 | Pins 2 & 3 | Pins 2 & 3 | Pins 2 & 3 |
| 2C8 | Pins 1 & 2 | Pins 2 & 3 | Pins 1 & 2 | Pins 1 & 2 | Pins 2 & 3 | Pins 2 & 3 | Pins 1 & 2 |
| 2D0 | Pins 1 & 2 | Pins 2 & 3 | Pins 1 & 2 | Pins 1 & 2 | Pins 2 & 3 | Pins 1 & 2 | Pins 2 & 3 |
| 2D8 | Pins 1 & 2 | Pins 2 & 3 | Pins 1 & 2 | Pins 1 & 2 | Pins 2 & 3 | Pins 1 & 2 | Pins 1 & 2 |
| 2E0 | Pins 1 & 2 | Pins 2 & 3 | Pins 1 & 2 | Pins 1 & 2 | Pins 1 & 2 | Pins 2 & 3 | Pins 2 & 3 |
| 2E8 | Pins 1 & 2 | Pins 2 & 3 | Pins 1 & 2 | Pins 1 & 2 | Pins 1 & 2 | Pins 2 & 3 | Pins 1 & 2 |
| 2F0 | Pins 1 & 2 | Pins 2 & 3 | Pins 1 & 2 | Pins 1 & 2 | Pins 1 & 2 | Pins 1 & 2 | Pins 2 & 3 |
| 2F8 | Pins 1 & 2 | Pins 2 & 3 | Pins 1 & 2 | Pins 1 & 2 | Pins 1 & 2 | Pins 1 & 2 | Pins 1 & 2 |
| 300 | Pins 1 & 2 | Pins 1 & 2 | Pins 2 & 3 | Pins 2 & 3 | Pins 2 & 3 | Pins 2 & 3 | Pins 2 & 3 |
| 308 | Pins 1 & 2 | Pins 1 & 2 | Pins 2 & 3 | Pins 2 & 3 | Pins 2 & 3 | Pins 2 & 3 | Pins 1 & 2 |
| 310 | Pins 1 & 2 | Pins 1 & 2 | Pins 2 & 3 | Pins 2 & 3 | Pins 2 & 3 | Pins 1 & 2 | Pins 2 & 3 |
| 318 | Pins 1 & 2 | Pins 1 & 2 | Pins 2 & 3 | Pins 2 & 3 | Pins 2 & 3 | Pins 1 & 2 | Pins 1 & 2 |
| 320 | Pins 1 & 2 | Pins 1 & 2 | Pins 2 & 3 | Pins 2 & 3 | Pins 1 & 2 | Pins 2 & 3 | Pins 2 & 3 |
| 328 | Pins 1 & 2 | Pins 1 & 2 | Pins 2 & 3 | Pins 2 & 3 | Pins 1 & 2 | Pins 2 & 3 | Pins 1 & 2 |
| 330 | Pins 1 & 2 | Pins 1 & 2 | Pins 2 & 3 | Pins 2 & 3 | Pins 1 & 2 | Pins 1 & 2 | Pins 2 & 3 |
| 338 | Pins 1 & 2 | Pins 1 & 2 | Pins 2 & 3 | Pins 2 & 3 | Pins 1 & 2 | Pins 1 & 2 | Pins 1 & 2 |
| 340 | Pins 1 & 2 | Pins 1 & 2 | Pins 2 & 3 | Pins 1 & 2 | Pins 2 & 3 | Pins 2 & 3 | Pins 2 & 3 |
| 348 | Pins 1 & 2 | Pins 1 & 2 | Pins 2 & 3 | Pins 1 & 2 | Pins 2 & 3 | Pins 2 & 3 | Pins 1 & 2 |
| 350 | Pins 1 & 2 | Pins 1 & 2 | Pins 2 & 3 | Pins 1 & 2 | Pins 2 & 3 | Pins 1 & 2 | Pins 2 & 3 |
| 358 | Pins 1 & 2 | Pins 1 & 2 | Pins 2 & 3 | Pins 1 & 2 | Pins 2 & 3 | Pins 1 & 2 | Pins 1 & 2 |
| 360 | Pins 1 & 2 | Pins 1 & 2 | Pins 2 & 3 | Pins 1 & 2 | Pins 1 & 2 | Pins 2 & 3 | Pins 2 & 3 |
| 368 | Pins 1 & 2 | Pins 1 & 2 | Pins 2 & 3 | Pins 1 & 2 | Pins 1 & 2 | Pins 2 & 3 | Pins 1 & 2 |
| 370 | Pins 1 & 2 | Pins 1 & 2 | Pins 2 & 3 | Pins 1 & 2 | Pins 1 & 2 | Pins 1 & 2 | Pins 2 & 3 |
| 378 | Pins 1 & 2 | Pins 1 & 2 | Pins 2 & 3 | Pins 1 & 2 | Pins 1 & 2 | Pins 1 & 2 | Pins 1 & 2 |
| 380 | Pins 1 & 2 | Pins 1 & 2 | Pins 1 & 2 | Pins 2 & 3 | Pins 2 & 3 | Pins 2 & 3 | Pins 2 & 3 |
| 388 | Pins 1 & 2 | Pins 1 & 2 | Pins 1 & 2 | Pins 2 & 3 | Pins 2 & 3 | Pins 2 & 3 | Pins 1 & 2 |
| 390 | Pins 1 & 2 | Pins 1 & 2 | Pins 1 & 2 | Pins 2 & 3 | Pins 2 & 3 | Pins 1 & 2 | Pins 2 & 3 |
| 398 | Pins 1 & 2 | Pins 1 & 2 | Pins 1 & 2 | Pins 2 & 3 | Pins 2 & 3 | Pins 1 & 2 | Pins 1 & 2 |
| 3A0 | Pins 1 & 2 | Pins 1 & 2 | Pins 1 & 2 | Pins 2 & 3 | Pins 1 & 2 | Pins 2 & 3 | Pins 2 & 3 |
| 3A8 | Pins 1 & 2 | Pins 1 & 2 | Pins 1 & 2 | Pins 2 & 3 | Pins 1 & 2 | Pins 2 & 3 | Pins 1 & 2 |
| 3B0 | Pins 1 & 2 | Pins 1 & 2 | Pins 1 & 2 | Pins 2 & 3 | Pins 1 & 2 | Pins 1 & 2 | Pins 2 & 3 |
| 3B8 | Pins 1 & 2 | Pins 1 & 2 | Pins 1 & 2 | Pins 2 & 3 | Pins 1 & 2 | Pins 1 & 2 | Pins 1 & 2 |
| 3C0 | Pins 1 & 2 | Pins 1 & 2 | Pins 1 & 2 | Pins 1 & 2 | Pins 2 & 3 | Pins 2 & 3 | Pins 2 & 3 |
| 3C8 | Pins 1 & 2 | Pins 1 & 2 | Pins 1 & 2 | Pins 1 & 2 | Pins 2 & 3 | Pins 2 & 3 | Pins 1 & 2 |

Note: Pins designated are in the closed position.

## RACAL INTERLAN, INC.
## NP600/XL (NP600/14)

### I/O ADDRESS

| Address | JP1 | JP2 | JP3 | JP4 | JP5 | JP6 | JP7 |
|---------|-----|-----|-----|-----|-----|-----|-----|
| 3D0 | Pins 1 & 2 | Pins 1 & 2 | Pins 1 & 2 | Pins 1 & 2 | Pins 2 & 3 | Pins 1 & 2 | Pins 2 & 3 |
| 3D8 | Pins 1 & 2 | Pins 1 & 2 | Pins 1 & 2 | Pins 1 & 2 | Pins 2 & 3 | Pins 1 & 2 | Pins 1 & 2 |
| 3E0 | Pins 1 & 2 | Pins 1 & 2 | Pins 1 & 2 | Pins 1 & 2 | Pins 1 & 2 | Pins 2 & 3 | Pins 2 & 3 |
| 3E8 | Pins 1 & 2 | Pins 1 & 2 | Pins 1 & 2 | Pins 1 & 2 | Pins 1 & 2 | Pins 2 & 3 | Pins 1 & 2 |
| 3F0 | Pins 1 & 2 | Pins 1 & 2 | Pins 1 & 2 | Pins 1 & 2 | Pins 1 & 2 | Pins 1 & 2 | Pins 2 & 3 |
| 3F8 | Pins 1 & 2 | Pins 1 & 2 | Pins 1 & 2 | Pins 1 & 2 | Pins 1 & 2 | Pins 1 & 2 | Pins 1 & 2 |

Note: Pins designated are in the closed position.

### INTERRUPT SELECT

| IRQ | JP20 | JP21 | JP22 | JP23 | JP24 |
|-----|------|------|------|------|------|
| IRQ3 | Closed | Open | Open | Open | Open |
| IRQ5 | Open | Closed | Open | Open | Open |
| IRQ9 | Open | Open | Closed | Open | Open |
| IRQ11 | Open | Open | Open | Closed | Open |
| IRQ15 | Open | Open | Open | Open | Closed |

### DMA CHANNEL SELECT

| Channel | JP12 | JP13 | JP14 | JP15 | JP16 | JP17 | JP18 | JP19 |
|---------|------|------|------|------|------|------|------|------|
| 3 | Closed | Closed | Open | Open | Open | Open | Open | Open |
| 1 | Open | Open | Closed | Closed | Open | Open | Open | Open |
| 5 | Open | Open | Open | Open | Closed | Closed | Open | Open |
| 7 | Open | Open | Open | Open | Open | Open | Closed | Closed |

### CABLE TYPE SELECT

| Cable | SW1 |
|-------|-----|
| Standard Ethernet | On |
| Thin Ethernet | Off |

### FACTORY CONFIGURED SETTINGS

| Function | Jumper | Setting |
|----------|--------|---------|
| Factory configured - do not alter | JP8 | Pins 1 & 2 closed |
| Factory configured - do not alter | JP9 | Pins 1 & 2 closed |
| Factory configured - do not alter | JP10 | Pins 1 & 2 closed |
| Factory configured - do not alter | JP11 | Pins 1 & 2 closed |

## RACORE COMPUTER PRODUCTS, INC.
## M8110 (REV. 1.2)

**NIC Type:** Token-Ring
**Transfer Rate:** 4Mbps
**Data Bus:** 16-bit ISA
**Topology:** Ring
**Wiring Type:** Shielded twisted pair
**Boot ROM:** Available

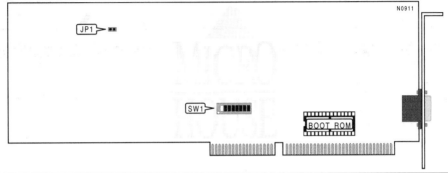

| LOGICAL LINK CONTROL FEATURE ||
| Setting | JP1 |
|---|---|
| Disabled | Closed |
| Enabled | Open |

Note: The Logical Link Control feature must have the appropriate firmware installed to function.

| I/O ADDRESS |||
|---|---|---|
| Address | SW1/1 | SW1/2 |
| A00h | On | On |
| A20h | On | Off |
| A40h | Off | On |
| A60h | Off | Off |

| INTERRUPT SELECT |||
|---|---|---|
| IRQ | SW1/3 | SW1/4 |
| IRQ3 | On | On |
| IRQ9 | On | Off |
| IRQ10 | Off | On |
| IRQ11 | Off | Off |

## RACORE COMPUTER PRODUCTS, INC.
M8110 (REV. 1.2)

| BOOT ROM ADDRESS | | |
|---|---|---|
| Address | SW1/5 | SW1/6 |
| CC000h | On | On |
| D0000h | On | Off |
| D8000h | Off | On |
| DC000h | Off | Off |

| DMA SELECT | | |
|---|---|---|
| Channel | SW1/7 | SW1/8 |
| DMA0 | On | On |
| DMA5 | On | Off |
| DMA6 | Off | On |
| DMA7 | Off | Off |

## RACORE COMPUTER PRODUCTS, INC.
## M 8 1 1 7

| | |
|---|---|
| **NIC Type** | Token-Ring |
| **Transfer Rate** | 4/16Mbps |
| **Data Bus** | 8/16-bit ISA |
| **Topology** | Ring |
| **Wiring Type** | Shielded twisted pair |
| | Unshielded twisted pair(with external media filter) |
| | 62.5/125 or 50/125 µm Fiber optic cable |
| **Boot ROM** | Available |

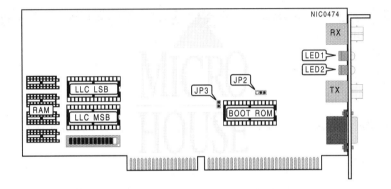

| MAU TYPE ||
|---|---|
| **Setting** | **JP2** |
| ⇨802.5J | Pins 2 & 3 closed |
| Alternate | Pins 1 & 2 closed |

| ROM MEMORY ENABLE ||
|---|---|
| **Setting** | **JP3** |
| ⇨Enabled | Closed |
| Disabled | Open |

Note: The boot ROM socket can contain a BIA ROM or a boot ROM. The BIA ROM provides the node address if a boot ROM is not installed. If the ROM memory is disabled, the node address must be set using the RTR16ASI.SYS device driver.

| I/O BASE ADDRESS |||
|---|---|---|
| **Address** | **SW1/1** | **SW1/2** |
| ⇨A00 - A0Fh | On | On |
| A20 - A2Fh | Off | On |
| A40 - A4Fh | On | Off |
| A60 - A6Fh | Off | Off |

## RACORE COMPUTER PRODUCTS, INC.
## M 8117

### INTERRUPT REQUEST

| IRQ | SW1/3 | SW1/4 |
|---|---|---|
| 2/9 | Off | On |
| ⇨3 | On | On |
| 10 | On | Off |
| 11 | Off | Off |

Note: If using the card in an 8-bit slot, do not use IRQ10 or IRQ11.

### BOOT ROM ADDRESS

| Address | SW1/5 | SW1/6 |
|---|---|---|
| CC000 - CDFFFh | On | On |
| D0000 - D1FFFh | Off | On |
| D8000 - D9FFFh | On | Off |
| ⇨DC000 - DDFFFh | Off | Off |

### DMA CHANNEL

| Channel | SW1/7 | SW1/8 |
|---|---|---|
| DMA0 | On | On |
| DMA5 | Off | On |
| ⇨DMA6 | On | Off |
| DMA7 | Off | Off |

### BOOT ROM

| Setting | SW1/9 |
|---|---|
| ⇨Disabled | On |
| Enabled | Off |

Note: To install the boot ROM remove the BIA ROM (if installed) and replace it with the boot ROM. JP3 must be closed for the boot ROM to function.

### NETWORK SEGMENT SPEED

| Speed | SW1/10 |
|---|---|
| ⇨16Mbps | On |
| 4Mbps | Off |

Note: All cards on the network segment must have this option set the same.

# RACORE COMPUTER PRODUCTS, INC.
## M 8 1 1 7

| LOGIC LINK CONTROL (LLC) LOCATION ||
|---|---|
| Speed | SW1/11 |
| ⇨ROM based | On |
| RAM based | Off |

Note: LLC allows the card to access networks using the ISO and SNA protocols. (e.g. IBM PC LAN, IBM 3270 terminal emulation, and AS400)
The card has built-in LLC support on the two ROMs above SW1. To allow for future compatibility and upgrades you can load the LLC code into the onboard RAM.

| CABLE TYPE & MAXIMUM CABLE LENTGH |||
|---|---|---|
| Type | Max. Length | SW1/12 |
| ⇨Fiber optic cable | 75m | On |
| Shielded twisted pair | 2000m | Off |
| Unshielded twisted pair with media-filter | 100m | Off |

| DIAGNOSTIC LED(S) |||
|---|---|---|
| LED | Status | Condition |
| LED1 | On | Fiber optic cable connection is good |
| LED1 | Off | Fiber optic cable connection is broken |
| LED2 | On | Card is inserted into the ring |
| LED2 | Off | Card is not inserted into the ring |

Note: If the MAU type is set to 802.2J (JP2 - Pins 2 & 3 closed) LED1 will not light until the adapter attempts insertion into the network.

## RACORE COMPUTER PRODUCTS, INC.
## M 8 1 1 9

| | |
|---|---|
| **NIC Type** | Token-Ring |
| **Transfer Rate** | 4/16Mbps |
| **Data Bus** | 8/16-bit ISA |
| **Topology** | Ring |
| **Wiring Type** | Shielded twisted pair |
| | Unshielded twisted pair |
| **Boot ROM** | Available (Built-in Remote Program Load) |

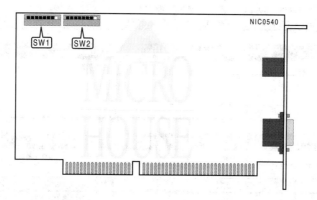

| I/O BASE ADDRESS | | |
|---|---|---|
| Address | SW1/1 | SW1/2 |
| ⇨A00 - A0Fh | On | On |
| A20 - A2Fh | Off | On |
| A40 - A4Fh | On | Off |
| A60 - A6Fh | Off | Off |

| INTERRUPT REQUEST | | |
|---|---|---|
| IRQ | SW1/3 | SW1/4 |
| 2/9 | Off | On |
| ⇨3 | On | On |
| 10 | On | Off |
| 11 | Off | Off |

Note: If using the card in an 8-bit slot IRQ10 becomes IRQ3 and IRQ11 becomes IRQ2.

| FLASH ROM ADDRESS | | |
|---|---|---|
| Address | SW1/5 | SW1/6 |
| CC000 - CDFFFh | On | On |
| D0000 - D1FFFh | Off | On |
| ⇨D8000 - D9FFFh | On | Off |
| DC000 - DDFFFh | Off | Off |

## RACORE COMPUTER PRODUCTS, INC.
## M 8119

### DMA CHANNEL

| Channel | SW1/7 | SW1/8 |
|---|---|---|
| ⇨DMA0 | On | On |
| DMA5 | Off | On |
| DMA6 | On | Off |
| DMA7 | Off | Off |

### NETWORK SEGMENT SPEED

| Speed | SW2/2 |
|---|---|
| ⇨16Mbps | On |
| 4Mbps | Off |

Note: All cards on the network segment must have this option set the same.

### CABLE TYPE

| Type | SW2/3 |
|---|---|
| Shielded twisted pair | On |
| Unshielded twisted pair | Off |

### FLASH ROM ENABLE

| Setting | SW2/4 |
|---|---|
| ⇨Enabled | Off |
| Disabled | On |

Note: The flash ROM stores the ROM boot programs, RPL (Remote Program Load) programs, LLC, and node address. If the flash ROM is disabled, the node address must be passed to the installed driver and the file RTR16LLC.COD must be present.

### DMA BURST SPEED

| Setting | SW2/6 |
|---|---|
| ⇨Minimum | On |
| Maximum | Off |

### CLOCK SOURCE

| Setting | SW2/7 |
|---|---|
| ⇨Onboard 8MHz clock | On |
| I/O bus clock | Off |

Note: If your system has an I/O bus speed greater than 8MHz you can improve system performance by selecting the I/O bus clock as the clock signal (SW2/7 Off) but the manufacturer warns that this setting will not work in all systems. If using the I/O bus clock setting, you should test the card thoroughly before attaching it to your network.

### FACTORY CONFIGURED SETTINGS

| Jumper | Setting |
|---|---|
| SW2/1 | Off |
| SW2/5 | On |

# SIEMENS NIXDORF INFORMATIONSSYSTEME AG
# S26361-D633-V2

| | |
|---|---|
| **NIC Type** | Ethernet |
| **Transfer Rate** | 10Mbps |
| **Data Bus** | 16-bit ISA |
| **Topology** | Linear Bus/Star |
| **Wiring Type** | Unshielded twisted pair |
| | AUI Transceiver via DB-15 port |
| **Boot ROM** | Available |

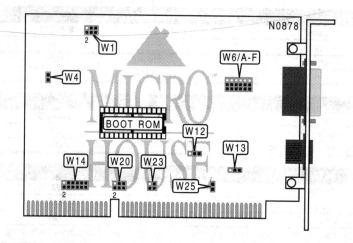

| CABLE TYPE SELECTION | | |
|---|---|---|
| **Cable Type** | **W6/A-F** | **W12** |
| ⇨ Unshielded twisted pair | Pins 2 & 3 closed | Pins 2 & 3 closed |
| AUI Transceiver via DB-15 port | Pins 1 & 2 closed | Pins 1 & 2 closed |

| BOOT ROM SELECTION | |
|---|---|
| **Mode** | **W1** |
| ⇨ Disabled | Pins 3 & 4 open |
| Enabled | Pins 3 & 4 closed |

| BASE I/O AND BOOT ROM ADDRESS SELECTION | | |
|---|---|---|
| **Base I/O Address** | **Boot ROM Address** | **W1** |
| ⇨ 300h | C8000h | Pins 1 & 2, 5 & 6 closed |
| 320h | CC000h | Pins 1 & 2 closed |
| 340h | D0000h | Pins 5 & 6 closed |
| 360h | D4000h | Pins 1 & 2, 5 & 6 open |

# SIEMENS NIXDORF INFORMATIONSSYSTEME AG
## S26361-D633-V2

| INTERRUPT SELECTION | | |
|---|---|---|
| IRQ | W20 | W25 |
| None | Open | Open |
| IRQ2/9 | Open | Closed |
| ⇨ IRQ3 | Pins 1 & 2 closed | Open |
| IRQ4 | Pins 3 & 4 closed | Open |
| IRQ5 | Pins 5 & 6 closed | Open |

| DMA CHANNEL SELECTION | | |
|---|---|---|
| Channel | W14 | W23 |
| 3 | Open | Pins 1 & 2, 3 & 4 closed |
| ⇨ 5 | Pins 9 & 10, 11 & 12 closed | Open |
| 6 | Pins 5 & 6, 7 & 8 closed | Open |
| 7 | Pins 1 & 2, 3 & 4 closed | Open |

| WAIT STATE CONFIGURATION | |
|---|---|
| Wait states | W4 |
| ⇨ 0 | Closed |
| 1 | Open |

| SIGNAL QUALITY ERROR TEST CONFIGURATION | |
|---|---|
| Mode | W12 |
| ⇨ Enabled | Pins 2 & 3 closed |
| Disabled | Pins 1 & 2 closed |

## SIEMENS NIXDORF INFORMATIONSSYSTEME AG
## S26361-D640-V1

| | |
|---|---|
| **NIC Type** | Token Ring |
| **Transfer Rate** | 4Mbps |
| | 16Mbps |
| **Data Bus** | 16-bit ISA |
| **Topology** | Ring |
| **Wiring Type** | Shielded twisted pair |
| | Unshielded twisted pair |
| **Boot ROM** | Available |

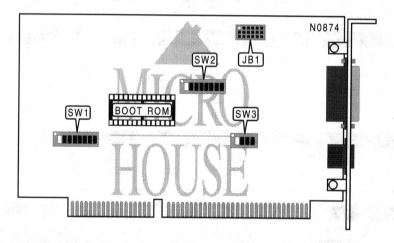

| CABLE TYPE ||
|---|---|
| JB1 ||
| UTP | STP |
| ↑ | ↓ |

| INTERRUPT SELECTION ||
|---|---|
| IRQ | SW1 |
| 2/9 | SW1/8 on |
| ⇨ 3 | SW1/7 on |
| 5 | SW1/6 on |
| 7 | SW1/5 on |
| 10 | SW1/4 on |
| 11 | SW1/3 on |
| 12 | SW1/2 on |
| 15 | SW1/1 on |

## SIEMENS NIXDORF INFORMATIONSSYSTEME AG
## S26361-D640-V1

### BASE I/O AND ROM ADDRESS SELECTION

| I/O Address | ROM Address | SW2/1 | SW2/2 |
|---|---|---|---|
| ⇨ 0A20h | CC000h | On | On |
| 1A20h | DC000h | Off | On |
| 2A20h | CE000h | On | Off |
| 3A20h | DE000h | Off | Off |

### BOOT ROM CONFIGURATION

| Mode | SW2/3 |
|---|---|
| ⇨ Disabled | Off |
| Enabled | On |

### DMA MODE SELECTION

| Mode | SW2/4 |
|---|---|
| Enabled | On |
| Disabled | Off |

### DMA CHANNEL SELECTION

| Channel | SW3 |
|---|---|
| 1 | SW3/1 on |
| 3 | SW3/2 on |
| ⇨ 5 | SW3/3 on |
| 6 | SW3/4 on |

### DATA BUS SIZE SELECTION

| Size | SW2/5 |
|---|---|
| ⇨ 8-bit | On |
| 16-bit | Off |

### BUS TIMING CONFIGURATION

| Machine type | SW2/6 |
|---|---|
| ⇨ Fully compatible AT | On |
| Near-compatible AT | Off |

### BUS SPEED SELECTION

| Speed | SW2/7 |
|---|---|
| ⇨ >10MHz | Off |
| <=10MHz | On |

### NETWORK SPEED SELECTION

| Speed | SW2/8 |
|---|---|
| ⇨ 4Mbps | On |
| 16Mbps | Off |

# SILCOM MANUFACTURING TECHNOLOGY, INC.
## TR DIRECT/16

|  |  |
|---|---|
| **NIC Type** | Token Ring |
| **Transfer Rate** | 4/16Mbps |
| **Data Bus** | 16-bit ISA |
| **Topology** | Ring |
| **Wiring Type** | AUI transceiver via DB-15 port |
|  | Shielded twisted pair |
|  | Unshielded twisted pair |
| **Boot ROM** | Available |

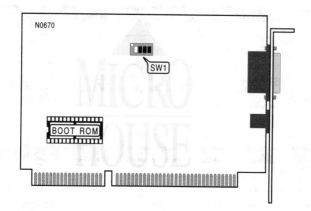

| BOOT ROM | | |
|---|---|---|
| **Setting** | **SW1/3** | **SW1/4** |
| ⇨ C8000h | On | On |
| D0000h | On | Off |
| E0000h | Off | On |
| Disabled | Off | Off |

| PORT ADDRESS SELECTION | | |
|---|---|---|
| **Address** | **SW1/1** | **SW1/2** |
| ⇨ A20h | Off | Off |
| 1A20h | Off | On |
| 2A20 | On | Off |
| 3A20 | On | On |

## SILCOM MANUFACTURING TECHNOLOGY, INC.
## TR DIRECT/BLUE ISA

| | |
|---|---|
| **NIC Type** | Token Ring |
| **Transfer Rate** | 4/16Mbps |
| **Data Bus** | 16-bit ISA |
| **Topology** | Ring |
| **Wiring Type** | AUI transceiver via DB-15 port |
| | Shielded twisted pair |
| | Unshielded twisted pair |
| **Boot ROM** | Available |

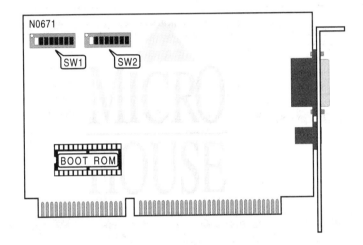

| ADAPTER SELECT | |
|---|---|
| **Setting** | **SW2/6** |
| ⇨ Primary A20 | Off |
| Secondary A24 | On |

| BUS WIDTH CONFIGURATION | |
|---|---|
| **BUS Width** | **SW2/7** |
| ⇨ Auto sense mode | Off |
| Force 8-bit | On |

| FACTORY CONFIGURED SETTINGS | |
|---|---|
| **Switch** | **Setting** |
| SW1/7 | Off |
| SW1/8 | Off |
| SW2/8 | Off |

# SILCOM MANUFACTURING TECHNOLOGY, INC.
## TR DIRECT/BLUE ISA

| INTERRUPT SELECTION | | |
|---|---|---|
| IRQ | SW2/1 | SW2/2 |
| ⇨ IRQ2 | Off | Off |
| IRQ3 | Off | On |
| IRQ6 (diskless only) | On | Off |
| IRQ7 | On | On |

| NETWORK SPEED | |
|---|---|
| Speed | SW2/5 |
| ⇨ 4Mbps | Off |
| 16Mbps | On |

Note: All cards on a segment must have this option set the same.

| RAM PAGE SIZE | | |
|---|---|---|
| Page Size | SW2/3 | SW2/3 |
| ⇨ 16KB | Off | On |
| 8KB | Off | Off |
| 32KB | On | Off |
| 64KB | On | On |

| ROM ADDRESS | | | | | | |
|---|---|---|---|---|---|---|
| Address | SW1/1 | SW1/2 | SW1/3 | SW1/4 | SW1/5 | SW1/6 |
| ⇨ CC00h | On | Off | Off | On | On | Off |
| C000h | On | Off | Off | Off | Off | Off |
| C200h | On | Off | Off | Off | Off | On |
| C400h | On | Off | Off | Off | On | Off |
| C600h | On | Off | Off | Off | On | On |
| C800h | On | Off | Off | On | Off | Off |
| CA00h | On | Off | Off | On | Off | On |
| CE00h | On | Off | Off | On | On | On |
| D000h | On | Off | On | Off | Off | Off |
| DC00h | On | Off | On | On | On | Off |

# STANDARD MICROSYSTEMS CORPORATION
# ARCNET PS310

| | |
|---|---|
| **NIC Type** | ARCnet |
| **Transfer Rate** | 2.5Mbps |
| **Data Bus** | 16-bit MCA |
| **Topology** | Star |
| | Linear Bus |
| **Wiring Type** | Unshielded twisted pair |
| | RG-62A/U 93ohm coaxial |
| **Boot ROM** | Available |

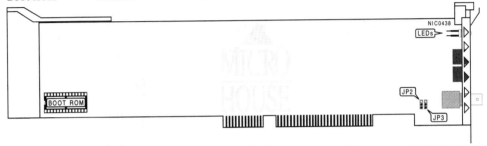

| TOPOLOGY SELECT ||
|---|---|
| **Topology** | **JP2** |
| ▷Star | Pins 1 & 2 closed |
| Linear Bus | Pins 2 & 3 closed |

| CABLE TYPE ||
|---|---|
| **Type** | **JP3** |
| ▷RG-62A/U 93ohm coaxial | Pins 1 & 2 closed |
| Unshielded twisted pair | Pins 2 & 3 closed |

Note: If the topology is star and the cable type is unshielded twisted pair, the card is self-terminating. No external terminator is needed in this case.

| DIAGNOSTIC LED(S) | | | |
|---|---|---|---|
| **LED** | **Color** | **Status** | **Condition** |
| LED1 | Red | On | Card is being accessed by host computer |
| LED1 | Red | Off | Card is not being accessed by host computer |
| LED2 | Green | On (Flickering) | Normal operation |
| LED2 | Green | Off | Card is malfuntioning or node I/D is set to zero |
| LED2 | Green | Blinking | Card is reconfiguring |

# STANDARD MICROSYSTEMS CORPORATION
## ELITE10T

**NIC Type** Ethernet
**Transfer Rate** 10Mbps
**Data Bus** 16-bit ISA
**Topology** Linear Bus
**Wiring Type** Unshielded twisted pair
AUI transceiver via DB-15 port
**Boot ROM** Available

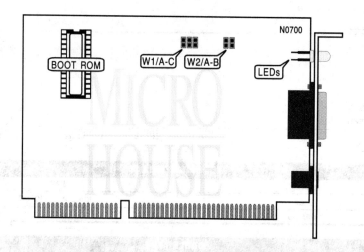

### ADDRESS SELECT

| Base Address | Memory Address | IRQ | W1/A | W1/B | W1/C |
|---|---|---|---|---|---|
| ◊ 280h | D0000h | IRQ3 | Open | Closed | Open |
| 240h | CC000h | IRQ2 disabled | Closed | Open | Open |
| 300h | CA000h | IRQ5 | Open | Open | Closed |

### BOOT ROM

| Setting | W2/A | W]2/B |
|---|---|---|
| Disabled | Closed | Open |
| Enabled | Open | Closed |

## STANDARD MICROSYSTEMS CORPORATION
## ELITE10T

| DIAGNOSTIC LED | | | |
|---|---|---|---|
| **LED** | **Color** | **Status** | **Condition** |
| LED1 | Green | On | Network connection is good |
| LED1 | Green | Off | Network connection is broken |
| LED2 | Yellow | On | Data is being transmitted |
| LED2 | Yellow | Off | Data is not being transmitted |
| LED3 | Green | On | Data is being received |
| LED3 | Green | Off | Data is not being received |

# STANDARD MICROSYSTEMS CORPORATION
## ELITE16 COMBO

| | |
|---|---|
| **NIC Type** | Ethernet |
| **Transfer Rate** | 10Mbps |
| **Data Bus** | 16-bit ISA |
| **Topology** | Linear Bus |
| **Wiring Type** | Unshielded twisted pair |
| | AUI transceiver via DB-15 port |
| | RG58A/U 50ohm coaxial |
| **Boot ROM** | Available |

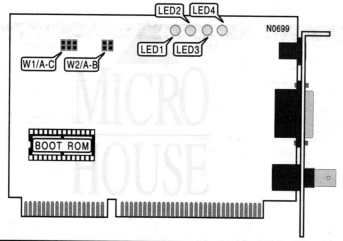

| ADDRESS SELECT | | | | | |
|---|---|---|---|---|---|
| I/O Address | Memory Address | IRQ | W1/A | W1/B | W1/C |
| ⇨ 280h | D000h | IRQ3 | Open | Closed | Open |
| 240h | Software configured | Software configured | Closed | Open | Open |
| 300h | CC00h | IRQ10 | Open | Open | Closed |

| BOOT ROM | | |
|---|---|---|
| Setting | W1/A | W1/B |
| ⇨ Disabled | Closed | Open |
| Enabled | Open | Closed |

| DIAGNOSTIC LED | | | |
|---|---|---|---|
| LED | Color | Status | Condition |
| LED1 | Orange | On | Data is being transmitted |
| LED1 | Orange | Off | Data is not being transmitted |
| LED2 | Orange | On | Data is being received |
| LED2 | Orange | Off | Data is not being received |
| LED3 | Green | On | Polarity is reversed and corrected |
| LED4 | Green | On | Link Integrity established |

# STANDARD MICROSYSTEMS CORPORATION
## ETHERCARD ELITE16C ULTRA ADAPTER

**NIC Type**　　　　Ethernet
**Transfer Rate**　　10Mbps
**Data Bus**　　　　16-bit ISA
**Topology**　　　　Linear Bus, Star
**Wiring Type**　　　AUI transceiver via DB-15 port
　　　　　　　　　　RG58A/U 50ohm coaxial
　　　　　　　　　　Unshielded twisted pair
**Boot ROM**　　　　Available

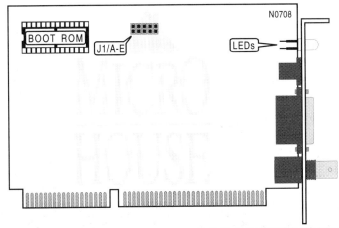

| ADDRESS/INTERRUPT CONFIGURATION | | | | | | | | |
|---|---|---|---|---|---|---|---|---|
| I/O Address | RAM Address | ROM Address | IRQ | W1/A | W1/B | W1/C | W1/D | W1/E |
| Software | Software | Software | Software | Closed | Open | Open | Open | Open |
| 280h | D000h | None | IRQ3 | Open | Closed | Open | Open | Open |
| 300h | CC00h | None | IRQ10 | Open | Open | Closed | Open | Open |
| 280h | D000h | D800h | IRQ3 | Open | Open | Open | Closed | Open |
| 300h | CC00h | D800h | IRQ10 | Open | Open | Open | Open | Closed |
| Note: The factory default is software configured address and interrupt selection. | | | | | | | | |

| DIAGNOSTIC LEDs | | | |
|---|---|---|---|
| LED | Color | Status | Condition |
| LED1 | Green | Off | Network connection is broken |
| LED1 | Green | On | Network connection is good |
| LED2 | Yellow | Off | Data is not being transmitted/received |
| LED2 | Yellow | On | Data is being transmitted/received |

# STANDARD MICROSYSTEMS CORPORATION
## TOKENCARD ELITE SMC8115T

**NIC Type**        Token Ring
**Transfer Rate**   4/16Mbps
**Data Bus**        16-bit ISA
**Topology**        Ring
**Wiring Type**     Shielded twisted pair via DB-9 port
                    Unshielded twisted pair
**Boot ROM**        Available

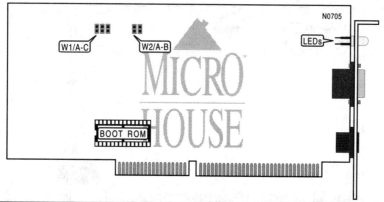

| ADDRESS/INTERRUPT CONFIGURATION | | | | | |
|---|---|---|---|---|---|
| I/O Address | Memory Address | IRQ | W1/A | W1/B | W1/C |
| ⇨ 240h | CC00h | Disabled | Closed | Open | Open |
| 280h | D000h | IRQ3 | Open | Closed | Open |
| 300h | CC00h | IRQ10 | Open | Open | Closed |

| BOOT ROM | | |
|---|---|---|
| Setting | W1/A | W1/B |
| ⇨ Disabled | Closed | Open |
| Enabled | Open | Closed |

| DIAGNOSTIC LED(S) | | | |
|---|---|---|---|
| LED | Color | Status | Condition |
| LED1 | Green | On | Loopback testing is in progress |
| LED1 | Green | Off | Loopback testing is not in progress |
| LED2 | Amber | On | Network connection is good |
| LED2 | Amber | Off | Network connection is broken |

## SVEC COMPUTER CORPORATION
## FD0490P

| | |
|---|---|
| **NIC Type:** | Ethernet |
| **Transfer Rate:** | 10Mbps |
| **Data Bus:** | 16-bit ISA |
| **Topology:** | Star |
| **Wiring Type** | AUI transceiver via DB-15 port |
| | Shielded/Unshielded twisted pair |
| **Boot ROM** | Available |

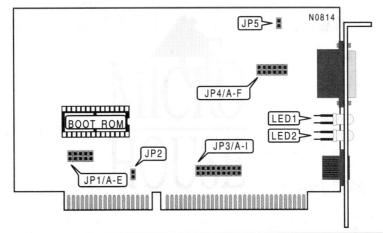

| I/O ADDRESS ||| 
|---|---|---|
| Address | JP1/Jumper A | JP1/Jumper B |
| 300h | Closed | Closed |
| 320h | Closed | Open |
| 340h | Open | Closed |
| 360h | Open | Open |

| BOOT ROM ADDRESS ||||
|---|---|---|---|
| Address | JP1/Jumper C | JP1/Jumper D | JP1/Jumper E |
| C800h | Closed | Open | Closed |
| C000h | Closed | Closed | Closed |
| C400h | Open | Closed | Closed |
| CC00h | Open | Open | Closed |
| D000h | Closed | Closed | Open |
| D400h | Open | Closed | Open |
| D800h | Closed | Open | Open |
| DC00h | Open | Open | Open |

## SVEC COMPUTER CORPORATION
## FD0490P

| TIMING ||
|---|---|
| Setting | JP2 |
| Standard | Closed |
| Non-standard | Open |
| Note: If compatibility problems are encountered, try opening JP2. ||

| INTERRUPT REQUEST - JP3 ||||||||||
|---|---|---|---|---|---|---|---|---|---|
| IRQ | A | B | C | D | E | F | G | H | I |
| IRQ2 | Closed | Open | Open | Open | Open | Open | Open | Open | Open |
| IRQ3 | Open | Closed | Open | Open | Open | Open | Open | Open | Open |
| IRQ4 | Open | Open | Closed | Open | Open | Open | Open | Open | Open |
| IRQ5 | Open | Open | Open | Closed | Open | Open | Open | Open | Open |
| IRQ6 | Open | Open | Open | Open | Closed | Open | Open | Open | Open |
| IRQ7 | Open | Open | Open | Open | Open | Closed | Open | Open | Open |
| IRQ10 | Open | Open | Open | Open | Open | Open | Closed | Open | Open |
| IRQ11 | Open | Open | Open | Open | Open | Open | Open | Closed | Open |
| IRQ12 | Open | Open | Open | Open | Open | Open | Open | Open | Closed |

| CABLE TYPE ||
|---|---|
| Type | JP4 |
| AUI transceiver via DB-15 port | All jumpers open |
| Shielded/Unshielded twisted pair | All jumpers closed |

| CONNECTORS ||
|---|---|
| Purpose | Location |
| External LED connector | JP5 |

| DIAGNOSTIC LED |||
|---|---|---|
| LED | Status | Condition |
| LED1 | On | Data is being received |
| LED2 | On | Data is being transmitted |
| JP5 | On | Power on |
| JP5 | Blinking | Data is being transmitted or received |
| JP5 | Off | Connection is broken/power off |

# SYSKONNECT, INC.
## SK-NET G16

| | |
|---|---|
| **NIC Type** | Ethernet |
| **Transfer Rate** | 10Mbps |
| **Data Bus** | 8/16-bit ISA |
| **Topology** | Linear bus |
| | Star |
| **Wiring Type** | DIX transceiver via DB-15 port |
| | RG-58A/U 50ohm coaxial |
| **Boot ROM** | Available |

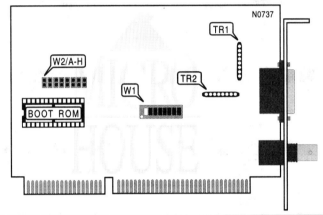

| BOOT ROM | |
|---|---|
| Setting | W1/2 |
| ⇨ Disabled | On |
| Enabled | Off |

| BASE MEMORY ADDRESS | | | |
|---|---|---|---|
| Address | W1/3 | W1/4 | W1/5 |
| ⇨ DC00h | Off | Off | Off |
| C000h | On | On | On |
| C400h | Off | On | On |
| C800h | On | Off | On |
| CC00h | Off | Off | On |
| D000h | On | On | Off |
| D400h | Off | On | Off |
| D800h | On | Off | Off |

| CABLE TYPE | | |
|---|---|---|
| Type | TR1 | TR2 |
| Thick Ethernet | Installed | Not installed |
| Thin Ethernet | Not installed | Installed |

## SYSKONNECT, INC.
## SK-NET G16

| DATA BUS SIZE(VERSION A) | | | | | | | | |
|---|---|---|---|---|---|---|---|---|
| Size | W2/A | W2/B | W2/C | W2/D | W2/E | W2/F | W2/G | W2/H |
| ⇨ 16-bit | Closed | Closed | Closed | Closed | Closed | Closed | Closed | Closed |
| 8-bit | Closed | Closed | Closed | Closed | Closed | Closed | Closed | Closed |

| DATA BUS SIZE(VERSION B) | | | | | | | | |
|---|---|---|---|---|---|---|---|---|
| Size | W2/A | W2/B | W2/C | W2/D | W2/E | W2/F | W2/G | W2/H |
| ⇨ 16-bit | Closed | Closed | Closed | Closed | Closed | Closed | Closed | Open |
| 8-bit | Open | Closed | Closed | Closed | Closed | Closed | Closed | Closed |

| INTERRUPT CONFIGURATION | |
|---|---|
| IRQ3 | W1/1 |
| Enabled | On |
| Disabled | Off |

| I/O ADDRESS | | | |
|---|---|---|---|
| Address | W1/6 | W1/7 | W1/8 |
| ⇨ 390h | Off | Off | Off |
| 100h | On | On | On |
| 180h | Off | On | On |
| 208h | On | Off | On |
| 220h | On | On | Off |
| 288h | Off | On | Off |
| 320h | Off | Off | On |
| 328h | On | Off | Off |

# SYSKONNECT, INC.
## SK-NET G32+

**NIC Type** Ethernet
**Transfer Rate** 10Mbps
**Data Bus** 16-bit EISA
**Topology** Linear bus
Star
**Wiring Type** Unshielded twisted pair
RG-58/U 50ohm coaxial
Unshielded twisted pair
**Boot ROM** Not available

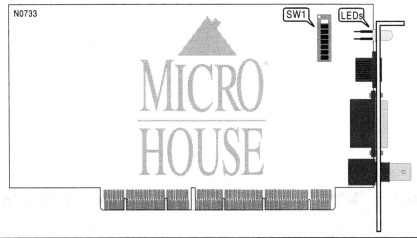

| CABLE TYPE | |
|---|---|
| **Type** | **SW1/switches 1 - 8** |
| ▷All types other than IBM type 1 | On |
| IBM type 1 | Off |

| DIAGNOSTIC LEDs | | |
|---|---|---|
| **LED** | **Status** | **Condition** |
| LED1 | Off | Polarity normal |
| LED1 | On | Polarity reversed |
| LED2 | Off | Network connection is broken |
| LED2 | On | Network connection is good |

# SYSKONNECT, INC.
## SK-NET TR4/16+

| | |
|---|---|
| **NIC Type** | Token Ring |
| **Transfer Rate** | 4/16Mbps |
| **Data Bus** | 16-bit ISA |
| **Topology** | Ring |
| **Wiring Type** | Shielded twisted pair via DB-9 port |
| | Unshielded twisted pair |
| **Boot ROM** | Available (Location unknown) |

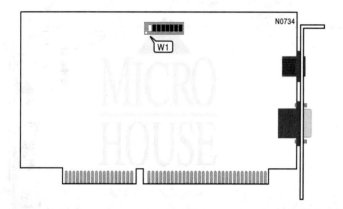

| BOOT ROM ||
|---|---|
| **Setting** | **W1/2** |
| ⇨ Disabled | Off |
| Enabled | On |

| BASE MEMORY ADDRESS SELECT ||||
|---|---|---|---|
| **Address** | **W1/3** | **W1/4** | **W1/5** |
| C0000h | On | On | On |
| C4000h | On | On | Off |
| C8000h | On | Off | On |
| CC000h | On | Off | Off |
| D0000h | Off | On | On |
| D4000h | Off | On | Off |
| D8000h | Off | Off | On |
| DC000h | Off | Off | Off |

## SYSKONNECT, INC.
## SK-NET TR4/16+

| I/O ADDRESS SELECT ||||
|---|---|---|---|
| Address | W1/6 | W1/7 | W1/8 |
| 0A20h | Off | Off | Off |
| 1A20h | Off | Off | On |
| 0B20h | Off | On | Off |
| 1B20h | Off | On | On |
| 0980h | On | Off | Off |
| 1980h | On | Off | On |
| 0900h | On | On | Off |
| 1900h | On | On | On |

| FACTORY CONFIGURED SETTINGS ||
|---|---|
| Switch | Setting |
| W1/1 | On |
| Note: This setting should be altered only if the system's motherboard uses address lines SA17 to SA19. ||

## THOMAS-CONRAD CORPORATION
## TC4035

| | |
|---|---|
| **NIC Type** | Token-Ring |
| **Transfer Rate** | 4Mbps |
| **Data Bus** | 8/16-bit ISA |
| **Topology** | Ring |
| **Wiring Type** | Shielded twisted pair |
| | Unshielded twisted pair |
| **Boot ROM** | Available |

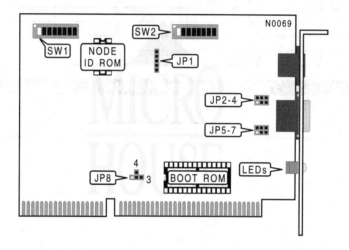

| CABLE TYPE | | | | | | |
|---|---|---|---|---|---|---|
| Type | JP2 | JP3 | JP4 | JP5 | JP6 | JP7 |
| ⇨Shielded twisted pair | Open | Closed | Closed | Open | Closed | Closed |
| Unshielded twisted pair | Closed | Closed | Open | Closed | Closed | Open |

| CARD SPEED | |
|---|---|
| Speed | JP8 |
| ⇨System clock speed | Pins 2 & 3 Closed |
| 8MHz | Pins 2 & 4 closed |
| 16MHz | Pins 1 & 2 Closed |

# THOMAS-CONRAD CORPORATION
## TC4035

| I/O BASE ADDRESS | | | |
|---|---|---|---|
| Address | SW1/1 | SW1/2 | SW1/3 |
| ↪1A20h | Off | Off | Off |
| 2A20h | On | Off | Off |
| 3A20h | Off | On | Off |
| 3A40h | On | On | Off |
| 3A60h | Off | Off | On |
| 3A80h | On | Off | On |
| 3AA0h | Off | On | On |
| 4AE0h | On | On | On |

| BOOT ROM ADDRESS | | | |
|---|---|---|---|
| Address | SW1/4 | SW1/5 | SW1/6 |
| ↪Disabled | Off | Off | Off |
| C400h | On | Off | Off |
| C800h | Off | On | Off |
| CC00h | On | On | Off |
| D000h | Off | Off | On |
| D400h | On | Off | On |
| D800h | Off | On | On |
| DC00h | On | On | On |

| WAIT STATE | |
|---|---|
| Setting | SW1/7 |
| ↪Zero wait states enabled | Off |
| Zero wait states disabled | On |

| DMA BURST SIZE | |
|---|---|
| Mode | SW1/8 |
| ↪Long burst (14 microseconds) | Off |
| Short burst (10 microseconds) | On |

| DMA CHANNEL | | |
|---|---|---|
| Channel | SW2/1 | SW2/2 |
| ↪DMA5 | Off | On |
| DMA1 | On | On |
| DMA6 | On | Off |
| DMA7 | Off | Off |

# Hardware Settings

## THOMAS-CONRAD CORPORATION
## TC4035

### INTERRUPT REQUEST

| IRQ | SW2/3 | SW2/4 | SW2/5 |
|---|---|---|---|
| 2/9 | Off | Off | Off |
| 3 | On | Off | Off |
| 5 | Off | Off | On |
| 6 | Off | On | Off |
| 7 | On | On | Off |
| 10 | On | Off | On |
| 11 | Off | On | On |
| 12 | On | On | On |

### FACTORY CONFIGURED SETTINGS

| Jumper | Setting |
|---|---|
| JP1 | Unused |
| SW2/6 | Unused |
| SW2/7 | Unused |
| SW2/8 | Unused |

Note: JP1 is not included on later versions of this card.

### LEDS

| LED | Color | Status | Condition |
|---|---|---|---|
| LED1 | Red | On | Data is being transmitted or received |
| LED1 | Red | Off | Data is being transmitted or received |
| LED2 | Green | On | Network connection is good |
| LED2 | Green | Off | Network connection is broken |

## THOMAS-CONRAD CORPORATION
## TC4043 TROPIC

**NIC Type**      Token Ring
**Transfer Rate**      4/16Mbps
**Data Bus**      16-bit ISA
**Topology**      Ring
**Wiring Type**      Shielded twisted pair via DB-9 port
**Boot ROM**      In firmware

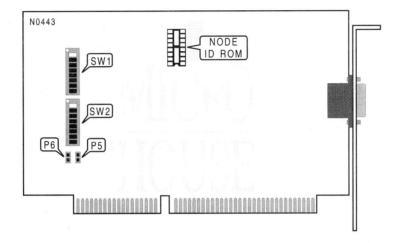

| BASE I/O ADDRESS ||
| --- | --- |
| Address | SW2/switch 1 |
| ⇨ 0A20h | Off |
| 0A24h | On |

| BLOCK DECODE SIZE ||
| --- | --- |
| Size | SW1/switch 8 |
| ⇨ 16KB | Off |
| 128KB | On |
| Note: This setting should only be changed if the TC4043 is installed in a IBM Personal System/2 Model 30-286 or if the adapter fails to initialize when set to the default 16KB block decode size. ||

| BOOT ROM ||
| --- | --- |
| Setting | P5 |
| ⇨ Disabled | Open |
| Enabled | Closed |

| BUS WIDTH SELECTION ||
| --- | --- |
| Setting | P6 |
| ⇨ Force 8-bit operation | Closed |
| Auto slot sense enabled | Open |

# THOMAS-CONRAD CORPORATION
## TC4043 TROPIC

### FACTORY CONFIGURED SETTINGS

| Switch | Setting |
|---|---|
| SW2/switch 5 | Off |
| SW2/switch 6 | Off |
| SW2/switch 7 | Off |

### INTERRUPT REQUEST

| IRQ | SW1/switch 7 | SW1/switch 8 |
|---|---|---|
| ⇨ IRQ2 | On | Off |
| IRQ3 | On | On |
| IRQ6 | Off | On |
| IRQ7 | Off | Off |

### NETWORK SPEED

| Speed | SW2/switch 4 |
|---|---|
| ⇨ 16Mbps | On |
| 4Mbps | Off |

### ROM MEMORY MAPPED I/O ADDRESS

| Address | SW1/1 | SW1/2 | SW1/3 | SW1/4 | SW1/5 | SW1/6 |
|---|---|---|---|---|---|---|
| ⇨ CC000h | Off | On | On | Off | Off | On |
| DC000h | Off | On | Off | Off | Off | On |
| C0000h | Off | On | On | On | On | On |
| C2000h | Off | On | On | On | On | Off |
| C4000h | Off | On | On | On | Off | On |
| C6000h | Off | On | On | On | Off | Off |
| C8000h | Off | On | On | Off | On | On |
| CA000h | Off | On | On | Off | On | Off |
| CC000h | Off | On | On | Off | Off | On |
| CE000h | Off | On | On | Off | Off | Off |
| D0000h | Off | On | Off | On | On | On |
| D2000h | Off | On | Off | On | On | Off |
| D4000h | Off | On | Off | On | Off | On |
| D6000h | Off | On | Off | On | Off | Off |
| D8000h | Off | On | Off | Off | On | On |
| DA000h | Off | On | Off | Off | On | Off |
| DC000h | Off | On | Off | Off | Off | On |
| DE000h | Off | On | Off | Off | Off | Off |

### SHARED RAM SIZE

| Size | SW2/switch 2 | SW2/switch 3 |
|---|---|---|
| ⇨ 16KB | Off | On |
| 64KB | Off | Off |

# THOMAS-CONRAD CORPORATION
## TC 4045

| | |
|---|---|
| **NIC Type** | Token-Ring |
| **Transfer Rate** | 4/16Mbps |
| **Data Bus** | 8/16-bit ISA |
| **Topology** | Ring |
| **Wiring Type** | Shielded |
| | Unshielded twisted pair |
| **Boot ROM** | Available |

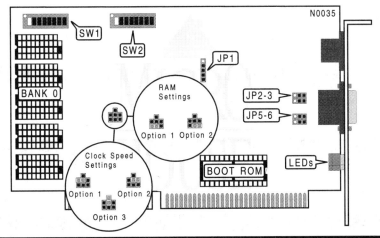

### FACTORY CONFIGURED SETTINGS

| Jumper | Setting |
|---|---|
| JP1 | Unused |

Note: JP1 is not included on later versions of the card.

### CABLE TYPE

| Type | JP2 | JP3 | JP5 | JP6 |
|---|---|---|---|---|
| ⇨Shielded twisted pair (DB9 port) | Open | Open | Open | Open |
| Shielded/Unshielded twisted pair (RJ-45 jack) | Closed | Closed | Closed | Closed |

### I/O BASE ADDRESS

| Address | SW1/1 | SW1/2 | SW1/3 |
|---|---|---|---|
| ⇨1A20h | Off | Off | Off |
| 2A20h | On | Off | Off |
| 3A20h | Off | On | Off |
| 3A40h | On | On | Off |
| 3A60h | Off | Off | On |
| 3A80h | On | Off | On |
| 3AA0h | Off | On | On |
| 4AE0h | On | On | On |

# THOMAS-CONRAD CORPORATION
## TC4045

### BOOT ROM ADDRESS

| Address | SW1/4 | SW1/5 | SW1/6 |
|---|---|---|---|
| ▷Disabled | Off | Off | Off |
| C400h | On | Off | Off |
| C800h | Off | On | Off |
| CC00h | On | On | Off |
| D000h | Off | Off | On |
| D400h | On | Off | On |
| D800h | Off | On | On |
| DC00h | On | On | On |

### WAIT STATE

| Setting | SW1/7 |
|---|---|
| ▷Zero wait states enabled | Off |
| Zero wait states disabled | On |

### DMA BURST SIZE

| Size | SW1/8 |
|---|---|
| ▷Long burst (14 microseconds) | Off |
| Short burst (10 microseconds) | On |

### DMA CHANNEL

| Channel | SW2/1 | SW2/2 |
|---|---|---|
| ▷5 | Off | On |
| 1 | On | On |
| 6 | On | Off |
| 7 | Off | Off |

### INTERRUPT REQUEST

| IRQ | SW2/3 | SW2/4 | SW2/5 |
|---|---|---|---|
| ▷2/9 | Off | Off | Off |
| 3 | On | Off | Off |
| 5 | Off | Off | On |
| 6 | Off | On | Off |
| 7 | On | On | Off |
| 10 | On | Off | On |
| 11 | Off | On | On |
| 12 | On | On | On |

# THOMAS-CONRAD CORPORATION
## TC4045

| NETWORK SPEED ||
|---|---|
| **Speed** | **SW2/6** |
| ⇨16Mbps | On |
| 4Mbps | Off |
| Note: All cards on the network must have this option set the same. ||

| DRAM CONFIGURATION |||
|---|---|---|
| **Size** | **RAM settings** | **Bank0** |
| 128KB | Option 1 | (5) 4464 |
| 512KB | Option 2 | (5) 44256 |
| 2MB | Option 2 | (5) 441000 |

| CLOCK SPEED ||
|---|---|
| **Speed** | **Clock speed settings** |
| ⇨System clock | Option 2 |
| 8MHz | Option 3 |
| 16MHz | Option 1 |

| DIAGNOSTIC LED(S) ||||
|---|---|---|---|
| **LED** | **Color** | **Status** | **Condition** |
| 1 | Green | On | Network connection is good |
| 1 | Green | Off | Network connection is broken |
| 2 | Red | On | Data is being transmitted or received |
| 2 | Red | Off | Data is not being transmitted or received |

## THOMAS-CONRAD CORPORATION
## TC4047

| | |
|---|---|
| NIC Type | Token Ring |
| Transfer Rate | 4/16Mbps |
| Data Bus | 16-bit EISA |
| Topology | Ring |
| Wiring Type | Shielded twisted pair |
| | Unshielded twisted pair |
| Boot ROM | In Firmware |

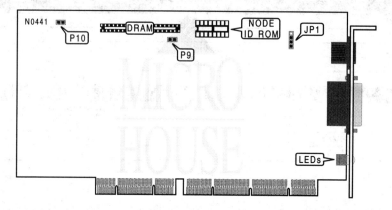

| FACTORY CONFIGURED SETTINGS ||
|---|---|
| Jumper | Setting |
| JP1 | Open |
| P10 | Open |

| DRAM UPGRADE CONFIGURATION ||
|---|---|
| 512KB Upgrade | P9 |
| ⇨ Installed | Open |
| Not installed | Closed |

| DIAGNOSTIC LED(S) ||||
|---|---|---|---|
| LED | Color | Status | Condition |
| LED1 | Green | On | Card is inserted into the ring |
| LED1 | Green | Off | Card is not inserted into the ring |
| LED2 | Red | On | Data is being transmitted |
| LED2 | Red | Off | Data is not being transmitted |

# THOMAS-CONRAD CORPORATION
## TC5043-2

| | |
|---|---|
| **NIC Type** | Ethernet |
| **Transfer Rate** | 10Mbps |
| **Data Bus** | 8/16-bit ISA |
| **Topology** | Star |
| **Wiring Type** | RG-58A/U 50ohm coaxial |
| | AUI transceiver via DB-15 port |
| **Boot ROM** | Available |

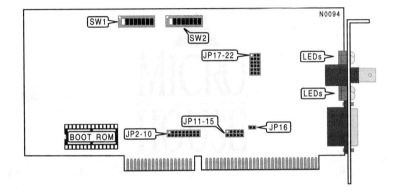

| INTERRUPT REQUEST | | | | | | | | | |
|---|---|---|---|---|---|---|---|---|---|
| IRQ | JP2 | JP3 | JP4 | JP5 | JP6 | JP7 | JP8 | JP9 | JP10 |
| ᴼ3 | Open | Open | Open | Open | Open | Open | Open | Closed | Open |
| 2/9 | Open | Open | Open | Open | Open | Open | Open | Open | Closed |
| 4 | Open | Open | Open | Open | Open | Open | Closed | Open | Open |
| 5 | Open | Open | Open | Open | Open | Closed | Open | Open | Open |
| 7 | Open | Open | Open | Open | Closed | Open | Open | Open | Open |
| 10 | Open | Open | Open | Closed | Open | Open | Open | Open | Open |
| 11 | Open | Open | Closed | Open | Open | Open | Open | Open | Open |
| 12 | Open | Closed | Open | Open | Open | Open | Open | Open | Open |
| 15 | Closed | Open | Open | Open | Open | Open | Open | Open | Open |

| I/O BASE ADDRESS | | | |
|---|---|---|---|
| I/O Address | SW2/1 | SW2/2 | SW2/3 |
| ᴼ300h | On | Off | Off |
| 100h | On | On | On |
| 120h | On | On | Off |
| 140h | On | Off | On |
| 320h | Off | On | On |
| 340h | Off | On | Off |
| 360h | Off | Off | On |

# THOMAS-CONRAD CORPORATION
## TC5043-2

| DRIVER TYPE ||
|---|---|
| **Driver** | **SW1/1** |
| ⇨NE2000 | On |
| NE1000 | Off |

| TIMING COMPATIBILITY ||
|---|---|
| **Setting** | **SW1/2** |
| ⇨Normal | On |
| C & T | Off |

Note: The Compaq portable 286, IBM PS/2 Model 30-286 & any computer utilizing a C & T chipset, must be set to C & T timing compatibility.

| BOOT ROM ||
|---|---|
| **Setting** | **SW2/7** |
| ⇨Disabled | Off |
| Enabled | On |

| BASE MEMORY ADDRESS ||||
|---|---|---|---|
| **Memory Address** | **SW2/4** | **SW2/5** | **SW2/6** |
| ⇨D000h | Off | On | On |
| C400h | On | On | Off |
| C800h | On | Off | On |
| CC00h | On | Off | Off |
| D400h | Off | On | Off |
| D800h | Off | Off | On |
| DC00h | Off | Off | Off |

| CABLE TYPE |||||||||
|---|---|---|---|---|---|---|---|---|
| **Type** | **JP16** | **JP17** | **JP18** | **JP19** | **JP20** | **JP21** | **JP22** ||
| RG58A/U 50ohm | Closed | Pins 1 & 2 | Pins 1 & 2 | Pins 1 & 2 | Pins 1 & 2 | Pins 1 & 2 | Pins 1 & 2 ||
| AUI transceiver | Open | Pins 2 & 3 | Pins 2 & 3 | Pins 2 & 3 | Pins 2 & 3 | Pins 2 & 3 | Pins 2 & 3 ||

Note: Pins designated should be in the closed position.

## THOMAS-CONRAD CORPORATION
## TC5043-2

| DIAGNOSTIC LED(S) ||||
|---|---|---|---|
| **LED** | **Color** | **Status** | **Condition** |
| LED1 | Yellow | On | Data is being transmitted |
| LED1 | Yellow | Off | Data is not being transmitted |
| LED2 | Green | On | Data is being received |
| LED2 | Green | Off | Data is not being received |
| LED3 | Yellow | On | Collision detected on network |
| LED3 | Yellow | Off | No collisions detected on network |
| LED4 | Green | On | 10BASE-T link integrity exists |
| LED4 | Green | On | 10BASE-T link integrity broken |

| BUS SPEED ||||
|---|---|---|---|
| **Speed** | **SW1/3** | **SW1/4** | **SW1/5** |
| ➪Normal | Off | Off | Off |
| 6/8MHz | Off | Off | On |
| 10MHz | Off | On | Off |
| 12/16MHz | On | Off | Off |

| FACTORY CONFIGURED SETTINGS - DO NOT ALTER ||
|---|---|
| **Jumper** | **Setting** |
| JP11 | Unidentified |
| JP12 | Unidentified |
| JP13 | Unidentified |
| JP14 | Unidentified |
| JP15 | Unidentified |
| SW2/8 | On |

# THOMAS-CONRAD CORPORATION
## TC5143-T / TC5143-2

| | |
|---|---|
| **NIC Type** | Ethernet |
| **Transfer Rate** | 10Mbps |
| **Data Bus** | 16-bit ISA |
| **Topology** | Star |
| **Wiring Type** | Unshielded twisted pair (TC5143-T) |
| | RG58A/U 50ohm coaxial (TC5143-2) |
| | AUI transceiver via DB-15 port (TC5143-2) |
| **Boot ROM** | Available |

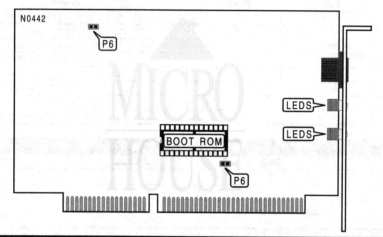

| BOOT ROM ||
|---|---|
| **Setting** | **P5** |
| ⇨ Disabled | Open |
| Enabled | Closed |

| CONFIGURATION OVERRIDE SETTING ||
|---|---|
| **Setting** | **P6** |
| ⇨ Normal operation | Open |
| Reset configuration to default | Closed |

## THOMAS-CONRAD CORPORATION
## TC5143-2

| LED | Color | Status | Condition |
|---|---|---|---|
| **DIAGNOSTIC LED(S)** | | | |
| LED1 | Yellow | Off | Data is not being transmitted |
| LED1 | Yellow | Blinking | Data is being transmitted |
| LED2 | Green | On | Data is not being received |
| LED2 | Green | Blinking | Data is being received |
| LED3 | Yellow | On | Collision detected on network |
| LED3 | Yellow | Off | No collisions detected on network |
| LED4 | Green | On | 10BaseT link integrity exists |
| LED4 | Green | Off | 10BaseT link integrity broken |
| Note: LED4 is only functional when link integrity is enabled, and when installed on a 10BaseT network. | | | |

## TIARA COMPUTER SYSTEMS, INC.
## ETHERNET LANCARD E2000

**NIC Type** Ethernet
**Transfer Rate** 10Mbps
**Data Bus** 16-bit ISA
**Topology** Linear Bus
**Wiring Type** RG-58A/U 50ohm coaxial
AUI transceiver via DB-15 port
**Boot ROM** Available

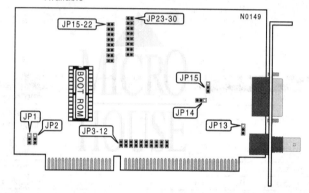

| Address | JP23 | JP24 | JP25 | JP26 | JP27 | JP28 | JP29 | JP30 |
|---|---|---|---|---|---|---|---|---|
| **BOOT ROM BASE ADDRESS** | | | | | | | | |
| Disabled | Closed | Open | Open | Open | Open | Open | Open | Open |
| C400h | Open | Open | Open | Open | Open | Open | Open | Closed |
| C800h | Open | Closed | Open | Open | Open | Open | Open | Open |
| CC00h | Open | Open | Closed | Open | Open | Open | Open | Open |
| D000h | Open | Open | Open | Closed | Open | Open | Open | Open |
| D400h | Open | Open | Open | Open | Closed | Open | Open | Open |
| D800h | Open | Open | Open | Open | Open | Closed | Open | Open |
| DC00h | Open | Open | Open | Open | Open | Open | Closed | Open |

| IRQ | JP3 | JP4 | JP5 | JP6 | JP7 | JP8 | JP9 | JP10 | JP11 | JP12 |
|---|---|---|---|---|---|---|---|---|---|---|
| **INTERRUPT REQUEST** | | | | | | | | | | |
| 2 | Open | Open | Open | Open | Open | Open | Open | Open | Open | Closed |
| 3 | Open | Open | Open | Open | Open | Open | Open | Open | Closed | Open |
| 4 | Open | Open | Open | Open | Open | Open | Open | Closed | Open | Open |
| 5 | Open | Open | Open | Open | Open | Open | Closed | Open | Open | Open |
| 7 | Open | Open | Open | Open | Open | Closed | Open | Open | Open | Open |
| 9 | Open | Open | Open | Open | Closed | Open | Open | Open | Open | Open |
| 10 | Open | Open | Open | Closed | Open | Open | Open | Open | Open | Open |
| 11 | Open | Open | Closed | Open | Open | Open | Open | Open | Open | Open |
| 12 | Open | Closed | Open | Open | Open | Open | Open | Open | Open | Open |
| 15 | Closed | Open | Open | Open | Open | Open | Open | Open | Open | Open |

## TIARA COMPUTER SYSTEMS, INC.
## ETHERNET LANCARD E2000

### SEGMENT LENGTH

| Maximum | JP13 |
|---|---|
| ⇨185 meters | Pins 1 & 2 Closed |
| 305 meters | Pins 2 & 3 Closed |

Note: All cards on the network must have this jumper set the same.

### CABLE TYPE

| Type | JP14 |
|---|---|
| ⇨RG-58A/U 50ohm coaxial | Pins 1 & 2 Closed |
| AUI transceiver via DB-15 port | Pins 2 & 3 Closed |

### I/O BASE ADDRESS

| Address | JP15 | JP16 | JP17 | JP18 | JP19 | JP20 | JP21 | JP22 |
|---|---|---|---|---|---|---|---|---|
| 280h | Open | Open | Open | Open | Closed | Open | Open | Open |
| 2A0h | Open | Open | Open | Open | Open | Closed | Open | Open |
| 2C0h | Open | Open | Open | Open | Open | Open | Closed | Open |
| 2E0h | Open | Open | Open | Open | Open | Open | Open | Closed |
| ⇨300h | Closed | Open | Open | Open | Open | Open | Open | Open |
| 320h | Open | Closed | Open | Open | Open | Open | Open | Open |
| 340h | Open | Open | Closed | Open | Open | Open | Open | Open |
| 360h | Open | Open | Open | Closed | Open | Open | Open | Open |

### COMPATIBILITY MODE

| Compatibility | JP1 | JP2 |
|---|---|---|
| ⇨Standard | Pins 1 & 2 Closed | Pins 1 & 2 Closed |
| Special | Pins 2 & 3 Closed | Pins 2 & 3 Closed |

Note: Use the special mode on the PS/2 Model 30-286 if the computer hangs after running IPX.

### DB-15 PORT GROUND CONFIGURATION

| Ground | JP15 |
|---|---|
| AUI (logic ground) | Pins 2 & 3 Closed |
| DIX (chassis ground) | Pins 1 & 2 Closed |

Note: The older, pre-IEEE 802.3 Ethernet Digital Equipment/Intel/Xerox (DIX) cables use a different type of grounding than the IEEE 802.3 Ethernet standards.

## TIARA COMPUTER SYSTEMS, INC.
## LANCARD/E * AT TP

| | |
|---|---|
| **NIC Type** | Ethernet |
| **Transfer Rate** | 10Mbps |
| **Data Bus** | 16-bit ISA |
| **Topology** | Linear Bus |
| **Wiring Type** | Shielded twisted pair |
| | Unshielded twisted pair |
| | AUI transceiver via DB-15 port |
| **Boot ROM** | Available |

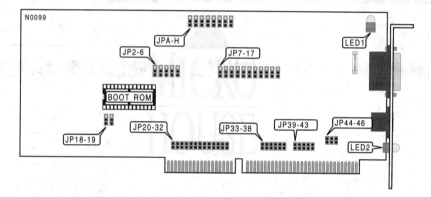

| CABLE TYPE | |
|---|---|
| **Jumper** | **JPA - JPH** |
| Shielded/Unshielded twisted pair | Pins 1 & 2 closed |
| AUI transceiver via DB-15 port | Pins 2 & 3 closed |

| DIAGNOSTIC LED(S) | | |
|---|---|---|
| **LED** | **Status** | **Condition** |
| LED1 | On | Power on |
| LED1 | Off | Card is not receiving power, the fuse is blown, or cable type is twisted pair. |
| LED2 | On | Twisted pair network connection is good |
| LED2 | Off | Twisted pair network connection is broken |
| Note: LED2 will only function is the cable type is twisted pair. | | |

## TIARA COMPUTER SYSTEMS, INC.
LANCARD/E * AT TP

### I/O BASE ADDRESS

| Address | JP2 | JP3 | JP4 | JP5 | JP6 |
|---|---|---|---|---|---|
| 300h | Pins 2 & 3 | Pins 2 & 3 | Pins 2 & 3 | Pins 1 & 2 | Pins 1 & 2 |
| 000h | Pins 2 & 3 | Pins 2 & 3 | Pins 2 & 3 | Pins 2 & 3 | Pins 2 & 3 |
| 020h | Pins 1 & 2 | Pins 2 & 3 | Pins 2 & 3 | Pins 2 & 3 | Pins 2 & 3 |
| 040h | Pins 2 & 3 | Pins 1 & 2 | Pins 2 & 3 | Pins 2 & 3 | Pins 2 & 3 |
| 060h | Pins 1 & 2 | Pins 1 & 2 | Pins 2 & 3 | Pins 2 & 3 | Pins 2 & 3 |
| 080h | Pins 2 & 3 | Pins 2 & 3 | Pins 1 & 2 | Pins 2 & 3 | Pins 2 & 3 |
| 0A0h | Pins 1 & 2 | Pins 2 & 3 | Pins 1 & 2 | Pins 2 & 3 | Pins 2 & 3 |
| 0C0h | Pins 2 & 3 | Pins 1 & 2 | Pins 1 & 2 | Pins 2 & 3 | Pins 2 & 3 |
| 0E0h | Pins 1 & 2 | Pins 1 & 2 | Pins 1 & 2 | Pins 2 & 3 | Pins 2 & 3 |
| 100h | Pins 2 & 3 | Pins 2 & 3 | Pins 2 & 3 | Pins 1 & 2 | Pins 2 & 3 |
| 120h | Pins 1 & 2 | Pins 2 & 3 | Pins 2 & 3 | Pins 1 & 2 | Pins 2 & 3 |
| 140h | Pins 2 & 3 | Pins 1 & 2 | Pins 2 & 3 | Pins 1 & 2 | Pins 2 & 3 |
| 160h | Pins 1 & 2 | Pins 1 & 2 | Pins 2 & 3 | Pins 1 & 2 | Pins 2 & 3 |
| 180h | Pins 2 & 3 | Pins 2 & 3 | Pins 1 & 2 | Pins 1 & 2 | Pins 2 & 3 |
| 1A0h | Pins 1 & 2 | Pins 2 & 3 | Pins 1 & 2 | Pins 1 & 2 | Pins 2 & 3 |
| 1C0h | Pins 2 & 3 | Pins 1 & 2 | Pins 1 & 2 | Pins 1 & 2 | Pins 2 & 3 |
| 1E0h | Pins 1 & 2 | Pins 1 & 2 | Pins 1 & 2 | Pins 1 & 2 | Pins 2 & 3 |
| 200h | Pins 2 & 3 | Pins 2 & 3 | Pins 2 & 3 | Pins 2 & 3 | Pins 1 & 2 |
| 220h | Pins 1 & 2 | Pins 2 & 3 | Pins 2 & 3 | Pins 2 & 3 | Pins 1 & 2 |
| 240h | Pins 2 & 3 | Pins 1 & 2 | Pins 2 & 3 | Pins 2 & 3 | Pins 1 & 2 |
| 260h | Pins 1 & 2 | Pins 1 & 2 | Pins 2 & 3 | Pins 2 & 3 | Pins 1 & 2 |
| 280h | Pins 2 & 3 | Pins 2 & 3 | Pins 1 & 2 | Pins 2 & 3 | Pins 1 & 2 |
| 2A0h | Pins 1 & 2 | Pins 2 & 3 | Pins 1 & 2 | Pins 2 & 3 | Pins 1 & 2 |
| 2C0h | Pins 2 & 3 | Pins 1 & 2 | Pins 1 & 2 | Pins 2 & 3 | Pins 1 & 2 |
| 2E0h | Pins 1 & 2 | Pins 1 & 2 | Pins 1 & 2 | Pins 2 & 3 | Pins 1 & 2 |
| 320h | Pins 1 & 2 | Pins 2 & 3 | Pins 2 & 3 | Pins 1 & 2 | Pins 1 & 2 |
| 340h | Pins 2 & 3 | Pins 1 & 2 | Pins 2 & 3 | Pins 1 & 2 | Pins 1 & 2 |
| 360h | Pins 1 & 2 | Pins 1 & 2 | Pins 2 & 3 | Pins 1 & 2 | Pins 1 & 2 |
| 380h | Pins 2 & 3 | Pins 2 & 3 | Pins 1 & 2 | Pins 1 & 2 | Pins 1 & 2 |
| 3A0h | Pins 1 & 2 | Pins 2 & 3 | Pins 1 & 2 | Pins 1 & 2 | Pins 1 & 2 |
| 3C0h | Pins 2 & 3 | Pins 1 & 2 | Pins 1 & 2 | Pins 1 & 2 | Pins 1 & 2 |
| 3E0h | Pins 1 & 2 | Pins 1 & 2 | Pins 1 & 2 | Pins 1 & 2 | Pins 1 & 2 |

Note: Pins designated should be in the closed position.

### BOOT ROM

| Setting | JP7 |
|---|---|
| Disabled | Pins 2 & 3 closed |
| Enabled | Pins 1 & 2 closed |

# TIARA COMPUTER SYSTEMS, INC.
## LANCARD/E * AT TP

### BASE MEMORY ADDRESS - FIRST DIGIT

| Address Segment | JP8 | JP9 | JP10 | JP11 |
|---|---|---|---|---|
| 0h | Pins 2 & 3 closed | Pins 2 & 3 closed | Pins 2 & 3 closed | Pins 2 & 3 closed |
| 1h | Pins 2 & 3 closed | Pins 2 & 3 closed | Pins 2 & 3 closed | Pins 1 & 2 closed |
| 2h | Pins 2 & 3 closed | Pins 2 & 3 closed | Pins 1 & 2 closed | Pins 2 & 3 closed |
| 3h | Pins 2 & 3 closed | Pins 2 & 3 closed | Pins 1 & 2 closed | Pins 1 & 2 closed |
| 4h | Pins 2 & 3 closed | Pins 1 & 2 closed | Pins 2 & 3 closed | Pins 2 & 3 closed |
| 5h | Pins 2 & 3 closed | Pins 1 & 2 closed | Pins 2 & 3 closed | Pins 1 & 2 closed |
| 6h | Pins 2 & 3 closed | Pins 1 & 2 closed | Pins 1 & 2 closed | Pins 2 & 3 closed |
| 7h | Pins 2 & 3 closed | Pins 1 & 2 closed | Pins 1 & 2 closed | Pins 1 & 2 closed |
| 8h | Pins 1 & 2 closed | Pins 2 & 3 closed | Pins 2 & 3 closed | Pins 2 & 3 closed |
| 9h | Pins 1 & 2 closed | Pins 2 & 3 closed | Pins 2 & 3 closed | Pins 1 & 2 closed |
| Ah | Pins 1 & 2 closed | Pins 2 & 3 closed | Pins 1 & 2 closed | Pins 2 & 3 closed |
| Bh | Pins 1 & 2 closed | Pins 2 & 3 closed | Pins 1 & 2 closed | Pins 1 & 2 closed |
| Ch | Pins 1 & 2 closed | Pins 1 & 2 closed | Pins 2 & 3 closed | Pins 2 & 3 closed |
| Dh | Pins 1 & 2 closed | Pins 1 & 2 closed | Pins 2 & 3 closed | Pins 1 & 2 closed |
| Eh | Pins 1 & 2 closed | Pins 1 & 2 closed | Pins 1 & 2 closed | Pins 2 & 3 closed |
| Fh | Pins 1 & 2 closed | Pins 1 & 2 closed | Pins 1 & 2 closed | Pins 1 & 2 closed |

Note: The Address Segment is the first digit in the Base Memory Address. Refer to the next table for the remaining three digits in the address. Note that not all configurable addresses are valid on all systems.

### BASE MEMORY ADDRESS - LAST THREE DIGITS

| Address | ROM Size | JP11 | JP12 | JP30 |
|---|---|---|---|---|
| x000h | 8K x 8 | Pins 2 & 3 closed | Pins 2 & 3 closed | Pins 2 & 3 closed |
| x200h | 8K x 8 | Pins 2 & 3 closed | Pins 2 & 3 closed | Pins 1 & 2 closed |
| x400h | 8K x 8 | Pins 2 & 3 closed | Pins 1 & 2 closed | Pins 2 & 3 closed |
| x600h | 8K x 8 | Pins 2 & 3 closed | Pins 1 & 2 closed | Pins 1 & 2 closed |
| x800h | 8K x 8 | Pins 1 & 2 closed | Pins 2 & 3 closed | Pins 2 & 3 closed |
| xA00h | 8K x 8 | Pins 1 & 2 closed | Pins 2 & 3 closed | Pins 1 & 2 closed |
| xC00h | 8K x 8 | Pins 1 & 2 closed | Pins 1 & 2 closed | Pins 2 & 3 closed |
| xE00h | 8K x 8 | Pins 1 & 2 closed | Pins 1 & 2 closed | Pins 1 & 2 closed |
| x000h | 16K x 8 | Pins 2 & 3 closed | Pins 2 & 3 closed | Pins 2 & 3 closed |
| x400h | 16K x 8 | Pins 2 & 3 closed | Pins 1 & 2 closed | Pins 2 & 3 closed |
| x800h | 16K x 8 | Pins 1 & 2 closed | Pins 2 & 3 closed | Pins 2 & 3 closed |
| xC00h | 16K x 8 | Pins 1 & 2 closed | Pins 1 & 2 closed | Pins 2 & 3 closed |
| x000h | 32K x 8 | Pins 2 & 3 closed | Pins 2 & 3 closed | Pins 2 & 3 closed |
| x800h | 32K x 8 | Pins 1 & 2 closed | Pins 2 & 3 closed | Pins 2 & 3 closed |
| x000h | 64K x 8 | Pins 2 & 3 closed | Pins 2 & 3 closed | Pins 2 & 3 closed |

Note: Place the three digit address given here behind the single digit given in the previous table to get the complete Base Memory Address.

## TIARA COMPUTER SYSTEMS, INC.
## LANCARD/E * AT TP

| | | | BOOT ROM SIZE | | |
|---|---|---|---|---|---|
| Size | JP15 | JP16 | JP17 | JP18 | JP19 |
| 2764 (8KB) | Pins 1 & 2 | Pins 1 & 2 | Pins 1 & 2 | Pins 2 & 3 | Pins 2 & 3 |
| 27128 (16KB) | Pins 2 & 3 | Pins 1 & 2 | Pins 1 & 2 | Pins 2 & 3 | Pins 1 & 2 |
| 27256 (32KB) | Pins 2 & 3 | Pins 2 & 3 | Pins 1 & 2 | Pins 1 & 2 | Pins 2 & 3 |
| 27512 (64KB) | Pins 2 & 3 | Pins 2 & 3 | Pins 2 & 3 | Pins 1 & 2 | Pins 1 & 2 |

Note: Pins designated should be in the closed position.

| | | | DMA REQUEST CHANNEL | | | | | |
|---|---|---|---|---|---|---|---|---|
| Channel | JP20 | JP22 | JP24 | JP26 | JP39 | JP41 | JP44 | JP46 |
| Disabled | Open | Open | Open | Open | Open | Open | Open | Closed |
| DRQ0 | Open | Open | Open | Closed | Open | Open | Open | Open |
| DRQ1 | Open | Open | Open | Open | Closed | Open | Open | Open |
| DRQ2 | Open | Open | Open | Open | Open | Open | Closed | Open |
| DRQ3 | Open | Open | Open | Open | Open | Closed | Open | Open |
| DRQ5 | Open | Open | Closed | Open | Open | Open | Open | Open |
| DRQ6 | Open | Closed | Open | Open | Open | Open | Open | Open |
| DRQ7 | Closed | Open | Open | Open | Open | Open | Open | Open |

| | | | DMA ACKNOWLEDGE CHANNEL | | | | | |
|---|---|---|---|---|---|---|---|---|
| Channel | JP21 | JP23 | JP25 | JP27 | JP33 | JP40 | JP42 | JP43 |
| Disabled | Open | Open | Open | Open | Open | Open | Open | Closed |
| DACK0 | Open | Open | Open | Closed | Open | Open | Open | Open |
| DACK1 | Open | Open | Open | Open | Closed | Open | Open | Open |
| DACK2 | Open | Open | Open | Open | Open | Open | Closed | Open |
| DACK3 | Open | Open | Open | Open | Open | Closed | Open | Open |
| DACK5 | Open | Open | Closed | Open | Open | Open | Open | Open |
| DACK6 | Open | Closed | Open | Open | Open | Open | Open | Open |
| DACK7 | Closed | Open | Open | Open | Open | Open | Open | Open |

Note: DMA acknowledge channel must be set to the same number as the DMA request.

| | | | | INTERRUPT REQUEST | | | | | | |
|---|---|---|---|---|---|---|---|---|---|---|
| IRQ | JP28 | JP29 | JP30 | JP31 | JP32 | JP34 | JP35 | JP36 | JP37 | JP38 | JP45 |
| 3 | Open | Open | Open | Open | Open | Closed | Open | Open | Open | Open | Open |
| 2 | Open | Open | Open | Open | Open | Open | Open | Open | Open | Open | Closed |
| 4 | Open | Open | Open | Open | Open | Open | Closed | Open | Open | Open | Open |
| 5 | Open | Open | Open | Open | Open | Open | Open | Closed | Open | Open | Open |
| 6 | Open | Open | Open | Open | Open | Open | Open | Open | Closed | Open | Open |
| 7 | Open | Open | Open | Open | Open | Open | Open | Open | Open | Closed | Open |
| 10 | Open | Open | Open | Open | Closed | Open | Open | Open | Open | Open | Open |
| 11 | Open | Open | Open | Closed | Open | Open | Open | Open | Open | Open | Open |
| 12 | Open | Open | Closed | Open | Open | Open | Open | Open | Open | Open | Open |
| 14 | Closed | Open | Open | Open | Open | Open | Open | Open | Open | Open | Open |
| 15 | Open | Closed | Open | Open | Open | Open | Open | Open | Open | Open | Open |

# TIARA COMPUTER SYSTEMS, INC.
## LANCARD/E*PC10BT

| | |
|---|---|
| **NIC Type** | Ethernet |
| **Transfer Rate** | 10Mbps |
| **Data Bus** | 8-bit ISA |
| **Topology** | Star |
| **Wiring Type** | Shielded/Unshielded twisted pair |
| | AUI transceiver via DB-15 port |
| **Boot ROM** | Available |

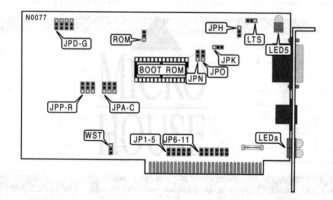

| INTERRUPT REQUEST | | | | | |
|---|---|---|---|---|---|
| **IRQ** | **JP1** | **JP2** | **JP3** | **JP4** | **JP5** |
| 2 | Open | Open | Open | Open | Closed |
| ⇨3 | Closed | Open | Open | Open | Open |
| 4 | Open | Closed | Open | Open | Open |
| 5 | Open | Open | Closed | Open | Open |
| 7 | Open | Open | Open | Closed | Open |

| DMA CHANNEL | | | | | | |
|---|---|---|---|---|---|---|
| **Channel** | **JP6** | **JP7** | **JP8** | **JP9** | **JP10** | **JP11** |
| ⇨Disabled | Open | Open | Closed | Open | Open | Closed |
| DMA1 | Closed | Open | Open | Closed | Open | Open |
| DMA3 | Open | Closed | Open | Open | Closed | Open |

## TIARA COMPUTER SYSTEMS, INC.
## LANCARD/E*PC10BT

| I/O BASE ADDRESS | | | |
|---|---|---|---|
| I/O Address | JPA | JPB | JPC |
| 200h | Pins 1 & 2 Closed | Pins 1 & 2 Closed | Pins 1 & 2 Closed |
| 240h | Pins 1 & 2 Closed | Pins 1 & 2 Closed | Pins 2 & 3 Closed |
| 280h | Pins 1 & 2 Closed | Pins 2 & 3 Closed | Pins 1 & 2 Closed |
| 2C0h | Pins 1 & 2 Closed | Pins 2 & 3 Closed | Pins 2 & 3 Closed |
| ▷300h | Pins 2 & 3 Closed | Pins 1 & 2 Closed | Pins 1 & 2 Closed |
| 340h | Pins 2 & 3 Closed | Pins 1 & 2 Closed | Pins 2 & 3 Closed |
| 380h | Pins 2 & 3 Closed | Pins 2 & 3 Closed | Pins 1 & 2 Closed |
| 3C0h | Pins 2 & 3 Closed | Pins 2 & 3 Closed | Pins 2 & 3 Closed |

| CABLE TYPE | | | | |
|---|---|---|---|---|
| Setting | JPD | JPE | JPF | JPG |
| ▷Shielded/Unshielded twisted pair | Pins 1 & 2 | Pins 1 & 2 | Pins 1 & 2 | Pins 1 & 2 |
| AUI transceiver via DB-15 port | Pins 2 & 3 | Pins 2 & 3 | Pins 2 & 3 | Pins 2 & 3 |

Note: Pins designated should be in the closed position.

| BASE MEMORY ADDRESS | | | |
|---|---|---|---|
| Address | JPP | JPQ | JPR |
| C0000h | Pins 2 & 3 Closed | Pins 2 & 3 Closed | Pins 2 & 3 Closed |
| C4000h | Pins 2 & 3 Closed | Pins 1 & 2 Closed | Pins 2 & 3 Closed |
| C8000h | Pins 1 & 2 Closed | Pins 2 & 3 Closed | Pins 2 & 3 Closed |
| CC000h | Pins 1 & 2 Closed | Pins 1 & 2 Closed | Pins 2 & 3 Closed |
| ▷D0000h | Pins 2 & 3 Closed | Pins 2 & 3 Closed | Pins 1 & 2 Closed |
| D4000h | Pins 2 & 3 Closed | Pins 1 & 2 Closed | Pins 1 & 2 Closed |
| D8000h | Pins 1 & 2 Closed | Pins 2 & 3 Closed | Pins 1 & 2 Closed |
| DC000h | Pins 1 & 2 Closed | Pins 1 & 2 Closed | Pins 1 & 2 Closed |

| 10BASE-T CONFIGURATION | | | |
|---|---|---|---|
| Setting | LTS | JPK | JPH |
| ▷Link Test Pulse enabled (10Base-T receive level standard) | Pins 1 & 2 | Pins 1 & 2 | Pins 2 & 3 |
| Link Test Pulse disabled (10Base-T receive level reduced by 4.5dB) | Pins 2 & 3 | Pins 2 & 3 | Pins 1 & 2 |

Note: Pins designated should be in the closed position.

| BOOT ROM | |
|---|---|
| Setting | Jumper ROM |
| ▷Disabled | Pins 2 & 3 Closed |
| Enabled | Pins 1 & 2 Closed |

## TIARA COMPUTER SYSTEMS, INC.
# LANCARD/E*PC10BT

| WAIT STATE ||
|---|---|
| Setting | Jumper WST |
| ➪Zero wait states disabled | Pins 2 & 3 Closed |
| Zero wait states enabled | Pins 1 & 2 Closed |

| DIAGNOSTIC LED(S) ||||
|---|---|---|---|
| LED | Color | Status | Condition |
| 1 | Green | On | Data is being transmitted |
| 1 | Green | Off | Data is not being transmitted |
| 2 | Green | On | Data is being received |
| 2 | Green | Off | Data is not being received |
| 3 | Green | On | Network connection is good |
| 3 | Green | Off | Network connection is broken |
| 4 | Amber | On | Collision is detected |
| 4 | Amber | Off | No collisions are detected |
| LED5 | Red | On | Data is being received power |
| LED5 | Red | Off | Data is not being received power |

| FACTORY CONFIGURED SETTINGS ||
|---|---|
| Jumper | Setting |
| JPN | Pins 1 & 2 Closed |
| JPO | Pins 1 & 2 Closed |

## TOP MICROSYSTEMS, INC.
## TE 2003

| | |
|---|---|
| **NIC Type** | Ethernet |
| **Transfer Rate** | 10Mbps |
| **Data Bus** | 16-bit ISA |
| **Topology** | Linear bus |
| **Wiring Type** | Unshielded twisted pair |
| | RG58A/U 50ohm coaxial |
| | AUI transceiver via DB-15 |
| **Boot ROM** | Available |

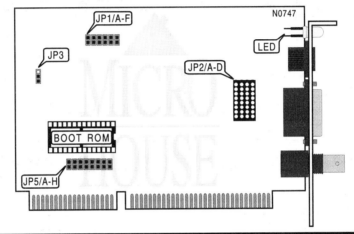

| BOOT ROM | |
|---|---|
| **Setting** | **JP1/A** |
| ⇨ Disabled | Open |
| Enabled | Closed |

| BOOT ROM ADDRESS SELECT | | | |
|---|---|---|---|
| **Address** | **JP1/B** | **JP1/C** | **JP1/D** |
| ⇨ C800h | Open | Closed | Open |
| C000h | Open | Open | Open |
| C400h | Open | Open | Closed |
| CC00h | Open | Closed | Closed |
| D000h | Closed | Open | Open |
| D400h | Closed | Open | Closed |
| D800h | Closed | Closed | Open |
| DC00h | Closed | Closed | Closed |

# TOP MICROSYSTEMS, INC.
# TE 2003

| CABLE TYPE | | | | |
|---|---|---|---|---|
| Type | JP2/A | JP2/B | JP2/C | JP2/D |
| ⇨ Thin Ethernet | Not installed | Not installed | Installed | Installed |
| Thick Ethernet | Not installed | Installed | Installed | Not installed |
| Unshielded twisted pair | Installed | Installed | Not installed | Not installed |

| INTERRUPT SELECT | | | | | | | | |
|---|---|---|---|---|---|---|---|---|
| ⇨ IRQ | JP5/A | JP5/B | JP5/C | JP5/D | JP5/E | JP5/F | JP5/G | JP5/H |
| IRQ3 | Open | Closed | Open | Open | Open | Open | Open | Open |
| IRQ2 | Closed | Open | Open | Open | Open | Open | Open | Open |
| IRQ4 | Open | Open | Closed | Open | Open | Open | Open | Open |
| IRQ5 | Open | Open | Open | Closed | Open | Open | Open | Open |
| IRQ10 | Open | Open | Open | Open | Closed | Open | Open | Open |
| IRQ11 | Open | Open | Open | Open | Open | Closed | Open | Open |
| IRQ12 | Open | Open | Open | Open | Open | Open | Closed | Open |
| IRQ15 | Open | Open | Open | Open | Open | Open | Open | Closed |

| I/O ADDRESS SELECT | | |
|---|---|---|
| Address | JP1/E | JP1/F |
| ⇨ 300h | Open | Open |
| 320h | Open | Closed |
| 340h | Closed | Open |
| 360h | Closed | Closed |

| SYSTEM TYPE | |
|---|---|
| Type | JP3 |
| ⇨ AT compatible | Pins 1 & 2 closed |
| Non-AT compatible | Pins 2 & 3 closed |

| DIAGNOSTIC LED | |
|---|---|
| Status | Condition |
| On | Network connection is good |
| Off | Network connection is broken |

## TRANSITION ENGINEERING, INC.
### TNIC-1500TA2/TNIC-1500T2/TNIC-1500A2

**NIC Type**          Ethernet
**Transfer Rate**     10Mbps
**Data Bus**          16-bit ISA
**Topology**          Linear bus
**Wiring Type**       Unshielded twisted pair
                      RG58A/U 50ohm coaxial
                      AUI transceiver via DB15
**Boot ROM**          Available

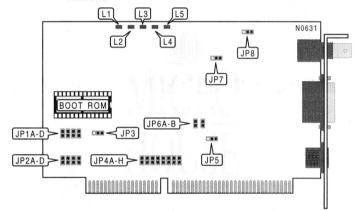

| | | | | DMA CHANNEL | | | | |
|---|---|---|---|---|---|---|---|---|
| Channel | JP1A | JP1B | JP1C | JP1D | JP2A | JP2B | JP2C | JP2D |
| DMA3 | Open | Open | Open | Closed | Open | Open | Open | Closed |
| DMA5 | Open | Open | Closed | Open | Open | Open | Closed | Open |
| DMA6 | Open | Closed | Open | Open | Open | Closed | Open | Open |
| DMA7 | Closed | Open | Open | Open | Closed | Open | Open | Open |

| ROM SIZE | |
|---|---|
| Size | JP3 |
| 64KB | Pins 1 & 2 closed |
| 128KB | Pins 1 & 2 closed |
| 256KB | Pins 2 & 3 closed |

## TRANSITION ENGINEERING, INC.
### TNIC-1500TA2/TNIC-1500T2/TNIC-1500A2

| INTERRUPT SETTINGS | | | | | | | | | |
|---|---|---|---|---|---|---|---|---|---|
| IRQ | JP4A | JP4B | JP4C | JP4D | JP4E | JP4H | JP4I | JP4J |
| 3 | Closed | Open | Open | Open | Open | Open | Open | Open |
| 4 | Open | Closed | Open | Open | Open | Open | Open | Open |
| 5 | Open | Open | Closed | Open | Open | Open | Open | Open |
| 9 | Open | Open | Open | Closed | Open | Open | Open | Open |
| 10 | Open | Open | Open | Open | Closed | Open | Open | Open |
| 11 | Open | Open | Open | Open | Open | Closed | Open | Open |
| 12 | Open | Open | Open | Open | Open | Open | Closed | Open |
| 15 | Open | Open | Open | Open | Open | Open | Open | Closed |

| I/O ADDRESS SELECTION | | | | |
|---|---|---|---|---|
| I/O Address | Boot ROM | JP5 | JP6A | JP6B |
| 310-31Fh | C8000-CBFFFh | Pins 1 & 2 closed | Closed | Closed |
| 320-32Fh | CC000-CFFFFh | Pins 1 & 2 closed | Closed | Open |
| 340-34Fh | D0000-D7FFFh | Pins 1 & 2 closed | Open | Closed |
| 360-36Fh | D4000-D7FFFh | Pins 1 & 2 closed | Open | Open |
| 300-30Fh | None | Pins 2 & 3 closed | Closed | Closed |
| 320-32Fh | None | Pins 2 & 3 closed | Closed | Open |
| 340-34Fh | None | Pins 2 & 3 closed | Open | Closed |
| 360-36Fh | None | Pins 2 & 3 closed | Open | Open |

| CABLE TYPE | | |
|---|---|---|
| Type | JP7 | JP8 |
| Unshielded twisted pair | Pins 1 & 2 closed | N/A |
| ➪RG58A/U 50ohm coaxial | Pins 2 & 3 closed | Pins 1 & 2 closed |
| AUI transceiver via DB-15 port | Pins 2 & 3 closed | Pins 2 & 3 closed |

| DIAGNOSTIC LED(S) | | |
|---|---|---|
| LED | Status | Condition |
| L1 | On | 10BaseT link integrity exists |
| L1 | Off | 10BaseT link integrity broken |
| L2 | On | Card receiving |
| L2 | Off | Card idle |
| L3 | On | Network polarity correct |
| L3 | Off | Network polarity reversed |
| L4 | On | Card transmitting |
| L4 | Off | Card idle |
| L5 | On | UTP disabled |
| L5 | Off | UTP enabled |

Note: L1 is only functional when card is installed on a 10BaseT network.

# TRANSITION ENGINEERING, INC.
## TNIC-1500AF/TNIC-1500TF

| | | |
|---|---|---|
| **NIC Type** | Ethernet | |
| **Transfer Rate** | 10Mbps | |
| **Data Bus** | 16-bit ISA | |
| **Topology** | Linear bus | |
| **Wiring Type** | Unshielded twisted pair | (TNIC-1500TF) |
| | Fiber optic cable | (TNIC-1500AF, TNIC-1500TF) |
| | AUI transceiver via DB15 | (TNIC-1500AF) |
| **Boot ROM** | Available | |

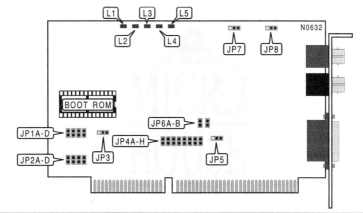

| DMA CHANNEL | | | | | | | | |
|---|---|---|---|---|---|---|---|---|
| Channel | JP1A | JP1B | JP1C | JP1D | JP2A | JP2B | JP2C | JP2D |
| DMA3 | Open | Open | Open | Closed | Open | Open | Open | Closed |
| DMA5 | Open | Open | Closed | Open | Open | Open | Closed | Open |
| DMA6 | Open | Closed | Open | Open | Open | Closed | Open | Open |
| DMA7 | Closed | Open | Open | Open | Closed | Open | Open | Open |

| ROM SIZE | |
|---|---|
| Size | JP3 |
| 64KB | Pins 1 & 2 closed |
| 128KB | Pins 1 & 2 closed |
| 256KB | Pins 2 & 3 closed |

# TRANSITION ENGINEERING, INC.
## TNIC-1500AF/TNIC-1500TF

### INTERRUPT SETTINGS

| IRQ | JP4A | JP4B | JP4C | JP4D | JP4E | JP4H | JP4I | JP4J |
|---|---|---|---|---|---|---|---|---|
| 3 | Closed | Open | Open | Open | Open | Open | Open | Open |
| 4 | Open | Closed | Open | Open | Open | Open | Open | Open |
| 5 | Open | Open | Closed | Open | Open | Open | Open | Open |
| 9 | Open | Open | Open | Closed | Open | Open | Open | Open |
| 10 | Open | Open | Open | Open | Closed | Open | Open | Open |
| 11 | Open | Open | Open | Open | Open | Closed | Open | Open |
| 12 | Open | Open | Open | Open | Open | Open | Closed | Open |
| 15 | Open | Open | Open | Open | Open | Open | Open | Closed |

### I/O ADDRESS SELECTION

| I/O Address | Boot ROM | JP5 | JP6A | JP6B |
|---|---|---|---|---|
| 310-31Fh | C8000-CBFFFh | Pins 1 & 2 closed | Closed | Closed |
| 320-32Fh | CC000-CFFFFh | Pins 1 & 2 closed | Closed | Open |
| 340-34Fh | D0000-D7FFFh | Pins 1 & 2 closed | Open | Closed |
| 360-36Fh | D4000-D7FFFh | Pins 1 & 2 closed | Open | Open |
| 300-30Fh | None | Pins 2 & 3 closed | Closed | Closed |
| 320-32Fh | None | Pins 2 & 3 closed | Closed | Open |
| 340-34Fh | None | Pins 2 & 3 closed | Open | Closed |
| 360-36Fh | None | Pins 2 & 3 closed | Open | Open |

### CABLE TYPE

| Type | JP7 | JP8 |
|---|---|---|
| Unshielded twisted pair | Pins 1 & 2 closed | N/A |
| Fiber optic cable | Pins 2 & 3 closed | Pins 1 & 2 closed |
| AUI transceiver via DB-15 port | Pins 2 & 3 closed | Pins 2 & 3 closed |

Note: JP7 should always be configured to Pins 2 & 3 closed on the TNIC-1500AF, and JP8 should always be configured to Pins 1 & 2 closed on the TNIC-1500TF.

### DIAGNOSTIC LED(S)

| LED | Status | Condition |
|---|---|---|
| L1 | On | 10BaseT link integrity exists |
| L1 | Off | 10BaseT link integrity broken |
| L2 | On | Card receiving |
| L2 | Off | Card idle |
| L3 | On | Network polarity correct |
| L3 | Off | Network polarity reversed |
| L4 | On | Card transmitting |
| L4 | Off | Card idle |
| L5 | On | UTP disabled |
| L5 | Off | UTP enabled |

Note: L1 is only functional when card is installed on a 10BaseT network.

## TTC COMPUTER PRODUCTS
### ET-001

| | |
|---|---|
| **NIC Type** | Ethernet |
| **Transfer Rate** | 10Mbps |
| **Data Bus** | 16-bit ISA |
| **Topology** | Star |
| **Wiring Type** | Shielded/Unshielded twisted pair |
| | AUI transceiver via DB-15 port |
| **Boot ROM** | Available |

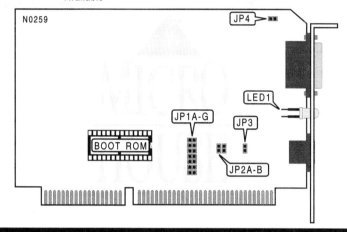

| INTERRUPT REQUEST | | | | | | | |
|---|---|---|---|---|---|---|---|
| IRQ | JP1A | JP1B | JP1C | JP1D | JP1E | JP1F | JP1G |
| 2 | Closed | Open | Open | Open | Open | Open | Open |
| ⇨3 | Open | Closed | Open | Open | Open | Open | Open |
| 4 | Open | Open | Closed | Open | Open | Open | Open |
| 5 | Open | Open | Open | Closed | Open | Open | Open |
| 7 | Open | Open | Open | Open | Closed | Open | Open |
| 10 | Open | Open | Open | Open | Open | Closed | Open |
| 11 | Open | Open | Open | Open | Open | Open | Closed |

| I/O BASE ADDRESS | | |
|---|---|---|
| Address | JP2A | JP2B |
| ⇨300-31Fh | Closed | Closed |
| 320-33Fh | Open | Closed |
| 340-35Fh | Closed | Open |
| 360-37Fh | Open | Open |

| BOOT ROM | |
|---|---|
| Setting | JP3 |
| ⇨Disabled | Closed |
| Enabled | Open |

*TTC COMPUTER PRODUCTS*

# ET-001

| CABLE TYPE ||
|---|---|
| Type | JP4 |
| ⇨Shielded/Unshielded twisted pair | Closed |
| AUI transceiver via DB-15 port | Open |

| DIAGNOSTIC LED(S) ||||
|---|---|---|---|
| LED | Color | Status | Condition |
| LED1 | Green | On | Network connection is good |
| LED1 | Green | Off | Network connection is broken |

## TULIP COMPUTERS
# TNCC-16

| | |
|---|---|
| **NIC Type** | Ethernet |
| **Transfer Rate** | 10Mbps |
| **Data Bus** | 16-bit ISA |
| **Topology** | Linear bus |
| **Wiring Type** | Shielded twisted pair |
| | Unshielded twisted pair |
| | RG58A/U 50ohm coaxial |
| | AUI transceiver via DB-15 |
| **Boot ROM** | Available |

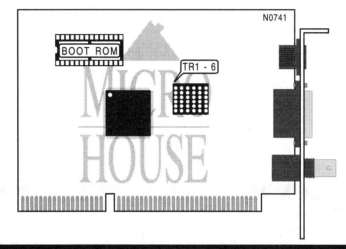

| CABLE TYPE | | | | | | |
|---|---|---|---|---|---|---|
| Type | TR1 | TR2 | TR3 | TR4 | TR5 | TR6 |
| Thick Ethernet | Not installed | Not installed | Not installed | Installed | Installed | Not installed |
| Thin Ethernet | Not installed | Not installed | Installed | Installed | Not installed | Not installed |
| Twisted pair | Not installed | Installed | Installed | Not installed | Not installed | Not installed |

## TULIP COMPUTERS
## TTCC-16

| | |
|---|---|
| **NIC Type** | Token Ring |
| **Transfer Rate** | 4/16 Mbps |
| **Data Bus** | 16-bit ISA |
| **Topology** | Ring |
| **Wiring Type** | Shielded twisted pair |
| | Unshielded twisted pair |
| **Boot ROM** | Not Available |

| ADDRESS CONFIGURATION | | | | |
|---|---|---|---|---|
| **Base Memory Address** | **I/O Address** | **SW1/1** | **SW1/2** | **SW1/3** |
| C0000h | 86A0h | On | On | On |
| C4000h | 96A0h | Off | On | On |
| C8000h | A6A0h | On | Off | On |
| CC000h | B6A0h | Off | Off | On |
| D0000h | C6A0h | On | On | Off |
| D4000h | D6A0h | Off | On | Off |
| D8000h | E6A0h | On | Off | Off |
| DC000h | F6A0h | Off | Off | Off |

| CABLE TYPE | |
|---|---|
| **Type** | **SW1/12** |
| Shielded twisted pair | On |
| Unshielded twisted pair | Off |

| DATA RATE SELECT | |
|---|---|
| **Rate** | **SW1/11** |
| 4Mbps | On |
| 16Mbps | Off |

# TULIP COMPUTERS
## TTCC-16

### DMA CONFIGURATION

| DMA | SW1/9 | SW1/10 |
|---|---|---|
| 5 | On | On |
| 6 | Off | On |
| 7 | On | Off |
| Disabled | Off | Off |

### INTERRUPT SELECT

| IRQ | SW1/4 | SW1/5 |
|---|---|---|
| 9 | On | On |
| 10 | Off | On |
| 11 | On | Off |
| 15 | Off | Off |

### SYSTEM SPEED

| Speed | SW1/6 |
|---|---|
| Normal | On |
| Fast | Off |

### RPL CONFIGURATION

| Setting | SW1/8 |
|---|---|
| Enabled | Off |
| Disabled | On |

### FACTORY CONFIGURED SETTINGS

| Switch | Setting |
|---|---|
| SW1/7 | On |

### MISCELLANEOUS TECHNICAL NOTES

Note: Unshielded twisted pair cable is attached to the TTCC-16 adapter via a proprietary adapter cable.

## UNGERMANN-BASS NETWORKS, INC.
## NIUPC/EOTP

| | |
|---|---|
| NIC Type | Ethernet |
| Transfer Rate | 10Mbps |
| Data Bus | 8-bit ISA/16-bit ISA |
| Topology | Linear bus |
| Wiring Type | Unshielded twisted pair |
| | AUI transceiver via DB15 |
| | DIX transceiver via DB15 |
| Boot ROM | Available |

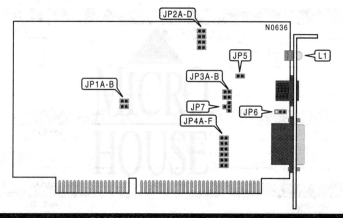

| FACTORY CONFIGURED SETTINGS ||
|---|---|
| Jumper | Setting |
| JP1A | Closed |
| JP1B | Closed |
| JP2C | Closed |
| JP2D | Open |
| JP3A | Open |
| JP3B | Open |

| BOOT ROM |||
|---|---|---|
| Setting | JP2C | JP2D |
| Disabled | Open | Open |
| Enabled with prompt | Open | Closed |
| Enabled without prompt | Closed | Closed |

| LINK INTEGRITY ||
|---|---|
| Function | JP4A |
| ⇨Link integrity enabled | Open |
| Link integrity disabled | Closed |

Note: If card is used in a pre-10BaseT cabling network, disable the link integrity function.

# UNGERMANN-BASS NETWORKS, INC.
## NIUPC/EOTP

| BOOT ROM ADDRESS | |
|---|---|
| Address | JP4B |
| ⇨C0000h | Open |
| D0000h | Closed |

| BASE I/O ADDRESS | | |
|---|---|---|
| Address | JP4C | JP4D |
| 350h | Closed | closed |
| 358h | Open | Closed |
| 360h | Closed | Open |
| ⇨ 368h | Open | Open |

| DATA BUS | | |
|---|---|---|
| Mode | JP4E | JP4F |
| 16-bit/8-bit auto detect | Closed | Closed |
| Force 8-bit data bus | Open | Open |

| SEGMENT LENGTH | |
|---|---|
| Length | JP5 |
| < 100 meters to hub | Open |
| > 100 meters to hub | Closed |

Note: A distance of more than 100 meters between the adapter and hub is not 10BaseT compliant.

| DB15 MODE CONFIGURATION | | |
|---|---|---|
| Setting | Ground Type | JP6 |
| DB15 operates as an AUI port | Logic | Pins 1 & 2 closed |
| DB15 operates as a DIX port | Chassis | Pins 2 & 3 closed |

| CABLE TYPE | |
|---|---|
| Mode | JP7 |
| Auto-switch between 10BaseT and DB15 | ⁃┋ |
| Force card to use 10BaseT | ⁃❗ |
| Force card to use DB15 port | ⁃┇ |

| DIAGNOSTIC LED(S) | | |
|---|---|---|
| LED | Status | Condition |
| L1 | On | 10BaseT link integrity exists |
| L1 | Off | 10BaseT link integrity broken |

# UNICOM ELECTRIC, INC.
## 3-IN-1 16-BIT ETHERNET ADAPTER CARD (ETP-4105-E)

**NIC Type**  Ethernet
**Transfer Rate**  10Mbps
**Data Bus**  16-bit ISA
**Topology**  Linear Bus, Star
**Wiring Type**  Unshielded twisted pair
AUI transceiver via DB-15 port
RG58A/U 50ohm coaxial
**Boot ROM**  Available

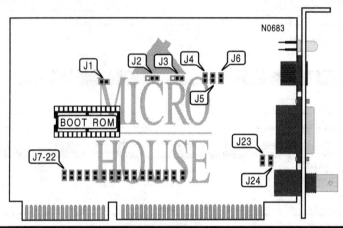

| BASE I/O ADDRESS SELECT | | | | |
|---|---|---|---|---|
| Address | J14 | J15 | J16 | J17 |
| ⇨ 300h | Open | Open | Open | Closed |
| 320h | Open | Open | Closed | Open |
| 340h | Open | Closed | Open | Open |
| 360h | Closed | Open | Open | Open |

| BASE MEMORY ADDRESS SELECT | | | | | |
|---|---|---|---|---|---|
| Address | J18 | J19 | J20 | J21 | J22 |
| ⇨ C8000h | Closed | Closed | Closed | Open | Closed |
| C0000h | Closed | Closed | Closed | Closed | Closed |
| C4000h | Closed | Closed | Closed | Closed | Open |
| CC000h | Closed | Closed | Closed | Open | Open |
| D0000h | Closed | Closed | Open | Closed | Closed |
| D4000h | Closed | Closed | Open | Closed | Open |
| D8000h | Closed | Closed | Open | Open | Closed |
| DC000h | Closed | Closed | Open | Open | Open |
| E0000h | Closed | Open | Closed | Closed | Closed |

## UNICOM ELECTRIC, INC.
## 3-IN-1 16-BIT ETHERNET ADAPTER CARD (ETP-4105-E)

| BOOTROM SELECT | |
|---|---|
| Type | J2 |
| 27128 | Pins 1 & 2 closed |
| 2764 | Pins 2 & 3 closed |

| CABLE TYPE | | | |
|---|---|---|---|
| Type | J4 | J5 | J6 |
| RG58A/U 50ohm coaxial | Open | Open | Closed |
| AUI transceiver via DB-15 port | Open | Closed | Open |
| Unshielded twisted pair | Closed | Open | Open |

| INTERRUPT SELECT | | | | | | | |
|---|---|---|---|---|---|---|---|
| IRQ | J7 | J8 | J9 | J10 | J11 | J12 | J13 |
| ⇨ IRQ3 | Open | Closed | Open | Open | Open | Open | Open |
| IRQ2 | Closed | Open | Open | Open | Open | Open | Open |
| IRQ4 | Open | Open | Closed | Open | Open | Open | Open |
| IRQ5 | Open | Open | Open | Closed | Open | Open | Open |
| IRQ10 | Open | Open | Open | Open | Closed | Open | Open |
| IRQ11 | Open | Open | Open | Open | Open | Closed | Open |
| IRQ12 | Open | Open | Open | Open | Open | Open | Closed |

| LINK TEST | |
|---|---|
| Setting | J3 |
| ⇨ Enabled | Pins 1 & 2 closed |
| Disabled | Pins 2 & 3 closed |

| RG58A TERMINATOR CONFIGURATION | |
|---|---|
| Setting | J23 |
| ⇨ RG58A/U 50ohm coaxial terminator disabled | Open |
| RG58A/U 50ohm coaxial terminator enabled | Closed |

| RG58A TRUNK SEGMENT LENGTH SELECT | |
|---|---|
| Cable Length | J24 |
| ⇨ 180M | Open |
| 300M | Closed |

| ZERO WAIT STATE | |
|---|---|
| Setting | J1 |
| ⇨ Enabled | Closed |
| Disabled | Open |

## UNICOM ELECTRIC, INC.
### ITR-1110

**NIC Type**          Token Ring
**Transfer Rate**     4/16Mbps
**Data Bus**          8-bit ISA
**Topology**          Token ring
**Wiring Type**       Shielded twisted pair
**Boot ROM**          Available

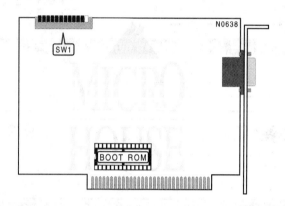

### BOOT ROM ADDRESS

| Address | SW1/1 | SW1/2 | SW1/3 | SW1/4 | SW1/5 | SW1/6 |
|---|---|---|---|---|---|---|
| C0000-C1FFFh | Off | On | On | On | On | On |
| C2000-C3FFFh | Off | On | On | On | On | Off |
| C4000-C5FFFh | Off | On | On | On | Off | On |
| C6000-C7FFFh | Off | On | On | On | Off | Off |
| C8000-C9FFFh | Off | On | On | Off | On | On |
| CA000-CBFFFh | Off | On | On | Off | On | Off |
| ↪CC000-CDFFFh | Off | On | On | Off | Off | On |
| CE000-CFFFFh | Off | On | On | Off | Off | Off |
| D0000-D1FFFh | Off | On | Off | On | On | On |
| D2000-D3FFFh | Off | On | Off | On | On | Off |
| D4000-D5FFFh | Off | On | Off | On | Off | On |
| D6000-D7FFFh | Off | On | Off | On | Off | Off |
| D8000-D9FFFh | Off | On | Off | Off | On | On |
| DA000-DBFFFh | Off | On | Off | Off | On | Off |
| ↪DC000-DDFFFh | Off | On | Off | Off | Off | On |
| DE000-DFFFFh | Off | On | Off | Off | Off | Off |
| E0000-E1FFFh | Off | Off | On | On | On | On |
| E2000-E3FFFh | Off | Off | On | On | On | Off |
| E4000-E5FFFh | Off | On | Off | On | Off | On |
| E6000-E7FFFh | Off | Off | On | On | Off | Off |

# UNICOM ELECTRIC, INC.
## 4/16MBPS TOKEN RING ADAPTER CARD(ITR-1110)

| BOOT ROM ADDRESS (CONTINUED) | | | | | | |
|---|---|---|---|---|---|---|
| Address | SW1/1 | SW1/2 | SW1/3 | SW1/4 | SW1/5 | SW1/6 |
| E8000-E9FFFh | Off | Off | On | Off | On | On |
| EA000-EBFFFh | Off | Off | On | Off | On | Off |
| EC000-EDFFFh | Off | Off | On | Off | Off | On |
| EE000-EFFFFh | Off | Off | On | Off | Off | Off |
| F0000-F1FFFh | Off | Off | Off | On | On | On |
| F2000-F3FFFh | Off | Off | Off | On | On | Off |
| F4000-F5FFFh | Off | Off | Off | On | Off | On |
| F6000-F7FFFh | Off | Off | Off | On | Off | Off |
| F8000-F9FFFh | Off | Off | Off | Off | On | On |
| FA000-FBFFFh | Off | Off | Off | Off | On | Off |
| FC000-FDFFFh | Off | Off | Off | Off | Off | On |
| FE000-FFFFFh | Off | Off | Off | Off | Off | Off |

Note: CC000h or DC000h are the recommended default settings.

| INTERRUPT REQUEST | | |
|---|---|---|
| IRQ | SW1/7 | SW1/8 |
| ⇨2 | On | On |
| 3 | On | Off |
| 6 | Off | On |
| 7 | Off | Off |

| I/O BASE ADDRESS | |
|---|---|
| Address | SW1/9 |
| ⇨Primary | On |
| Secondary | Off |

| SHARED RAM CONFIGURATION | | |
|---|---|---|
| Size | SW1/10 | SW1/11 |
| 8KB | On | On |
| ⇨16KB | Off | On |
| 32KB | On | Off |
| 64KB | Off | Off |

Note: This setting configures how much system memory the card will use.

| NETWORK SPEED | |
|---|---|
| Speed | SW1/12 |
| 4Mbps | On |
| ⇨16Mbps | Off |

Note: All cards on a segment must have this option set the same.

## XINETRON, INC.
## RE321

**NIC Type** Ethernet
**Transfer Rate** 10Mbps
**Data Bus** 16-bit ISA
**Topology** Linear bus
**Wiring Type** AUI transceiver via DB-15 port
RG-58A/U 50ohm coaxial
**Boot ROM** Available

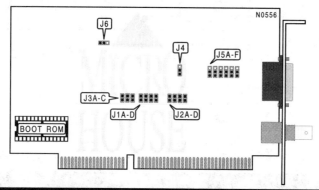

| INTERRUPT REQUEST | | | | |
|---|---|---|---|---|
| IRQ | J1A | J1B | J1C | J1D |
| 2/9 | Open | Open | Open | Closed |
| ↺3 | Open | Open | Closed | Open |
| 4 | Open | Closed | Open | Open |
| 5 | Closed | Open | Open | Open |

| I/O BASE ADDRESS | | | | |
|---|---|---|---|---|
| Address | J2A | J2B | J2C | J2D |
| ↺300 - 31Fh | Open | Closed | Closed | Closed |
| 320 - 33Fh | Open | Closed | Closed | Open |
| 340 - 35Fh | Open | Closed | Open | Closed |
| 360 - 37Fh | Open | Closed | Open | Open |

| BOOT ROM ADDRESS | | | |
|---|---|---|---|
| Address | J3A | J3B | J3C |
| C000h | Open | Open | Closed |
| CC00h | Open | Closed | Open |
| D000h | Closed | Open | Closed |

## XINETRON, INC.
## RE321

| CABLE TYPE | | |
|---|---|---|
| Type | J4 | J5A - J5F |
| ▷RG-58A/U 50ohm coaxial | Pins 1 & 2 closed | Pins 2 & 3 closed |
| AUI transceiver via DB-15 port | Pins 2 & 3 closed | Pins 1 & 2 closed |

| COMPATIBILITY MODE | |
|---|---|
| Setting | JP1 |
| ▷Disabled | Pins 2 & 3 closed |
| Enabled | Pins 1 & 2 closed |
| Note: On some systems, the data bus timing is not fully IBM PC/AT compatible. If the card is not initializing, enabling the compatibility mode may allow the card to operate normally. | |

## ZENITH DATA SYSTEMS
## LAN10E-MAT

| | |
|---|---|
| **NIC Type** | Ethernet |
| **Transfer Rate** | 10Mbps |
| **Data Bus** | 16-bit ISA |
| **Topology** | Star |
| | Linear Bus |
| **Wiring Type** | Shielded/Unshielded twisted pair |
| | AUI transceiver via DB-15 port |
| | RG–58A/U 50ohm coaxial |
| **Boot ROM** | Available |

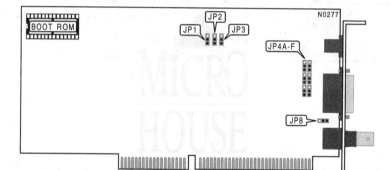

| BASE MEMORY ADDRESS ||
|---|---|
| **Address** | **JP1** |
| CC00h | Pins 1 & 2 Closed |
| D000h | Pins 2 & 3 Closed |

| INTERRUPT REQUEST ||
|---|---|
| **IRQ** | **JP2** |
| 5 | Pins 1 & 2 Closed |
| 10 | Pins 2 & 3 Closed |

| ROM ENABLE ||
|---|---|
| **Setting** | **JP3** |
| ➪Disabled | Pins 2 & 3 Closed |
| Enabled (at address D800h) | Pins 1 & 2 Closed |

| CABLE TYPE ||
|---|---|
| **Type** | **JP4A - JP4F** |
| Shielded/Unshielded twisted pair | Pins 1 & 2 Closed |
| RG–58A/U 50ohm or AUI transceiver via DB-15 | Pins 2 & 3 Closed |

| SIGNAL QUALITY ||
|---|---|
| **Setting** | **JP8** |
| ➪SQE test disabled | Pins 2 & 3 Closed |
| SQE test enabled | Pins 1 & 2 Closed |

## ZENITH DATA SYSTEMS
## LAN10FAT

| | |
|---|---|
| **NIC Type** | Ethernet |
| **Transfer Rate** | 10Mbps |
| **Data Bus** | 16-bit ISA |
| **Topology** | Linear Bus |
| **Wiring Type** | 50/62.5/80/100µ Fiber optic cable |
| **Boot ROM** | Available |

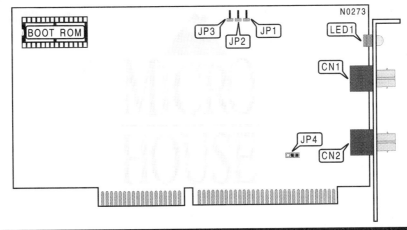

| CONNECTIONS ||
|---|---|
| **Purpose** | **Location** |
| SMA/ST Receive Connector (gray) | CN1 |
| SMA/ST Transmit Connector (black) | CN2 |

| BASE MEMORY ADDRESS ||
|---|---|
| **Address** | **JP1** |
| ⇨CC00h | Closed |
| D000h | Open |

| INTERRUPT REQUEST ||
|---|---|
| **IRQ** | **JP2** |
| ⇨5 | Closed |
| 10 | Open |

| ROM ENABLE ||
|---|---|
| **Setting** | **JP3** |
| ⇨Disabled | Open |
| Enabled (at address D800h) | Closed |

## ZENITH DATA SYSTEMS
## LAN10FAT

| CABLE LENGTH | |
|---|---|
| Cable length between two furthest nodes | JP4 |
| ▷Less than 1Km (normal distance) | Pins 1 & 2 Closed |
| More than 1Km (extended distance) | Pins 2 & 3 Closed |

| DIAGNOSTIC LED(S) | | |
|---|---|---|
| LED | Status | Condition |
| LED1 | On | Card is configured correctly and operating normally |
| LED1 | One blink | Data is not being received (receive link is down) |
| LED1 | Two blinks | Card is jabbering (transmitting continuously) |
| LED1 | Three blinks | Data is not being transmitted (transmit link is down) |
| LED1 | Four blinks | Card detected a remote fault |
| LED1 | Five blinks | Illegal data was input to the card |

## ZENITH DATA SYSTEMS
# LAN16TR-AT

| | |
|---|---|
| **NIC Type** | Token-Ring |
| **Transfer Rate** | 4/16Mbps |
| **Data Bus** | 16-bit ISA |
| **Topology** | Ring |
| **Wiring Type** | Unshielded twisted pair |
| | Shielded twisted pair |
| **Boot ROM** | Available |

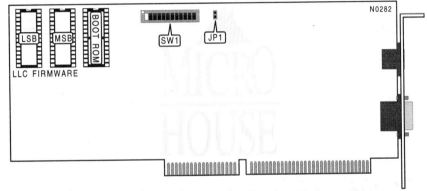

| FACTORY CONFIGURED SETTINGS ||
|---|---|
| **Jumper** | **Setting** |
| JP1 | Unused |

| I/O BASE ADDRESS |||
|---|---|---|
| **Address** | **SW1/1** | **SW1/2** |
| A00 - A0Fh | On | On |
| A20 - A2Fh | Off | On |
| A40 - A4Fh | On | Off |
| A60 - A6Fh | Off | Off |

| INTERRUPT REQUEST |||
|---|---|---|
| **IRQ** | **SW1/3** | **SW1/4** |
| 3 | On | On |
| 9 | Off | On |
| 10 | On | Off |
| 11 | Off | Off |

| BOOT ROM ADDRESS |||
|---|---|---|
| **Address** | **SW1/5** | **SW1/6** |
| CC000 - CFFFFh | On | On |
| D0000 - D3FFFh | Off | On |
| D8000 - D8FFFh | On | Off |
| DC000 - DFFFFh | Off | Off |

## ZENITH DATA SYSTEMS
## LAN16TR-AT

| DMA CHANNEL | | |
|---|---|---|
| Channel | SW1/7 | SW1/8 |
| DMA0 | On | On |
| DMA5 | Off | On |
| DMA6 | On | Off |
| DMA7 | Off | Off |

| BOOT ROM | |
|---|---|
| Setting | SW1/9 |
| ➪Disabled | On |
| Enabled | Off |

| NETWORK SEGMENT SPEED | |
|---|---|
| Speed | SW1/10 |
| 4Mbps | Off |
| 16Mbps | On |

Note: All cards on the network segment must have this option set the same.

| LLC EPROM CONFIGURATION | |
|---|---|
| Setting | SW1/11 |
| ➪ROM-based LLC | On |
| RAM-based LLC | Off |

Note: The LLC EPROMs contain the Logic Link Control program. With the LLC ROMs disabled, the LLC logic can be loaded into RAM. The card will function properly with the LLC chips disabled.

| CABLE TYPE | |
|---|---|
| Type | SW1/12 |
| Unshielded twisted pair (RJ-45 Jack) | Off |
| Shielded twisted pair (DB-9 Connector) | On |

## ZENITH DATA SYSTEMS
## LAN16TR-XT

| | |
|---|---|
| **NIC Type** | Token-Ring |
| **Transfer Rate** | 4/16Mbps |
| **Data Bus** | 16-bit ISA |
| **Topology** | Ring |
| **Wiring Type** | Unshielded twisted pair |
| | Shielded twisted pair |
| **Boot ROM** | Available |

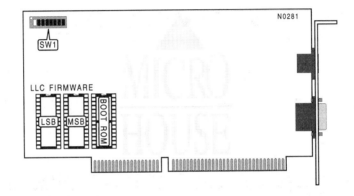

| I/O BASE ADDRESS ||
|---|---|
| Address | SW1/1 |
| A20 - A4Fh | On |
| A60 - A6Fh | Off |

| INTERRUPT REQUEST ||
|---|---|
| IRQ | SW1/2 |
| 2 | Off |
| 3 | On |

| BOOT ROM ADDRESS ||
|---|---|
| Address | SW1/3 |
| CC000 - CFFFFh | On |
| DC000 - DFFFFh | Off |

| BOOT ROM ||
|---|---|
| Setting | SW1/4 |
| ⇨Disabled | On |
| Enabled | Off |

## ZENITH DATA SYSTEMS
## LAN16TR-XT

| LLC EPROM ENABLED ||
|---|---|
| **Setting** | **SW1/5** |
| ⇨RAM-based LLC | On |
| ROM-based LLC | Off |

Note: The LLC EPROMs contain the Logic Link Control program. With the LLC ROMs disabled, the LLC logic can be loaded into RAM. The card will function properly with the LLC chips disabled.

| NETWORK SEGMENT SPEED ||
|---|---|
| **Speed** | **SW1/6** |
| ⇨16Mbps | On |
| 4Mbps | Off |

Note: All cards on the network segment must have this option set the same.

| CABLE TYPE ||
|---|---|
| **Type** | **SW1/7** |
| Unshielded twisted pair (RJ-45 Jack) | Off |
| Shielded twisted pair (DB-9 Connector) | On |

| FACTORY CONFIGURED SETTINGS ||
|---|---|
| **Switch** | **Setting** |
| SW1/8 | Unused |

# ZENITH DATA SYSTEMS
## LAN4TR-AT

| | |
|---|---|
| **NIC Type** | Token-Ring |
| **Transfer Rate** | 4Mbps |
| **Data Bus** | 16-bit ISA |
| **Topology** | Ring |
| **Wiring Type** | Shielded/Unshielded twisted pair |
| **Boot ROM** | Available |

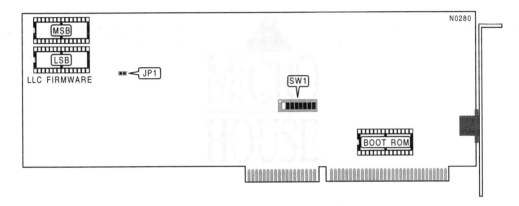

| LLC EPROM ENABLED ||
|---|---|
| Setting | JP1 |
| ▷External EPROMs disabled | Closed |
| External EPROMs enabled | Open |

Note: The LLC EPROMs contain the Logic Link Control program. The card will function properly with the LLC chips disabled.

| I/O BASE ADDRESS |||
|---|---|---|
| Address | SW1/1 | SW1/2 |
| A00 - A0Fh | On | On |
| A20 - A2Fh | On | Off |
| A40 - A4Fh | Off | On |
| A60 - A6Fh | Off | Off |

| INTERRUPT REQUEST |||
|---|---|---|
| IRQ | SW1/3 | SW1/4 |
| 3 | On | On |
| 9 | On | Off |
| 10 | Off | On |
| 11 | Off | Off |

## ZENITH DATA SYSTEMS
## LAN4TR-AT

| BOOT ROM ADDRESS | | | |
|---|---|---|---|
| Address | Boot ROM Size | SW1/5 | SW1/6 |
| CC000 - CCFFFh | 4KB | On | On |
| D0000 - D3FFFh | 16KB | On | Off |
| D8000 - D8FFFh | 4KB | Off | On |
| DC000 - DFFFFh | 16KB | Off | Off |

| DMA CHANNEL | | |
|---|---|---|
| Channel | SW1/7 | SW1/8 |
| DMA0 | On | On |
| DMA5 | On | Off |
| DMA6 | Off | On |
| DMA7 | Off | Off |

## ZERO ONE NETWORKING
## ZOT-N200E2/EC/ET

| | |
|---|---|
| **NIC Type** | Ethernet |
| **Transfer Rate** | 10Mbps |
| **Data Bus** | 16-bit ISA |
| **Topology** | Linear Bus, Star |
| **Wiring Type** | RG-58A/U 50ohm coaxial |
| | Unshielded twisted pair |
| **Boot ROM** | Available |

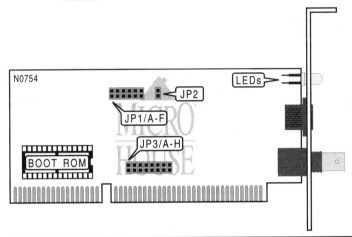

| BOOT ROM ||
|---|---|
| **Setting** | **JP1/A** |
| ⇨ Disabled | Open |
| Enabled | Closed |

| BASE MEMORY ADDRESS SELECT ||||
|---|---|---|---|
| **Address** | **JP1/B** | **JP1/C** | **JP1/D** |
| ⇨ C8000h | Open | Open | Open |
| C4000h | Closed | Closed | Open |
| CC000h | Open | Open | Closed |
| D0000h | Open | Closed | Open |
| D4000h | Open | Closed | Closed |
| D8000h | Closed | Open | Open |
| DC000h | Closed | Open | Closed |
| E0000h | Closed | Closed | Closed |

| I/O ADDRESS SELECT |||
|---|---|---|
| **Address** | **JP1/E** | **JP1/F** |
| ⇨ 300 - 31Fh | Open | Open |
| 320 - 33Fh | Open | Closed |
| 340 - 35Fh | Closed | Open |
| 360 - 37Fh | Closed | Closed |

## ZERO ONE NETWORKING
## ZOT-N200E2/EC/ET

| INTERRUPT SELECT | | | | | | | | | |
|---|---|---|---|---|---|---|---|---|---|
| IRQ | JP3/A | JP3/B | JP3/C | JP3/D | JP3/E | JP3/F | JP3/G | JP3/H | |
| ⇨ IRQ3 | Open | Closed | Open | Open | Open | Open | Open | Open | |
| IRQ2 | Closed | Open | Open | Open | Open | Open | Open | Open | |
| IRQ4 | Open | Open | Closed | Open | Open | Open | Open | Open | |
| IRQ5 | Open | Open | Open | Closed | Open | Open | Open | Open | |
| IRQ10 | Open | Open | Open | Open | Closed | Open | Open | Open | |
| IRQ11 | Open | Open | Open | Open | Open | Closed | Open | Open | |
| IRQ12 | Open | Open | Open | Open | Open | Open | Closed | Open | |
| IRQ15 | Open | Open | Open | Open | Open | Open | Open | Closed | |

| SYSTEM COMPATIBILITY | |
|---|---|
| System | JP2 |
| ⇨ Fully IBM compatible | Closed |
| Other | Open |

| DIAGNOSTIC LEDS | | | |
|---|---|---|---|
| LED | Color | Status | Condition |
| LED1 | Green | Off | Data is not being transmitted/received via BNC port |
| LED1 | Green | On | Data is being transmitted/received via BNC port |
| LED2 | Red | Off | Network UTP connection is broken |
| LED2 | Red | On | Network UTP connection is good |

**APPENDIX A**

# Node ID Quick Reference

## Appendix A Contents:

Part 1: Setting an ARCNET or Token Ring Node ID... 359

Part 2: Node ID Quick Reference Tables ...................... 360

# Setting an ARCNET or Token Ring Node ID

Each ARCNET/Token Ring card on the network must have its own unique node ID. Token Ring cards usually have the node ID configured at the factory, ARCNET cards do not.

The node ID is usually configured by a set of eight switches. These switches are a binary representation of the node ID. Some NICs require a switch to be off to enable the bit, some require it to be on to enable the bit. The table below specifies "O" and "X". Whether "O" or "X" signifies "ON" depends on your NIC.

Node ID 0 is normally reserved for messaging between nodes and is not a valid setting on most NICs. Node IDs 254 and 255 are reserved for network broadcast messages on Apple Computer networks. The nodes on a network do not need to be numbered sequentially (in order), but they do need to be unique.

# ARCNET Node ID Quick Reference Table

| Node ID | Switch 1 | Switch 2 | Switch 3 | Switch 4 | Switch 5 | Switch 6 | Switch 7 | Switch 8 |
|---|---|---|---|---|---|---|---|---|
| 1 | X | O | O | O | O | O | O | O |
| 2 | O | X | O | O | O | O | O | O |
| 3 | X | X | O | O | O | O | O | O |
| 4 | O | O | X | O | O | O | O | O |
| 5 | X | O | X | O | O | O | O | O |
| 6 | O | X | X | O | O | O | O | O |
| 7 | X | X | X | O | O | O | O | O |
| 8 | O | O | O | X | O | O | O | O |
| 9 | X | O | O | X | O | O | O | O |
| 10 | O | X | O | X | O | O | O | O |
| 11 | X | X | O | X | O | O | O | O |
| 12 | O | O | X | X | O | O | O | O |
| 13 | X | O | X | X | O | O | O | O |
| 14 | O | X | X | X | O | O | O | O |
| 15 | X | X | X | X | O | O | O | O |
| 16 | O | O | O | O | X | O | O | O |
| 17 | X | O | O | O | X | O | O | O |
| 18 | O | X | O | O | X | O | O | O |
| 19 | X | X | O | O | X | O | O | O |
| 20 | O | O | X | O | X | O | O | O |
| 21 | X | O | X | O | X | O | O | O |
| 22 | O | X | X | O | X | O | O | O |
| 23 | X | X | X | O | X | O | O | O |
| 24 | O | O | O | X | X | O | O | O |
| 25 | X | O | O | X | X | O | O | O |
| 26 | O | X | O | X | X | O | O | O |
| 27 | X | X | O | X | X | O | O | O |
| 28 | O | O | X | X | X | O | O | O |
| 29 | X | O | X | X | X | O | O | O |
| 30 | O | X | X | X | X | O | O | O |
| 31 | X | X | X | X | X | O | O | O |
| 32 | O | O | O | O | O | X | O | O |
| 33 | X | O | O | O | O | X | O | O |
| 34 | O | X | O | O | O | X | O | O |
| 35 | X | X | O | O | O | X | O | O |
| 36 | O | O | X | O | O | X | O | O |
| 37 | X | O | X | O | O | X | O | O |
| 38 | O | X | X | O | O | X | O | O |
| 39 | X | X | X | O | O | X | O | O |
| 40 | O | O | O | X | O | X | O | O |

## ARCNET NODE ID REFERENCE

| Node ID | Switch 1 | Switch 2 | Switch 3 | Switch 4 | Switch 5 | Switch 6 | Switch 7 | Switch 8 |
|---|---|---|---|---|---|---|---|---|
| 41 | X | O | O | X | O | X | O | O |
| 42 | O | X | O | X | O | X | O | O |
| 43 | X | X | O | X | O | X | O | O |
| 44 | O | O | X | X | O | X | O | O |
| 45 | X | O | X | X | O | X | O | O |
| 46 | O | X | X | X | O | X | O | O |
| 47 | X | X | X | X | O | X | O | O |
| 48 | O | O | O | O | X | X | O | O |
| 49 | X | O | O | O | X | X | O | O |
| 50 | O | X | O | O | X | X | O | O |
| 51 | X | X | O | O | X | X | O | O |
| 52 | O | O | X | O | X | X | O | O |
| 53 | X | O | X | O | X | X | O | O |
| 54 | O | X | X | O | X | X | O | O |
| 55 | X | X | X | O | X | X | O | O |
| 56 | O | O | O | X | X | X | O | O |
| 57 | X | O | O | X | X | X | O | O |
| 58 | O | X | O | X | X | X | O | O |
| 59 | X | X | O | X | X | X | O | O |
| 60 | O | O | X | X | X | X | O | O |
| 61 | X | O | X | X | X | X | O | O |
| 62 | O | X | X | X | X | X | O | O |
| 63 | X | X | X | X | X | X | O | O |
| 64 | O | O | O | O | O | O | X | O |
| 65 | X | O | O | O | O | O | X | O |
| 66 | O | X | O | O | O | O | X | O |
| 67 | X | X | O | O | O | O | X | O |
| 68 | O | O | X | O | O | O | X | O |
| 69 | X | O | X | O | O | O | X | O |
| 70 | O | X | X | O | O | O | X | O |
| 71 | X | X | X | O | O | O | X | O |
| 72 | O | O | O | X | O | O | X | O |
| 73 | X | O | O | X | O | O | X | O |
| 74 | O | X | O | X | O | O | X | O |
| 75 | X | X | O | X | O | O | X | O |
| 76 | O | O | X | X | O | O | X | O |
| 77 | X | O | X | X | O | O | X | O |
| 78 | O | X | X | X | O | O | X | O |
| 79 | X | X | X | X | O | O | X | O |
| 80 | O | O | O | O | X | O | X | O |
| 81 | X | O | O | O | X | O | X | O |
| 82 | O | X | O | O | X | O | X | O |
| 83 | X | X | O | O | X | O | X | O |
| 84 | O | O | X | O | X | O | X | O |

| Node ID | Switch 1 | Switch 2 | Switch 3 | Switch 4 | Switch 5 | Switch 6 | Switch 7 | Switch 8 |
|---|---|---|---|---|---|---|---|---|
| 85 | X | O | X | O | X | O | X | O |
| 86 | O | X | X | O | X | O | X | O |
| 87 | X | X | X | O | X | O | X | O |
| 88 | O | O | O | X | X | O | X | O |
| 89 | X | O | O | X | X | O | X | O |
| 90 | O | X | O | X | X | O | X | O |
| 91 | X | X | O | X | X | O | X | O |
| 92 | O | O | X | X | X | O | X | O |
| 93 | X | O | X | X | X | O | X | O |
| 94 | O | X | X | X | X | O | X | O |
| 95 | X | X | X | X | X | O | X | O |
| 96 | O | O | O | O | O | X | X | O |
| 97 | X | O | O | O | O | X | X | O |
| 98 | O | X | O | O | O | X | X | O |
| 99 | X | X | O | O | O | X | X | O |
| 100 | O | O | X | O | O | X | X | O |
| 101 | X | O | X | O | O | X | X | O |
| 102 | O | X | X | O | O | X | X | O |
| 103 | X | X | X | O | O | X | X | O |
| 104 | O | O | O | X | O | X | X | O |
| 105 | X | O | O | X | O | X | X | O |
| 106 | O | X | O | X | O | X | X | O |
| 107 | X | X | O | X | O | X | X | O |
| 108 | O | O | X | X | O | X | X | O |
| 109 | X | O | X | X | O | X | X | O |
| 110 | O | X | X | X | O | X | X | O |
| 111 | X | X | X | X | O | X | X | O |
| 112 | O | O | O | O | X | X | X | O |
| 113 | X | O | O | O | X | X | X | O |
| 114 | O | X | O | O | X | X | X | O |
| 115 | X | X | O | O | X | X | X | O |
| 116 | O | O | X | O | X | X | X | O |
| 117 | X | O | X | O | X | X | X | O |
| 118 | O | X | X | O | X | X | X | O |
| 119 | X | X | X | O | X | X | X | O |
| 120 | O | O | O | X | X | X | X | O |
| 121 | X | O | O | X | X | X | X | O |
| 122 | O | X | O | X | X | X | X | O |
| 123 | X | X | O | X | X | X | X | O |
| 124 | O | O | X | X | X | X | X | O |
| 125 | X | O | X | X | X | X | X | O |
| 126 | O | X | X | X | X | X | X | O |
| 127 | X | X | X | X | X | X | X | O |
| 128 | O | O | O | O | O | O | O | X |

## The Network Technical Guide 363

### ARCNET NODE ID REFERENCE

| Node ID | Switch 1 | Switch 2 | Switch 3 | Switch 4 | Switch 5 | Switch 6 | Switch 7 | Switch 8 |
|---|---|---|---|---|---|---|---|---|
| 129 | X | O | O | O | O | O | O | X |
| 130 | O | X | O | O | O | O | O | X |
| 131 | X | X | O | O | O | O | O | X |
| 132 | O | O | X | O | O | O | O | X |
| 133 | X | O | X | O | O | O | O | X |
| 134 | O | X | X | O | O | O | O | X |
| 135 | X | X | X | O | O | O | O | X |
| 136 | O | O | O | X | O | O | O | X |
| 137 | X | O | O | X | O | O | O | X |
| 138 | O | X | O | X | O | O | O | X |
| 139 | X | X | O | X | O | O | O | X |
| 140 | O | O | X | X | O | O | O | X |
| 141 | X | O | O | O | O | O | O | X |
| 142 | O | X | X | X | O | O | O | X |
| 143 | X | X | X | X | O | O | O | X |
| 144 | O | O | O | O | X | O | O | X |
| 145 | X | O | O | O | X | O | O | X |
| 146 | O | X | O | O | X | O | O | X |
| 147 | X | X | O | O | X | O | O | X |
| 148 | O | O | X | O | X | O | O | X |
| 149 | X | O | X | O | X | O | O | X |
| 150 | O | X | X | O | X | O | O | X |
| 151 | X | X | X | O | X | O | O | X |
| 152 | O | O | O | X | X | O | O | X |
| 153 | X | O | O | X | X | O | O | X |
| 154 | O | X | O | X | X | O | O | X |
| 155 | X | X | O | X | X | O | O | X |
| 156 | O | O | X | X | X | O | O | X |
| 157 | X | O | X | X | X | O | O | X |
| 158 | O | X | X | X | X | O | O | X |
| 159 | X | X | X | X | X | O | O | X |
| 160 | O | O | O | O | O | X | O | X |
| 161 | X | O | O | O | O | X | O | X |
| 162 | O | X | O | O | O | X | O | X |
| 163 | X | X | O | O | O | X | O | X |
| 164 | O | O | X | O | O | X | O | X |
| 165 | X | O | X | O | O | X | O | X |
| 166 | O | X | X | O | O | X | O | X |
| 167 | X | X | X | O | O | X | O | X |
| 168 | O | O | O | X | O | X | O | X |
| 169 | X | O | O | X | O | X | O | X |
| 170 | O | X | O | X | O | X | O | X |
| 171 | X | X | O | X | O | X | O | X |
| 172 | O | O | X | X | O | X | O | X |

## ARCNET NODE ID REFERENCE

| Node ID | Switch 1 | Switch 2 | Switch 3 | Switch 4 | Switch 5 | Switch 6 | Switch 7 | Switch 8 |
|---|---|---|---|---|---|---|---|---|
| 173 | X | O | X | X | O | X | O | X |
| 174 | O | X | X | X | O | X | O | X |
| 175 | X | X | X | X | O | X | O | X |
| 176 | O | O | O | O | X | X | O | X |
| 177 | X | O | O | O | X | X | O | X |
| 178 | O | X | O | O | X | X | O | X |
| 179 | X | X | O | O | X | X | O | X |
| 180 | O | O | X | O | X | X | O | X |
| 181 | X | O | X | O | X | X | O | X |
| 182 | O | X | X | O | X | X | O | X |
| 183 | X | X | X | O | X | X | O | X |
| 184 | O | O | O | X | X | X | O | X |
| 185 | X | O | O | X | X | X | O | X |
| 186 | O | X | O | X | X | X | O | X |
| 187 | X | X | O | X | X | X | O | X |
| 188 | O | O | X | X | X | X | O | X |
| 189 | X | O | X | X | X | X | O | X |
| 190 | O | X | X | X | X | X | O | X |
| 191 | X | X | X | X | X | X | O | X |
| 192 | O | O | O | O | O | O | X | X |
| 193 | X | O | O | O | O | O | X | X |
| 194 | O | X | O | O | O | O | X | X |
| 195 | X | X | O | O | O | O | X | X |
| 196 | O | O | X | O | O | O | X | X |
| 197 | X | O | X | O | O | O | X | X |
| 198 | O | X | X | O | O | O | X | X |
| 199 | X | X | X | O | O | O | X | X |
| 200 | O | O | O | X | O | O | X | X |
| 201 | X | O | O | X | O | O | X | X |
| 202 | O | X | O | X | O | O | X | X |
| 203 | X | X | O | X | O | O | X | X |
| 204 | O | O | X | X | O | O | X | X |
| 205 | X | O | X | X | O | O | X | X |
| 206 | O | X | X | X | O | O | X | X |
| 207 | X | X | X | X | O | O | X | X |
| 208 | O | O | O | O | X | O | X | X |
| 209 | X | O | O | O | X | O | X | X |
| 210 | O | X | O | O | X | O | X | X |
| 211 | X | X | O | O | X | O | X | X |
| 212 | O | O | X | O | X | O | X | X |
| 213 | X | O | X | O | X | O | X | X |
| 214 | O | X | X | O | X | O | X | X |
| 215 | X | X | X | O | X | O | X | X |
| 216 | O | O | O | X | X | O | X | X |

## ARCNET NODE ID REFERENCE

| Node ID | Switch 1 | Switch 2 | Switch 3 | Switch 4 | Switch 5 | Switch 6 | Switch 7 | Switch 8 |
|---|---|---|---|---|---|---|---|---|
| 217 | X | O | O | X | X | O | X | X |
| 218 | O | X | O | X | X | O | X | X |
| 219 | X | X | O | X | X | O | X | X |
| 220 | O | O | X | X | X | O | X | X |
| 221 | X | O | X | X | X | O | X | X |
| 222 | O | X | X | X | X | O | X | X |
| 223 | X | X | X | X | X | O | X | X |
| 224 | O | O | O | O | O | X | X | X |
| 225 | X | O | O | O | O | X | X | X |
| 226 | O | X | O | O | O | X | X | X |
| 227 | X | X | O | O | O | X | X | X |
| 228 | O | O | X | O | O | X | X | X |
| 229 | X | O | X | O | O | X | X | X |
| 230 | O | X | X | O | O | X | X | X |
| 231 | X | X | X | O | O | X | X | X |
| 232 | O | O | O | X | O | X | X | X |
| 233 | X | O | O | X | O | X | X | X |
| 234 | O | X | O | X | O | X | X | X |
| 235 | X | X | O | X | O | X | X | X |
| 236 | O | O | X | X | O | X | X | X |
| 237 | X | O | X | X | O | X | X | X |
| 238 | O | X | X | X | O | X | X | X |
| 239 | X | X | X | X | O | X | X | X |
| 240 | O | O | O | O | X | X | X | X |
| 241 | X | O | O | O | X | X | X | X |
| 242 | O | X | O | O | X | X | X | X |
| 243 | X | X | O | O | X | X | X | X |
| 244 | O | O | X | O | X | X | X | X |
| 245 | X | O | X | O | X | X | X | X |
| 246 | O | X | X | O | X | X | X | X |
| 247 | X | X | X | O | X | X | X | X |
| 248 | O | O | O | X | X | X | X | X |
| 249 | X | O | O | X | X | X | X | X |
| 250 | O | X | O | X | X | X | X | X |
| 251 | X | X | O | X | X | X | X | X |
| 252 | O | O | X | X | X | X | X | X |
| 253 | X | O | X | X | X | X | X | X |
| 254 | O | X | X | X | X | X | X | X |
| 255 | X | X | X | X | X | X | X | X |

# APPENDIX B

# Network Specifications

## Appendix B Contents:

Part 1: Overview .................................................. 369
Part 2: Ethernet Specifications ........................... 369
Part 3: Token Ring Specifications ...................... 380
Part 4: FDDI Specifications ................................. 392
Part 5: ARCNET Specifications .......................... 393

# Overview

This appendix provides summaries for popular LAN standards as well as specifications for common cables and their pin assignments.

The actual pin assignments and cabling may vary depending on the manufacturer of the hardware.

# Ethernet Specifications

## General

*Table B-1: Ethernet Summary*

| | |
|---|---|
| **Applicable Standards** | IEEE 802.3 series |
| | 10Base2, 10Base5, 10BaseT, 10BaseF, 100BaseT |
| **Max Bandwidth** | 10Mbps-100Mbps |
| **MAC Type** | CSMA/CD (contention-type) |
| **Topology** | Linear Bus, Star |
| **Cabling** | UTP, TP, Coaxial, Fiber-Optic |
| **Fault Tolerance** | Not part of standard |

# Specifications for Ethernet Networks Over Thick Coaxial Cable (10Base5)

*Table B-2: 10Base5 Trunk Cable Specifications*

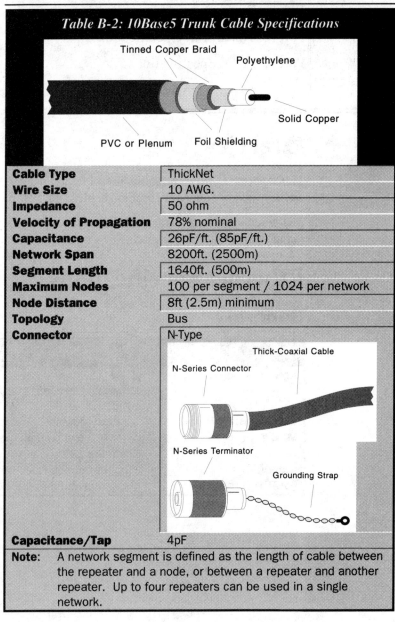

| | |
|---|---|
| Cable Type | ThickNet |
| Wire Size | 10 AWG. |
| Impedance | 50 ohm |
| Velocity of Propagation | 78% nominal |
| Capacitance | 26pF/ft. (85pF/ft.) |
| Network Span | 8200ft. (2500m) |
| Segment Length | 1640ft. (500m) |
| Maximum Nodes | 100 per segment / 1024 per network |
| Node Distance | 8ft (2.5m) minimum |
| Topology | Bus |
| Connector | N-Type |
| Capacitance/Tap | 4pF |

Note: A network segment is defined as the length of cable between the repeater and a node, or between a repeater and another repeater. Up to four repeaters can be used in a single network.

## Table B-3: 10Base5 AUI Drop Cable Specifications

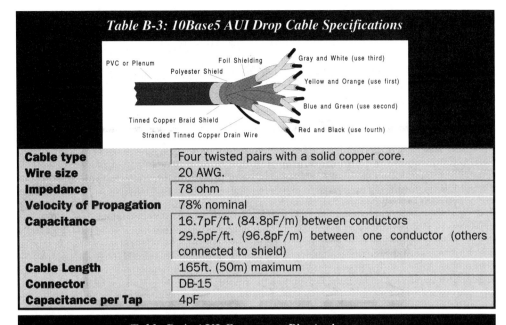

| | |
|---|---|
| **Cable type** | Four twisted pairs with a solid copper core. |
| **Wire size** | 20 AWG. |
| **Impedance** | 78 ohm |
| **Velocity of Propagation** | 78% nominal |
| **Capacitance** | 16.7pF/ft. (84.8pF/m) between conductors<br>29.5pF/ft. (96.8pF/m) between one conductor (others connected to shield) |
| **Cable Length** | 165ft. (50m) maximum |
| **Connector** | DB-15 |
| **Capacitance per Tap** | 4pF |

## Table B-4: AUI Connector Pin Assignments

| Pin | Circuit | Signal Name |
|---|---|---|
| 1 | CI-S | Control In - Circuit Shield |
| 2 | CI-A | Control In - Circuit A (+) |
| 3 | DO-A | Data Out - Circuit A (+) |
| 4 | DI-S | Data In - Circuit Shield |
| 5 | DI-A | Data In - Circuit A (+) |
| 6 | Vc | Voltage Common |
| 7 | N/C | Not Used |
| 8 | CO-S | Control Out - Circuit Shield |
| 9 | CI-B | Control In - Circuit B (-) |
| 10 | DO-B | Data Out - Circuit B (-) |
| 11 | DO-S | Data Out - Circuit Shield |
| 12 | DI-B | Data In - Circuit B (-) |
| 13 | VP | Voltage Plus (+) |
| 14 | VS | Voltage Shield |
| 15 | N/C | Not Used |
| Shell | PG | Protective Ground |

**Notes:** 1) VP and Vc use a single twisted pair in the AUI cable.
2) A variant of the AUI connector is the DIX connector. The two are identical except for the Pin 1 connection: The DIX connector attaches Pin 1 to the chassis ground while the AUI connector attaches Pin 1 to logic ground.

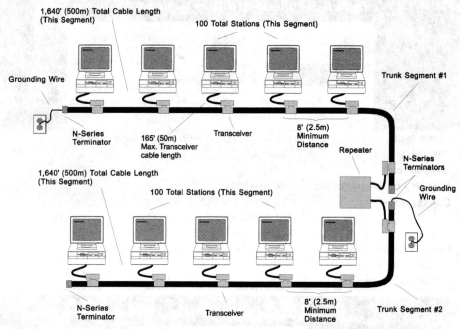

*Figure B-1: 10Base5 (Thick-Ethernet) Network*

- The arrangement must be in a linear bus configuration.
- Each end of the linear trunk cable must have an **N-series terminator** (see **Table B-2**) installed.
- Only one of the two terminators should be grounded. If you don't have a terminator with a grounding strap, then you can attach a piece of wire to it. Attach the other end of the grounding wire to the center screw on an electrical outlet cover plate.
- Maximum segment length is 1,640 feet (500 meters).
- Maximum total network length is 8,200 feet (2.5 kilometers).
- Maximum number of nodes per segment is 100 (includes workstations, file servers, and repeaters).
- Minimum distance between transceivers is 8 feet (2.5 meters).
- Maximum number of segments with no **inter-repeater links** is 2. Maximum number of segments with inter-repeater links is 5 (an inter-repeater link is a cable segment with no workstations attached).

- Thick and Thin-Ethernet networks may be linked together using special adapters.
- The maximum distance between the AUI port and the external transceiver is 164 feet (50m), there is no minimum distance (in some cases the transceiver may even plug directly into the AUI port.

## Specifications for Ethernet Networks Over Thin Coaxial Cable (10Base2)

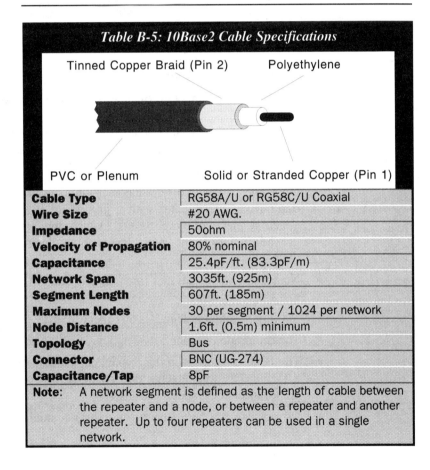

*Table B-5: 10Base2 Cable Specifications*

| | |
|---|---|
| **Cable Type** | RG58A/U or RG58C/U Coaxial |
| **Wire Size** | #20 AWG. |
| **Impedance** | 50ohm |
| **Velocity of Propagation** | 80% nominal |
| **Capacitance** | 25.4pF/ft. (83.3pF/m) |
| **Network Span** | 3035ft. (925m) |
| **Segment Length** | 607ft. (185m) |
| **Maximum Nodes** | 30 per segment / 1024 per network |
| **Node Distance** | 1.6ft. (0.5m) minimum |
| **Topology** | Bus |
| **Connector** | BNC (UG-274) |
| **Capacitance/Tap** | 8pF |

**Note:** A network segment is defined as the length of cable between the repeater and a node, or between a repeater and another repeater. Up to four repeaters can be used in a single network.

## Simple 10Base2 Guidelines

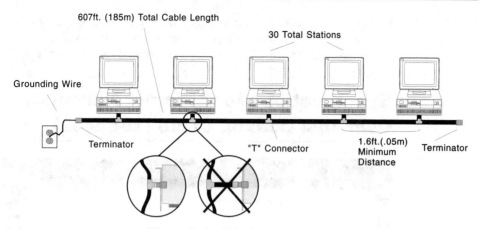

*Figure B-2: Simple 10Base2 Network Guidelines*

- The arrangement must be in a linear bus configuration.
- Each end of the linear cable must have a 50Ω terminator installed.
- Only one of the two terminators should be grounded. If you don't have a terminator with a grounding strap then you can attach a piece of wire to it. Attach the other end of the grounding wire to the center screw on an electrical outlet cover plate.
- No more than 30 devices (including repeaters) can be attached to a single segment.
- Maximum segment length is 607 feet (185m).
- The minimum distance between devices is 1.6 feet (.5m).
- The BNC connector on the coaxial cable must attach directly to a T-connector that attaches directly to the device; "drop cables" do not meet IEEE specifications.
- Do not use RG-58U (television cable) coaxial cable, use RG-58A/U.

## Complex 10Base2 Guidelines

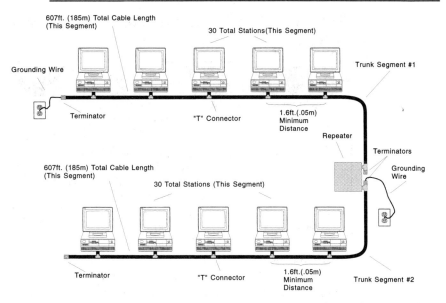

*Figure B-3: Complex 10Base2 Network Guidelines*

- Repeaters extend the allowable stations on a network and increase the total network span.
- Repeaters count as one device on each segment of the network to which it is attached.
- No more than 4 total repeaters allowed on the entire network.
- No more than 5 segments total connected by repeaters. Only 3 of these segments can contain workstations. The remaining two can only be used as inter-repeater links.
- The maximum cable distance of all segments can not exceed 3,035 feet (925 meters).

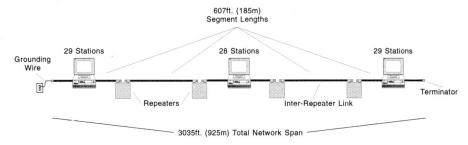

*Figure B-4: 10Base2 Network Spacing*

## Specifications for Ethernet Networks Over Unshielded Twisted Pair Cable (10BaseT)

*Table B-6: 10BaseT Cable Specifications*

| | |
|---|---|
| **Cable Type** | Two, three, four or six twisted pairs with a solid copper core. Four or six pairs are recommended for flexibility of the network layout. Do not use cable with more than six pairs. Cable twists should occur every foot, but two twists or more per foot is recommended to reduce interference. |
| **Wire Size** | #22, #24, or #26 AWG. |
| **Network Span** | 1640ft. (500 m) |
| **Segment Length** | 328ft. (100m) |
| **Maximum Nodes** | 1 per segment / 1024 per network |
| **Impedance** | 85 to 115ohms at 10MHz. |
| **Insertion Loss** | Maximum 11.5dB between 5 and 10Mhz |
| **Attenuation** | 8 to 10dB per 100m at 20° C (using PVC insulated cable) |
| **Topology** | Star |
| **Connector** | Modular RJ-45. This is an 8-pin connector, but only four pins are used by 10BASE-T. |
| **Segment Loss** | Maximum 11.5dB between 5 and 10MHz. |
| **Propagation Delay** | Maximum delay is 1000ns per segment. |
| **Notes:** | 1) A network segment is defined as the length of cable between the repeater and a node, or between a repeater and another repeater. Up to four repeaters can be used in a single network.<br>2) The telephone-type cable known as "silver satin" is not suitable for use with 10BASET networks. It can be identified by its flat shape and typically silver vinyl jacket. The use of "silver satin" cable can cause false data collisions on the network. |

The Network Technical Guide  377

### Table B-7: Connector Pin Assignments for Straight-Through Wiring

| From RJ-45 Pin | FROM Assignment | Color Code | TO Assignment | To RJ-45 Pin |
|---|---|---|---|---|
| 1 | Transmit (+) | White/Orange | Transmit (+) | 1 |
| 2 | Transmit (-) | Orange/White | Transmit (-) | 2 |
| 3 | Receive (+) | White/Green | Receive (+) | 3 |
| 4 | Not Used | Blue/White | Not Used | 4 |
| 5 | Not Used | White/Blue | Not Used | 5 |
| 6 | Receive (-) | Green/White | Receive (-) | 6 |
| 7 | Not Used | White/Brown | Not Used | 7 |
| 8 | Not Used | Brown/White | Not Used | 8 |

**Note:** The dashed line indicates pair number one, and the solid line indicates pair number two.

### Table B-8: Connector Pin Assignments for External Crossover Wiring

| From RJ-45 Pin | FROM Assignment | Color Code | TO Assignment | To RJ-45 Pin |
|---|---|---|---|---|
| 1 | Transmit (+) | White/Orange | Receive (+) | 1 |
| 2 | Transmit (-) | Orange/White | Receive (-) | 2 |
| 3 | Receive (+) | White/Green | Transmit (+) | 3 |
| 4 | Not Used | Blue/White | Not Used | 4 |
| 5 | Not Used | White/Blue | Not Used | 5 |
| 6 | Receive (-) | Green/White | Transmit (-) | 6 |
| 7 | Not Used | White/Brown | Not Used | 7 |
| 8 | Not Used | Brown/White | Not Used | 8 |

**Note:** The dashed line indicates pair number one, and the solid line indicates pair number two.

Two 10BaseT Ethernet devices can only communicate if the transmit pins on one device are connected to the receive pins on the other. This is called "crossover." The crossover can be implemented either in the wiring or in the device itself. The IEEE 10BaseT specification recommends that the crossover take place in the device. If both devices on each end of the cable implement crossover, then the cable must also use crossover. The only instance where a straight-through cable would be used is when only one of the devices is using crossover – not both devices.

Telephone cabling is often installed with crossover implemented in the wall plates. If existing telephone cabling is going to be used for a 10BaseT Ethernet network, keep this in mind.

- Stations connect to the network by way of a *concentrator*, a wiring center where all workstations connect to form a star topology. An external concentrator is shown below, but some specialized NICs have a number of RJ-45 jacks on the back – meaning the NIC has an internal concentrator.

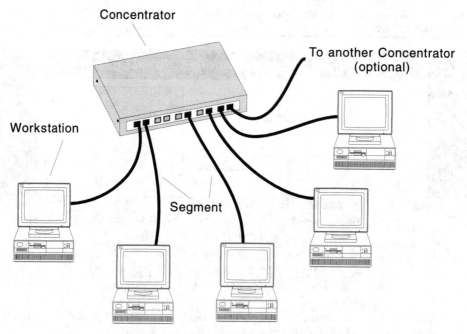

*Figure B-5: 10BaseT Network Guidelines*

- Maximum of 1,024 workstations on a network.
- Maximum cable distance between the concentrator/repeater and a workstation is 328 feet (100m).
- Minimum cable distance between the concentrator and a workstation is 2 feet (0.6m).
- Connections between twisted pair and thick or thin coaxial can be accomplished through the use of a special transceiver.
- Concentrators may be linked together to allow for additional workstations.

## Specifications for 100 Mbps Ethernet Networks Over Twisted Pair and Fiber-Optic Cable (100BaseT)

- Maximum number of workstations depends on the hardware vendor.
- 100BaseT is further divided into three types:
    - 100BaseTX applies to media that uses two twisted pairs.
    - 100BaseT4 applies to media that uses four twisted pairs
    - 100BaseFX applies to fiber-optic media
- Devices attach to the network via a Media Independent Interface (MII), which serves the same purpose as an AUI.
- 100BaseTX pin assignments are the same as 100BaseT, so no cabling changes are required for upgrade.
- Maximum twisted pair (TX and T4) cable distance between a hub and workstation is 328ft (100m).
- Maximum fiber-optic (FX) cable distance is 1312ft (400m).

# Token Ring Specifications

## General

*Table B-9: Token Ring Summary*

| | |
|---|---|
| **Applicable Standards** | IEEE 802.5 series, IBM specifications |
| **Max Bandwidth** | 4 or 16 Mbps |
| **MAC Type** | Token Passing |
| **Topology** | Ring |
| **Cabling** | UTP, TP, Coaxial, Fiber-Optic |
| **Fault Tolerance** | Back-up data path in intelligent MAUs |

- Stations attach to the network via MAUs (Multistation Access Units).

- The MAUs in a Token-Ring network must be connected in a ring configuration using cables. These cables are called "patch cables." The ring is formed by connecting the patch cable from the RO (Ring Out) connector of one MAU to the RI (Ring In) connector of the next MAU. This continues until the final patch cable is connected to the RI connector of the first MAU. If a network will only be using one MAU, its RO and RI connectors do <u>not</u> need to be connected with a patch cable.

- If a station attached to a MAU is disconnected or the cable brakes, the network will function normally. This is due to auto-sensing electronics contained in the MAU that close off the broken segment when a fault is detected.

- Maximum Total Ring Length: varies from vendor to vendor.

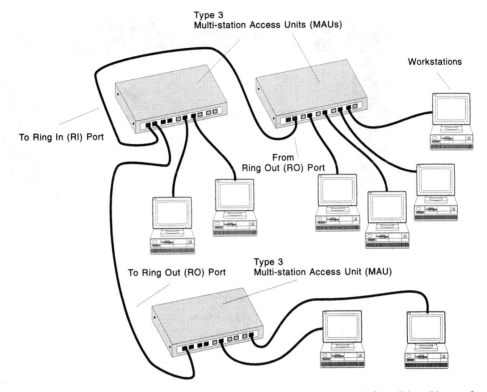

*Figure B-6: Connecting Multiple MAUs in a Token Ring Network*

## Standard Token Ring Network Requirements

Standard Token Ring configurations use IBM Type 1, 2, or 3 cabling.

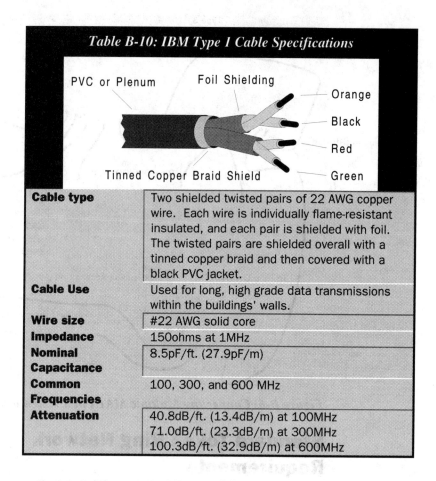

Table B-10: IBM Type 1 Cable Specifications

| | |
|---|---|
| **Cable type** | Two shielded twisted pairs of 22 AWG copper wire. Each wire is individually flame-resistant insulated, and each pair is shielded with foil. The twisted pairs are shielded overall with a tinned copper braid and then covered with a black PVC jacket. |
| **Cable Use** | Used for long, high grade data transmissions within the buildings' walls. |
| **Wire size** | #22 AWG solid core |
| **Impedance** | 150ohms at 1MHz |
| **Nominal Capacitance** | 8.5pF/ft. (27.9pF/m) |
| **Common Frequencies** | 100, 300, and 600 MHz |
| **Attenuation** | 40.8dB/ft. (13.4dB/m) at 100MHz<br>71.0dB/ft. (23.3dB/m) at 300MHz<br>100.3dB/ft. (32.9dB/m) at 600MHz |

### Table B-11: IBM Type 2 Cable Specifications

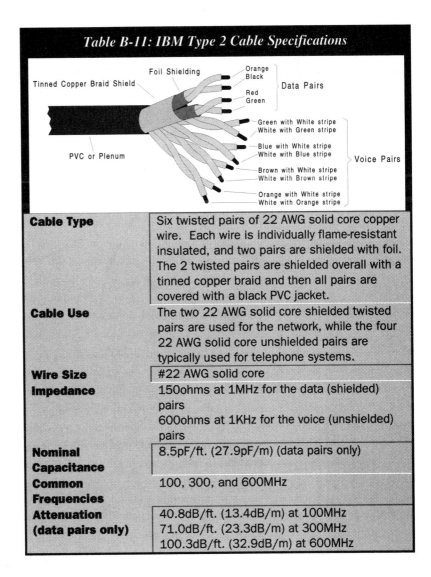

| | |
|---|---|
| **Cable Type** | Six twisted pairs of 22 AWG solid core copper wire. Each wire is individually flame-resistant insulated, and two pairs are shielded with foil. The 2 twisted pairs are shielded overall with a tinned copper braid and then all pairs are covered with a black PVC jacket. |
| **Cable Use** | The two 22 AWG solid core shielded twisted pairs are used for the network, while the four 22 AWG solid core unshielded pairs are typically used for telephone systems. |
| **Wire Size** | #22 AWG solid core |
| **Impedance** | 150ohms at 1MHz for the data (shielded) pairs<br>600ohms at 1KHz for the voice (unshielded) pairs |
| **Nominal Capacitance** | 8.5pF/ft. (27.9pF/m) (data pairs only) |
| **Common Frequencies** | 100, 300, and 600MHz |
| **Attenuation (data pairs only)** | 40.8dB/ft. (13.4dB/m) at 100MHz<br>71.0dB/ft. (23.3dB/m) at 300MHz<br>100.3dB/ft. (32.9dB/m) at 600MHz |

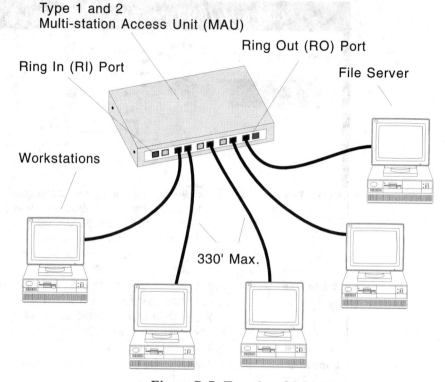

*Figure B-7: Type 1 and 2 MAU*

- Maximum number of MAU (Multistation Access Units) per ring (network) is 33.

- Maximum number of stations on one ring is 260.

- Maximum distance between a MAU and a station is 330 feet (100 meters).

- Maximum patch cable distance between two MAUs is 330 feet (100 meters).

- Maximum data rate: 4 or 16 Mbps. Maximum distances remain the same for 4 and 16 Mbps networks.

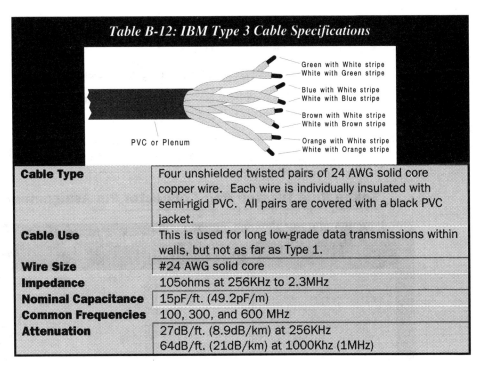

Table B-12: IBM Type 3 Cable Specifications

| | |
|---|---|
| **Cable Type** | Four unshielded twisted pairs of 24 AWG solid core copper wire. Each wire is individually insulated with semi-rigid PVC. All pairs are covered with a black PVC jacket. |
| **Cable Use** | This is used for long low-grade data transmissions within walls, but not as far as Type 1. |
| **Wire Size** | #24 AWG solid core |
| **Impedance** | 105ohms at 256KHz to 2.3MHz |
| **Nominal Capacitance** | 15pF/ft. (49.2pF/m) |
| **Common Frequencies** | 100, 300, and 600 MHz |
| **Attenuation** | 27dB/ft. (8.9dB/km) at 256KHz |
| | 64dB/ft. (21dB/km) at 1000Khz (1MHz) |

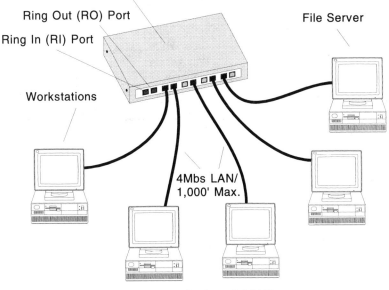

*Figure B-8: Type 3 MAU*

- Stations attach to the network via MAUs
- Maximum number of MAUs is 9.
- Maximum number of stations per ring is 72 (using 9 MAUs).
- Maximum distance between a MAU and a station is 1,000 feet (300 meters) for 4 Mbps network.
- Maximum data rate: 4 or 16 Mbps. Maximum distance between station and MAU is shorter when using 16 Mbps NICs in order to minimize line noise.

## Token Ring Twisted Pair Connector Pin Assignments

*Table B-13: Connector Pin Assignments/Straight-Through Wiring (RJ-45 Connector)*

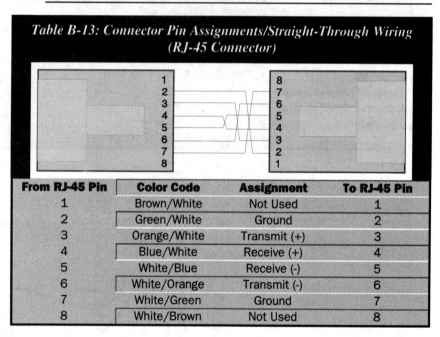

| From RJ-45 Pin | Color Code | Assignment | To RJ-45 Pin |
|---|---|---|---|
| 1 | Brown/White | Not Used | 1 |
| 2 | Green/White | Ground | 2 |
| 3 | Orange/White | Transmit (+) | 3 |
| 4 | Blue/White | Receive (+) | 4 |
| 5 | White/Blue | Receive (-) | 5 |
| 6 | White/Orange | Transmit (-) | 6 |
| 7 | White/Green | Ground | 7 |
| 8 | White/Brown | Not Used | 8 |

### Table B-14: Token Ring Pin Assignments/Crossover Wiring (RJ-45 Connector)

| From RJ-45 Pin | FROM Assignment | Color Code | TO Assignment | To RJ-45 Pin |
|---|---|---|---|---|
| 1 | Not Used | White/Orange | Not Used | 8 |
| 2 | Ground | Orange/White | Ground | 7 |
| 3 | Transmit (+) | White/Green | Transmit (-) | 6 |
| 4 | Receive (+) | White/Orange | Receive (-) | 5 |
| 5 | Receive (-) | Orange/White | Receive (+) | 4 |
| 6 | Transmit (-) | White/Green | Transmit (+) | 3 |
| 7 | Ground | White/Green | Ground | 2 |
| 8 | Not Used | Green/White | Not Used | 1 |

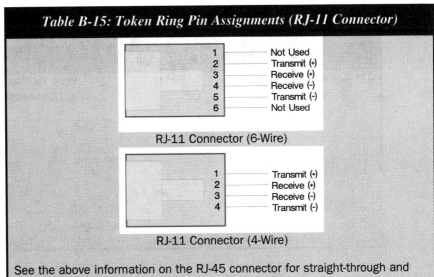

Table B-15: Token Ring Pin Assignments (RJ-11 Connector)

RJ-11 Connector (6-Wire)
1 — Not Used
2 — Transmit (+)
3 — Receive (+)
4 — Receive (-)
5 — Transmit (-)
6 — Not Used

RJ-11 Connector (4-Wire)
1 — Transmit (+)
2 — Receive (+)
3 — Receive (-)
4 — Transmit (-)

See the above information on the RJ-45 connector for straight-through and crossover wiring diagrams.
If the cable will carry data and power use an 8-position modular plug (RJ-45) with an 8-wire cable. If it will only be carrying data only, then a 6- (RJ-11) or 8-position (RJ-45) modular plug may be used.

### Table B-16: STP Connector Pin Assignments (DB-9 from NIC)

| DB-9 From NIC Pin Number | Color | Assignment | Wire Center Data Connector |
|---|---|---|---|
| 1 | Red | Receive (+) | R |
| 2 | | Not Used | |
| 3 | | Not Used | |
| 4 | | Not Used | |
| 5 | Black | Transmit (-) | B |
| 6 | Green | Receive (-) | G |
| 7 | | Not Used | |
| 8 | | Not Used | |
| 9 | Orange | Transmit (+) | O |

**Note:** Attach a DB-9 connector at the NIC end of the cable and an IBM MIC connector at the other.

### Table B-17: STP Connector Pin Assignments (From NIC DB-9 to IBM Wire Center)

| DB-15 Pin Number | Color | Wire Center Data Connector |
|---|---|---|
| 1 | Red | R |
| 2 | Not Used | |
| 3 | Not Used | |
| 4 | Not Used | |
| 5 | Black | B |
| 6 | Green | G |
| 7 | Not Used | |
| 8 | Not Used | |
| 9 | Orange | O |

## Table B-18: Token Ring Pin Assignments/Type 1, 2, or 6 IBM-Type Data Connector

| Pin Number | Color Code | Assignment |
|---|---|---|
| 1 | Red | R |
| 2 | Black | B |
| 3 | Green | G |
| 4 | Orange | O |

When the connector is not in use, shorting bars connect the red to the orange wire, and the black to the green wire.

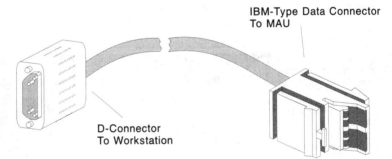

*Figure B-9: IBM Hermaphroditic Data Connector*

This is a proprietary connector used to attach the workstation to the MAU. When the connector is not plugged in, shorting bars connect the red to the orange wire and the black to the green wire.

# Token-Ring Cable Types 5, 6, 8, and 9

*Table B-19: IBM Type 5 Cable Specifications*

| | |
|---|---|
| **Cable Type** | Two color coded single fiber 100µm core graded index fiber optic cables with a Kevlar strength member covered by a flame resistant black polyurethane. |
| **Cable Use** | Trunk connections between MAUs or between repeaters. |
| **Numerical Aperture** | .290 nominal |
| **Bandwidth** | 150MHz at 850nm<br>500MHz at 1300nm |
| **Attenuation** | 6.0dB/km at 850nm<br>4.0dB/km at 1300nm |
| **Maximum Bend Radius** | 4.0in (6.0in for long term installations) |

*Table B-20: IBM Type 6 Cable Specifications*

| | |
|---|---|
| **Cable Type** | Two shielded twisted pairs of 26 AWG stranded core copper wire. Each wire is individually shielded. All pairs are then shielded with a tinned copper braid and covered with a black PVC jacket. |
| **Cable Use** | Commonly used to connect network devices to wall jacks. Is also used in patch panels and punch-down blocks. |
| **Wire Size** | #25 AWG stranded core |
| **Impedance** | 150ohms |
| **Nominal Capacitance** | 8.5pF/ft. (27.9pF/m) |
| **Common Frequencies** | 4, 16, 100 and 300MHz |
| **Attenuation** | 10dB/1000ft. (3.3dB/100m) at 4MHz<br>20dB/1000ft. (6.6dB/100m) at 16MHz<br>57dB/1000ft. (18.7dB/100m) at 100MHz<br>100dB/1000ft. (32.3dB/100m) at 300MHz |
| **Minimum Near End Crosstalk** | 52dB/1000ft.(305m) at 4MHz<br>44dB/1000ft.(305m) at 16MHz<br>33dB/1000ft.(305m) at 100MHz<br>25dB/1000ft.(305m) at 300MHz |

### Table B-21: IBM Type 8 Cable Specifications

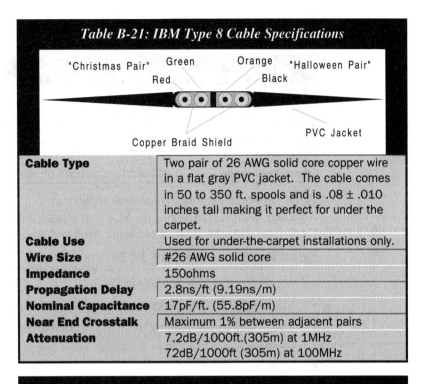

| | |
|---|---|
| **Cable Type** | Two pair of 26 AWG solid core copper wire in a flat gray PVC jacket. The cable comes in 50 to 350 ft. spools and is .08 ± .010 inches tall making it perfect for under the carpet. |
| **Cable Use** | Used for under-the-carpet installations only. |
| **Wire Size** | #26 AWG solid core |
| **Impedance** | 150ohms |
| **Propagation Delay** | 2.8ns/ft (9.19ns/m) |
| **Nominal Capacitance** | 17pF/ft. (55.8pF/m) |
| **Near End Crosstalk** | Maximum 1% between adjacent pairs |
| **Attenuation** | 7.2dB/1000ft.(305m) at 1MHz<br>72dB/1000ft (305m) at 100MHz |

### Table B-22: IBM Type 9 Cable Specifications

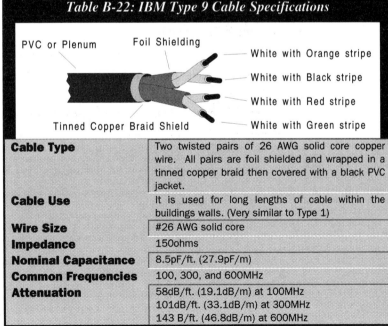

| | |
|---|---|
| **Cable Type** | Two twisted pairs of 26 AWG solid core copper wire. All pairs are foil shielded and wrapped in a tinned copper braid then covered with a black PVC jacket. |
| **Cable Use** | It is used for long lengths of cable within the buildings walls. (Very similar to Type 1) |
| **Wire Size** | #26 AWG solid core |
| **Impedance** | 150ohms |
| **Nominal Capacitance** | 8.5pF/ft. (27.9pF/m) |
| **Common Frequencies** | 100, 300, and 600MHz |
| **Attenuation** | 58dB/ft. (19.1dB/m) at 100MHz<br>101dB/ft. (33.1dB/m) at 300MHz<br>143 B/ft. (46.8dB/m) at 600MHz |

# FDDI Specifications

*Table B-23: FDDI Summary*

| | |
|---|---|
| **Applicable Standards** | ANSI X3T9.5 |
| **Max Bandwidth** | 100 Mbps |
| **MAC Type** | Token Passing |
| **Topology** | Ring |
| **Cabling** | Fiber-Optic |
| **Fault Tolerance** | Redundant Ring (DAS only) |

FDDI networks can use an optional redundant ring, known as Dual Attached Station (DAS) FDDI. FDDI networks that do not use the DAS option are categorized as Single Attached Station (SAS).

## Specifications for SAS FDDI

- Maximum data rate is 100 Mbps.
- Maximum of 1000 workstations on a network.
- Maximum cable distance between workstations is 2 km (1.24 miles).
- Maximum cable length (ring length) is 100 km for SAS.

## Specifications for DAS FDDI

- Maximum data rate is 100 Mbps.
- Maximum of 500 workstations on a network.
- Maximum cable distance between workstations is 2 km (1.24 miles).
- Maximum cable length is 200 km for DAS in an end-wrapped state.

# ARCNET Specifications

## General

*Table B-1: ARCNET Summary*

| | |
|---|---|
| **Applicable Standards** | Non-sponsored |
| **Max Bandwidth** | 2.5 Mbps or 20 Mbps |
| **MAC Type** | Token Passing |
| **Topology** | Star, Daisy-Chain (Linear Bus) |
| **Cabling** | Coaxial, TP, UTP |
| **Fault Tolerance** | Not part of standard |

## Specifications for ARCNET Networks Over Coaxial and Twisted Pair Cable

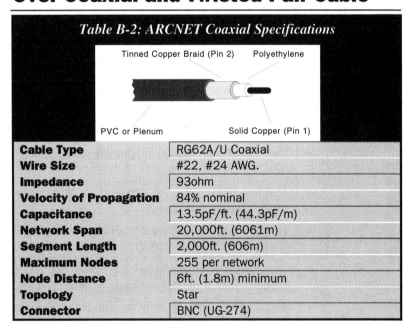

*Table B-2: ARCNET Coaxial Specifications*

| | |
|---|---|
| **Cable Type** | RG62A/U Coaxial |
| **Wire Size** | #22, #24 AWG. |
| **Impedance** | 93ohm |
| **Velocity of Propagation** | 84% nominal |
| **Capacitance** | 13.5pF/ft. (44.3pF/m) |
| **Network Span** | 20,000ft. (6061m) |
| **Segment Length** | 2,000ft. (606m) |
| **Maximum Nodes** | 255 per network |
| **Node Distance** | 6ft. (1.8m) minimum |
| **Topology** | Star |
| **Connector** | BNC (UG-274) |

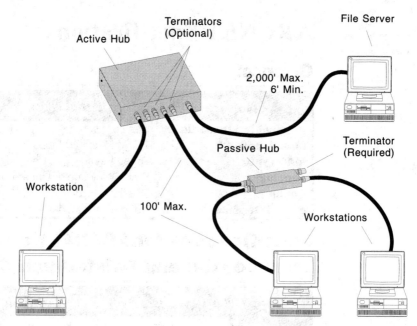

*Figure B-10: Typical Coaxial ARCNET Network*

- The coaxial cable used for ARCNET is RG-62A/U.
- The maximum distance between an active hub and a workstation is 2000ft. (606m).
- The maximum distance between two active hubs is 2000ft (606m).
- The maximum distance between an active and passive hub is 100ft. (30m).
- The maximum distance between a passive hub and a workstation is 100ft. (30m).
- The maximum distance between the workstations located at the farthest ends of the network must not exceed 20,000 ft (6060m).
- Workstations should only be connected to an active or passive hub (except in a two station or peer-to-peer network).
- File servers can be connected anywhere in the network layout.

## ARCNET Active Hubs

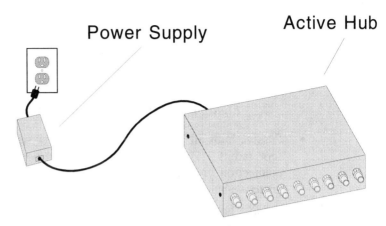

*Figure B-11: ARCNET Active Hub*

- Active hubs are used to condition and boost the network signal. They usually contain eight ports for workstation and additional hub connection.
- Active hubs, chained together, can achieve a maximum network length of 20,000 ft. (6060m).
- No more than 2000ft. (606m). of cable may separate two active hubs or an active hub and a workstation (1,000 feet on some implementations).
- No more than 100ft. (30m) of cable may separate an active hub from a passive hub.
- The ports may be used in any order to connect workstations or other active or passive hubs.
- Unused ports on active hubs do not need to be terminated, but it is recommended that you terminate them with 93ohm terminators.
- Do not connect two ports of an active hub together in a loop.

## ARCNET Passive Hubs

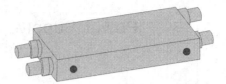

*Figure B-12: Typical ARCNET Passive Hub*

- Passive hubs are used to relay the network signals within a 100 foot radius. They usually contain four ports for workstation connection. No more than 100ft. (30m) of cable may separate an active hub from a passive hub.
- Passive hubs may only be used as an intermediate connection between workstations or between active hubs and workstations.
- Two passive hubs may not be connected together because they only split and relay the network signals. They do not boost the signal. If two passive hubs are connected together the signals will become too weak for the network to function properly.
- Unused ports on passive hubs do not need to be terminated, but it is recommended that you terminate them with 93ohm terminators.
- Do not connect two ports of a passive hub together in a loop.

# Specifications for ARCNET Networks Over Twisted Pair Cable

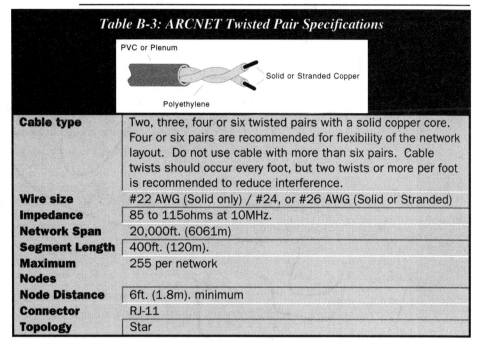

Table B-3: ARCNET Twisted Pair Specifications

| | |
|---|---|
| **Cable type** | Two, three, four or six twisted pairs with a solid copper core. Four or six pairs are recommended for flexibility of the network layout. Do not use cable with more than six pairs. Cable twists should occur every foot, but two twists or more per foot is recommended to reduce interference. |
| **Wire size** | #22 AWG (Solid only) / #24, or #26 AWG (Solid or Stranded) |
| **Impedance** | 85 to 115ohms at 10MHz. |
| **Network Span** | 20,000ft. (6061m) |
| **Segment Length** | 400ft. (120m). |
| **Maximum Nodes** | 255 per network |
| **Node Distance** | 6ft. (1.8m). minimum |
| **Connector** | RJ-11 |
| **Topology** | Star |

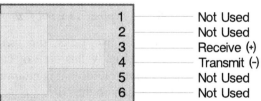

*Figure B-13: ARCNET RJ-11 Pin Assignments*

- ARCNET twisted pair networks operate with a single pair of unshielded wires. Unused pairs of existing telephone cables are often used, but must conform to the following specifications:
- A 22, 24 or 26 AWG solid copper core wire (RJ-11 plugs can only accommodate 24 or 26 AWG).
- A 24 or 26 AWG stranded core wire (RJ-11 plugs can only accommodate 26 AWG).
- Two twists per foot minimum.

## ARCNET Star Topology Network

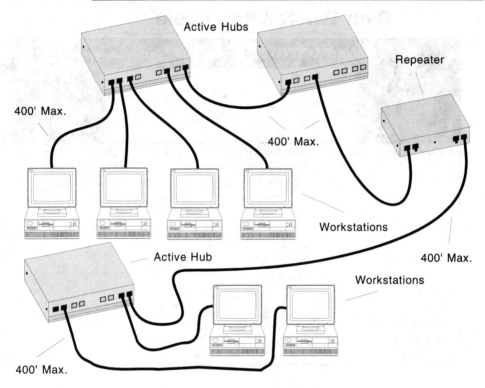

*Figure B-14: Typical Twisted Pair ARCNET Star Network*

- On NICs with two jacks, the jack not being used must be terminated. Usually this type of card comes standard with an internal terminator, all that is required for termination is to set a jumper or switch on the NIC.
- Maximum of 400ft. (120m) between any two network devices.
- Minimum of 6 ft. (1.8m) between any two network devices.

## ARCNET Twisted Pair Hubs and Repeaters

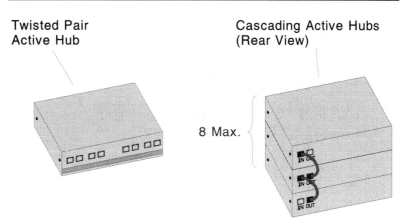

*Figure B-15: Single and Cascaded Active Hubs for Twisted Pair ARCNET*

- Hubs are the central wire center for workstation connections to the network.
- Open ports on the hub do not need to be terminated.
- No more than 8 hubs may be cascaded.

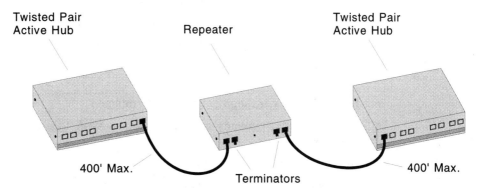

*Figure B-16: Repeaters and Hubs Twisted Pair ARCNET*

- Repeaters extend the distance between two hubs.
- Each port on the repeater usually has two jacks, the unused jacks must be terminated.

# ARCNET Daisy-Chain Topology Network

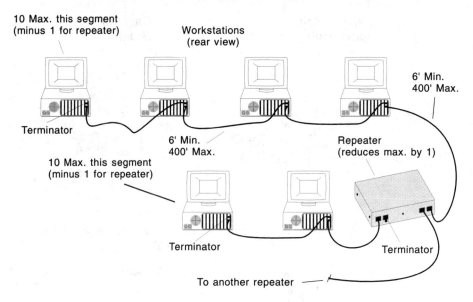

*Figure B-17: ARCNET Daisy-Chain Network*

- Only available on ARCNET NICs with two jacks.
- Maximum of 400ft. (120m) between any two network devices.
- Minimum of 6 ft. (1.8m) between any two network devices.
- Maximum number of 10 stations on a single daisy-chain wire segment.
- Each end of a wire segment must be terminated.
- Multiple wire segments can be combined with the use of a repeater.
- Adding a repeater reduces the number of stations by one.
- Each wire segment can contain no more than 15 attachments (twisted pair products and connectors, including modular couplers and modular wall jacks).
- Hubs may only be used to link into a star configured network, not for extending the daisy-chain.

# APPENDIX C

# Directory of Manufacturers

## Appendix C Contents:

Part 1: Disclaimer .......................................................... 403
Part 2: Contact Information ........................................... 403

# Disclaimer

Company profiles have been supplied by the companies themselves and do not necessarily reflect the views and opinions of Micro House International. Notes in parentheses are those of the editor.

# Contact Information

### 3COM CORPORATION

| | | |
|---|---|---|
| **Address** | 5400 Bayfront Plaza | |
| | Santa Clara, CA 95052-8145 | |
| **Phone** | (408) 764-5000 | Main |
| | (800) 638-3266 | Toll Free |
| | (800) 876-3266 | Tech Support |
| | (408) 764-5001 | Fax |
| | (408) 980-8204 | BBS |
| **E-Mail** | info@3com.com | |
| **WWW** | http://www.3com.com | |

### 3COM PRIMARY ACCESS

| | | |
|---|---|---|
| **Address** | 12230 World Trade Drive | |
| | San Diego, CA 92128-3765 | |
| **Phone** | (619) 675-4100 | Main |
| | (619) 674-8800 | Fax |
| **WWW** | http://www.3comnsp.com | |

### ACCTON TECHNOLOGY CORPORATION

| | | |
|---|---|---|
| **Address** | 1962 Zanker Road | |
| | San Jose, CA 95112 | |
| **Phone** | (408) 452-8900 | Main |
| | (800) 926-9288 | Toll Free |
| | (408) 452-8900 | Tech Support |
| | (408) 452-8988 | Fax |
| | (408) 452-8828 | BBS |
| **E-Mail** | support or sales@accton.com | |
| **WWW** | http://www.accton.com | |

## ACCULOGIC, INC.

| | | |
|---|---|---|
| **Address** | 7 Whatney<br>Irvine, CA  92718 | |
| **Phone** | (714) 454-2441 | Main |
| | (800) 234-7811 | Toll Free |
| | (714) 454-2441 | Tech Support |
| | (714) 454-8527 | Fax |
| | (714) 470-1759 | BBS |
| **E-Mail** | techsupport@acctechnology.com | |
| **WWW** | http://www.acctechnology.com | |

## ACEEX CORPORATION

| | | |
|---|---|---|
| **Address** | 1040a South Melrose Street<br>Placentia, CA  92670-7119 | |
| **Phone** | (714) 632-8889 | Main |
| | (714) 632-8876 | Tech Support |
| | (714) 632-8868 | Fax |
| | (714) 632-8853 | BBS |

## ACTIONTEC ELECTRONICS, INC.

| | | |
|---|---|---|
| **Address** | 750 North Mary Avenue<br>Sunnyvale, CA  94086 | |
| **Phone** | (408) 739-7000 | Main |
| | (800) 797-7001 | Toll Free |
| | (408) 739-7035 | Tech Support |
| | (408) 739-7001 | Fax |
| | (408) 739-7071 | BBS |
| **E-Mail** | techsupp@actiontec.com | |
| **WWW** | http://www.actiontec.com | |

## ACTIONTEC ELECTRONICS, INC.

| | | |
|---|---|---|
| **Address** | 17702 Mitchell North<br>Irvine, CA  92614 | |
| **Phone** | (714) 851-8242 | Main |
| | (800) Pmc-2277 | Toll Free |
| | (714) 851-8242 | Tech Support |
| | (714) 851-8249 | Fax |
| | (714) 851-1527 | BBS |

## ADAK COMMUNICATIONS CORPORATION
| | | |
|---|---|---|
| **Address** | 5840 Enterprise Drive<br>Lansing, MI 48911-4107 | |
| **Phone** | (517) 882-5191 | Main |
| | (517) 882-3194 | Fax |

## ADAPTEC, INC.
| | | |
|---|---|---|
| **Address** | 691 South Milpitas Blvd.<br>Milpitas, CA 95035 | |
| **Phone** | (408) 945-8600 | Main |
| | (800) 959-7274 | Toll Free |
| | (408) 934-7274 | Tech Support |
| | (408) 945-6776 | Fax |
| | (408) 945-7727 | BBS |
| **WWW** | http://www.adaptec.com | |

## ADC KENTROX
| | | |
|---|---|---|
| **Address** | 14375 Nw Science Park Drive<br>Portland, OR 97229 | |
| **Phone** | (503) 643-1681 | Main |
| | (800) 733-5511 | Toll Free |
| | (800) 733-5511 | Tech Support |
| | (503) 641-3341 | Fax |
| **WWW** | http://www.kentrox.com | |

## ADDTRON TECHNOLOGY CO., LTD.
| | | |
|---|---|---|
| **Address** | 46712 Fremont Blvd.<br>Fremont, CA 94538 | |
| **Phone** | (510) 770-0120 | Main |
| | (800) 998-4638 | Toll Free |
| | (800) 998-4646 | Tech Support |
| | (510) 770-0171 | Fax |
| | (510) 770-0272 | BBS |
| **E-Mail** | 73762.3706@compuserve.com | |
| **WWW** | http://www.addtron.com | |

## ADVANCED INTEGRATION RESEARCH, INC.
| | | |
|---|---|---|
| Address | 2188 Del Franco Street<br>San Jose, CA 95131 | |
| Phone | (408) 428-0800 | Main |
| | (800) 866-1945 | Toll Free |
| | (408) 428-1547 | Tech Support |
| | (408) 428-0950 | Fax |
| | (408) 428-1735 | BBS |
| E-Mail | air@ix.netcom.com | |
| WWW | http://www.airweb.com | |

## ADVANCED INTERLINK CORPORATION
| | | |
|---|---|---|
| Address | 15181 Springdale Street<br>Huntington Beach, CA 92649 | |
| Phone | (714) 894-1675 | Main |
| | (714) 893-1546 | Fax |
| | (714) 894-0893 | BBS |
| WWW | http://www.advancedinterlink.com | |

## ADVANCED LOGIC RESEARCH, INC.
| | | |
|---|---|---|
| Address | 9401 Jeronimo Road<br>Irvine, CA 92618 | |
| Phone | (800) 257-1230 | Toll Free |
| | (800) 257-1230 | Tech Support |
| | (714) 458-0532 | Fax |
| | (714) 458-6834 | BBS |

## ADVANCED TELECOMMUNICATIONS MODULES LTD.
| | | |
|---|---|---|
| Address | 1130 East Arques Avenue<br>Sunnyvale, CA 94086 | |
| Phone | (408) 523-1400 | Main |
| | (408) 523-1410 | Fax |
| WWW | http://www.atminc.com | |

## AIRGO COMMUNICATIONS, INC.
| | | |
|---|---|---|
| Address | Sorenson Research Park<br>849 West Levoy Drive<br>Salt Lake City, UT 84123 | |
| Phone | (801) 269-7200 | Main |

## AIRONET WIRELESS COMMUNICATIONS, INC.

| | | |
|---|---|---|
| **Address** | Po Box 5292 | |
| | Akron, OH  44334-0292 | |
| **Phone** | (216) 665-7900 | Main |
| | (216) 665-7922 | Fax |
| **E-Mail** | sales@aironet.com | |
| **WWW** | http://www.aironet.com | |

## ALFA NETCOM, INC.

| | | |
|---|---|---|
| **Address** | 477 Valley Way | |
| | Milpitas, CA  95035 | |
| **Phone** | (408) 934-3880 | Main |
| | (408) 934-3883 | Fax |
| **E-Mail** | Support@alfainc.com | |
| **WWW** | http://www.alfainc.com | |

## ALLIED TELESYN, INC.

| | | |
|---|---|---|
| **Address** | 950 Kifer Road | |
| | Sunnyvale, CA  94086 | |
| **Phone** | (408) 730-0950 | Main |
| | (800) 424-5012 | Toll Free |
| | (408) 428-4835 | Tech Support |
| | (408) 730-0950 | Fax |
| | (206) 483-7979 | BBS |
| **WWW** | http://www.alliedtelesyn.com | |

## ALTA RESEARCH CORPORATION

| | | |
|---|---|---|
| **Address** | 614 South Federal Highway | |
| | Deerfield Beach, FL  33441 | |
| **Phone** | (800) 423-8535 | Toll Free |
| | (800) 423-8535 | Tech Support |
| | (305) 428-8678 | Fax |
| | (305) 428-0616 | BBS |
| **E-Mail** | lan@altasrch.com | |

## AMERICAN DATA TECHNOLOGY

| | | |
|---|---|---|
| **Address** | 1825 S Peck Rd. Unit J | |
| | Monrovia, CA  91016 | |
| **Phone** | (818) 303-8567 | Main |
| | (818) 303-0610 | Fax |

## AMERICAN NATIONAL STANDARDS INSTITUTE (ANSI)
| | | |
|---|---|---|
| **Address** | 11 West 42nd Street<br>New York, New York 10036 | |
| **Phone** | (212) 642-4900 | Main |
| | (212) 398-0023 | Fax |
| **Web** | www.ansi.org | |

## AMERICAN RESEARCH CORPORATION
| | | |
|---|---|---|
| **Address** | 602 Monterey Pass Road<br>Monterey Park, CA 91754 | |
| **Phone** | (818) 284-1904 | Main |
| | (818) 284-4213 | Fax |

## AMKLY SYSTEMS, INC.
| | | |
|---|---|---|
| **Address** | 26796 Vista Terrace Drive<br>Lake Forest, CA 92630 | |
| **Phone** | (714) 768-3511 | Main |
| | (714) 768-3511 | Tech Support |
| | (714) 768-7885 | Fax |
| | (714) 597-0828 | BBS |
| **WWW** | http://www.amkly.com | |

## AMQUEST CORPORATION
| | | |
|---|---|---|
| **Address** | 1650 Manheim Pike<br>Lancaster, PA 17601 | |
| **Phone** | (717) 569-8030 | Main |
| | (717) 569-2054 | Tech Support |
| | (717) 569-8530 | Fax |
| **WWW** | http://www.amquestcorp.com | |

## ANCO TECHNOLOGY, INC.
| | | |
|---|---|---|
| **Address** | 140 North Palm Street<br>Brea, CA 92621 | |
| **Phone** | (714) 992-9000 | Main |
| | (800) 545-2626 | Toll Free |
| | (714) 992-9005 | Tech Support |
| | (714) 992-1672 | Fax |
| **WWW** | http://www.ancotech.com | |

### ANCOR COMMUNICATIONS, INC.

| | | |
|---|---|---|
| **Address** | 6130 Blue Circle Drive<br>Minnetonka, MN 55343 | |
| **Phone** | (612) 932-4000 | Main |
| | (800) 342-7379 | Toll Free |
| | (612) 932-4037 | Fax |
| **E-Mail** | info@ancor.com | |
| **WWW** | http://www.ancor.com | |

### ANDREW CORPORATION

| | | |
|---|---|---|
| **Address** | 23610 Telo Avenue<br>Torrance, CA 90505 | |
| **Phone** | (310) 784-8000 | Main |
| | (800) 826-3739 | Toll Free |
| | (310) 784-8090 | Fax |
| **WWW** | http://www.andrew.com | |

### ANGIA COMMUNICATIONS, INC.

| | | |
|---|---|---|
| **Address** | 441 E. Bay Blvd.<br>Provo, UT 84606 | |
| **Phone** | (801) 371-0488 | Main |
| | (800) 877-9159 | Toll Free |
| | (800) 998-0555 | Tech Support |
| | (801) 373-9847 | Fax |
| | (801) 373-7807 | BBS |
| **WWW** | http://www.angia.com | |

### ANSEL COMMUNICATIONS, INC.

| | | |
|---|---|---|
| **Address** | 8711 148th Avenue Northeast<br>Redmond, WA 98052 | |
| **Phone** | (206) 869-4928 | Main |
| | (800) 341-7978 | Toll Free |
| | (206) 869-2389 | Tech Support |
| | (206) 869-5015 | Fax |
| **E-Mail** | gomes@ansel.com | |
| **WWW** | http://www.ansel.com | |

### APACHE MICRO PERIPHERALS, INC.

| | | |
|---|---|---|
| **Address** | 17895 Skypark Circle, Unit F<br>Irvine, CA 92614 | |
| **Phone** | (714) 251-1818 | Main |
| | (714) 251-1877 | Fax |
| **WWW** | http://www.apache-micro.com | |

## ARCHTEK CORPORATION

| | | |
|---|---|---|
| Address | 18549 Gale Avenue<br>City Of Industry, CA 91748-1338 | |
| Phone | (818) 912-9800 | Main |
| | (818) 912-9700 | Fax |
| | (818) 912-3980 | BBS |
| E-Mail | smartlink@aol.com | |
| WWW | http://www.archtek.com | |

## ARCNET TRADE ASSOCIATION (ATA)

| | | |
|---|---|---|
| Address | 2460 Wisconsin Ave.<br>Downers Grove, Illinois USA 60515 | |
| Phone | (630) 960-5130 | Main |
| | (630) 963-2122 FAX | Fax |
| e-mail | ata@arcnet.com | |
| Web | www.arcnet.com | |

## ARK PC TECHNOLOGY

| | | |
|---|---|---|
| Address | 10 Hughes<br>Suite A-104<br>Irvine, CA 92618 | |
| Phone | (714) 452-0166 | Main |
| | (714) 452-0176 | Fax |
| E-Mail | tak6212@msn.com | |
| WWW | http://www.arkpc.com | |

## ARTISOFT, INC.

| | | |
|---|---|---|
| Address | 2202 North Forbes Blvd.<br>Tucson, AZ 85745 | |
| Phone | (520) 670-7100 | Main |
| | (800) 233-5564 | Toll Free |
| | (520) 670-7335 | Fax |
| WWW | http://www.artisoft.com | |

## ASANTE TECHNOLOGIES, INC.

| | | |
|---|---|---|
| Address | 821 Fox Lane<br>San Jose, CA 95131 | |
| Phone | (408) 435-8401 | Main |
| | (800) 662-9686 | Toll Free |
| | (408) 432-7511 | Fax |
| | (408) 432-1416 | BBS |
| E-Mail | sales@asante.com | |
| WWW | http://www.asante.com | |

## ASKEY COMPUTER CORPORATION
**Address**  162 Atlantic Street
Pamona, CA 91768

**Phone**  (909) 595-8851  Main
(909) 595-7282  Fax

## ASPEN TECHNOLOGIES, INC.
**Address**  400 Rogers Street
Princeton, WV 24740

**Phone**  (304) 425-1111  Main
(304) 425-Info  Tech Support
(304) 487-6111  Fax
(304) 487-0556  BBS

**WWW**  http://www.aspentek.com

## AST RESEARCH, INC.
**Address**  16215 Alton Parkway
Irvine, CA 92618-3618

**Phone**  (714) 727-4141  Main
(800) 727-1278  Tech Support
(714) 727-9355  Fax

**WWW**  http://www.ast.com

## ASUS COMPUTER INTERNATIONAL
**Address**  721 Charcot Avenue
San Jose, CA 95131

**Phone**  (408) 474-0567  Main
(408) 474-0568  Fax
(408) 474-0555  BBS

**WWW**  http://asustek.asus.com.tw

## AT&T, INC.
**Address**  32 Avenue Of The Americas
New York, NY 10013

**Phone**  (212) 387-5400  Main
**WWW**  http://att.com

## AT&T PARADYNE

| | | |
|---|---|---|
| **Address** | 8545 126th Avenue North<br>P.o. Box 2826<br>Largo, FL 34649-2826 | |
| **Phone** | (813) 530-2000 | Main |
| | (800) 237-0016 | Toll Free |
| | (800) 894-0596 | Tech Support |
| | (813) 530-2535 | Fax |
| | (813) 532-5254 | BBS |
| **WWW** | http://www.paradyne.att.com | |

## AT-LAN-TEC, INC.

| | | |
|---|---|---|
| **Address** | 7881 H Beachcraft Avenue<br>Gaithersburg, MD 20879 | |
| **Phone** | (301) 948-7070 | Main |
| | (301) 948-0658 | Fax |
| | (301) 990-6998 | BBS |
| **WWW** | http://www.atlantec.net | |

## AURORA TECHNOLOGIES, INC.

| | | |
|---|---|---|
| **Address** | 176 Second Avenue<br>Waltham, MA 02154 | |
| **Phone** | (617) 290-4800 | Main |
| | (617) 290-4844 | Fax |
| **E-Mail** | info@auratek.com | |
| **WWW** | http://www.auratek.com | |

## AZTECH LABS, INC.

| | | |
|---|---|---|
| **Address** | 47811 Warm Springs Blvd.<br>Fremont, CA 94539 | |
| **Phone** | (510) 623-8988 | Main |
| | (800) 886-8859 | Toll Free |
| | (800) 886-8879 | Tech Support |
| | (510) 623-8989 | Fax |
| | (510) 623-8933 | BBS |
| **WWW** | http://www.aztech.com | |

## BANYAN SYSTEMS, INC.

| | | |
|---|---|---|
| **Address** | 120 Flanders Rd. | |
| | P.O. Box 5103 | |
| | Westboro, MA  01581 | |
| **Phone** | (508) 898-1000 | Main |
| | (800) 222-6926 | Toll Free |
| | (508) 898-1755 | Fax |
| | (800) 932-9226 | Automated fax-back |
| **WWW** | http://www.banyan.com | |

## BAY NETWORKS, INC.

| | | |
|---|---|---|
| **Address** | 4401 Great America Parkway | |
| | Santa Clara, CA  95054 | |
| **Phone** | (408) 988-2400 | Main |
| | (408) 988-5525 | Fax |
| **WWW** | http://www.wellfleet.com | |

## BEC COMPUTER ENTERPRISE

| | | |
|---|---|---|
| **Address** | 425  Privet Road | |
| | Horsham, PA  19044 | |
| **Phone** | (800) 453-7630 | Toll Free |
| | (215) 672-5945 | Fax |

## BEST DATA PRODUCTS, INC.

| | | |
|---|---|---|
| **Address** | 21800 Nordhoff Street | |
| | Chatsworth, CA  91311 | |
| **Phone** | (818) 773-9600 | Main |
| | (818) 773-9619 | Fax |
| **WWW** | http://www.bestdata.com | |

## BLACK BOX CORPORATION

| | | |
|---|---|---|
| **Address** | 1000 Park Drive | |
| | Lawrence, PA  15055 | |
| **Phone** | (412) 746-5500 | Main |
| | (412) 746-0746 | Fax |
| | (412) 746-7120 | BBS |
| **WWW** | http://www.blackbox.com | |

## BOCA RESEARCH, INC.
| | | |
|---|---|---|
| **Address** | 1377 Clint Moore Road | |
| | Boca Raton, FL  33487-2722 | |
| **Phone** | (407) 997-6227 | Main |
| | (407) 997-5549 | Tech Support |
| | (407) 994-5848 | Fax |
| **WWW** | http://www.boca.org | |

## BROADTECH INTERNATIONAL COMPANY
| | | |
|---|---|---|
| **Address** | 451 W. Lambert Road, Suite 209 | |
| | Suite 209 | |
| | Brea, CA  92621 | |
| **Phone** | (714) 255-8898 | Main |
| | (714) 255-8899 | Fax |
| **WWW** | http://www.broadtech.com | |

## BROOKTROUT TECHNOLOGY, INC.
| | | |
|---|---|---|
| **Address** | 410 1st Avenue | |
| | Needham, MA  02194 | |
| **Phone** | (617) 449-4100 | Main |
| | (617) 449-9009 | Fax |
| | (617) 449-9279 | BBS |
| **E-Mail** | info@brooktrout.com | |
| **WWW** | http://www.brooktrout.com | |

## BUSLOGIC, INC.
| | | |
|---|---|---|
| **Address** | 4151 Burton Drive | |
| | Santa Clara, CA  95054 | |
| **Phone** | (408) 492-9090 | Main |
| | (408) 492-1542 | Fax |
| | (408) 492-1984 | BBS |
| **WWW** | http://www.buslogic.com | |

## CABLETRON SYSTEMS, INC.
| | | |
|---|---|---|
| **Address** | 35 Industrial Way | |
| | Rochester, NH  03867 | |
| **Phone** | (603) 332-9400 | Main |
| | (603) 332-9400 | Tech Support |
| | (603) 337-2444 | Fax |
| | (603) 335-3358 | BBS |
| **WWW** | http://www.ctron.com | |

## CARDINAL TECHNOLOGIES, INC.

| | | |
|---|---|---|
| **Address** | 1827 Freedom Road<br>Lancaster, PA  17601 | |
| **Phone** | (717) 293-3000 | Main |
| | (717) 293-3124 | Tech Support |
| | (717) 293-3055 | Fax |
| | (717) 293-3074 | BBS |
| **E-Mail** | prodinfo@cardtech.com | |
| **WWW** | http://www.cardtech.com/ | |

## CARDWARE LAB, INC.

| | | |
|---|---|---|
| **Address** | 1080 East Duane Avenue<br>Suite G<br>Sunnyvale, CA  94086 | |
| **Phone** | (408) 774-9101 | Main |
| | (408) 774-9102 | Fax |
| | (408) 774-9103 | BBS |
| **E-Mail** | tech@cwli.com | |
| **WWW** | http://www.cwli.com | |

## CASTELLE, INC.

| | | |
|---|---|---|
| **Address** | 3 Scott Blvd., Suite 3255-3<br>Santa Clara, CA  95054 | |
| **Phone** | (408) 496-0474 | Main |
| | (408) 496-0502 | Fax |
| **WWW** | http://www.castelle.com | |

## CELAN TECHNOLOGY, INC.

| | | |
|---|---|---|
| **Address** | 2323 Calle Del Mundo<br>Santa Clara, CA  95054 | |
| **Phone** | (408) 988-8288 | Main |
| | (800) 682-3526 | Toll Free |
| | (408) 988-8289 | Fax |
| | (408) 988-8388 | BBS |

## CHASE RESEARCH, INC.

| | | |
|---|---|---|
| **Address** | 545 Marriott Drive<br>Suite100<br>Nashville, TN  37214 | |
| **Phone** | (615) 872-0770 | Main |
| | (615) 872-0771 | Fax |
| **WWW** | http://www.chaser.com | |

## CISCO SYSTEMS, INC.
| | | |
|---|---|---|
| **Address** | 170 West Tasman Drive<br>P.o. Box 3075<br>San Jose, CA 95134 | |
| **Phone** | (408) 216-8000 | Main |
| | (408) 526-4100 | Fax |
| **WWW** | http://www.cisco.com | |

## CLAFLIN & CLAYTON, INC.
| | | |
|---|---|---|
| **Address** | 313 Boston Post Road West<br>Marlborough, MA 01752-4612 | |
| **Phone** | (508) 787-1000 | Main |
| | (508) 393-7979 | Tech Support |
| **WWW** | http://www.netphone.com | |

## CMD TECHNOLOGY, INC.
| | | |
|---|---|---|
| **Address** | 1 Vanderbilt<br>Irvine, CA 92618 | |
| **Phone** | (714) 454-0800 | Main |
| | (800) 426-3832 | Toll Free |
| | (714) 455-1656 | Fax |
| **WWW** | http://websvr.cmd.com | |

## CNET TECHNOLOGY, INC.
| | | |
|---|---|---|
| **Address** | 2199 Zanker Road<br>San Jose, CA 95131 | |
| **Phone** | (408) 954-8000 | Main |
| | (800) 486-2638 | Toll Free |
| | (408) 954-8800 | Tech Support |
| | (408) 954-8866 | Fax |
| | (408) 954-1787 | BBS |
| **WWW** | http://www.cnet.com.tw | |

## CODENOLL TECHNOLOGY CORPORATION
| | | |
|---|---|---|
| **Address** | 1086 North Broadway<br>Yonkers, NY 10701 | |
| **Phone** | (914) 965-6300 | Main |
| | (914) 965-6300 | Tech Support |
| | (914) 965-9811 | Fax |
| | (914) 965-1972 | BBS |
| **WWW** | http://www.codenoll.com | |

## COGENT DATA TECHNOLOGIES, INC.

| | | |
|---|---|---|
| **Address** | 640 Mullis Street<br>Friday Harbor, WA 98250 | |
| **Phone** | (360) 378-2929 | Main |
| | (800) 426-4368 | Toll Free |
| | (360) 378-2929 | Tech Support |
| | (360) 378-2882 | Fax |
| | (360) 378-5405 | BBS |
| **WWW** | http://www.cogentdata.com | |

## COGENT DATA TECHNOLOGIES, INC.

| | | |
|---|---|---|
| **Address** | 15375 SE 30th Place<br>Suite 310<br>Bellevue, WA 98007 | |
| **Phone** | (206) 603-0333 | Main |
| | (206) 603-9223 | Fax |
| **E-Mail** | sales@cogentdata.com | |

## COMPEX, INC.

| | | |
|---|---|---|
| **Address** | 4051 East La Palma Avenue<br>Anaheim, CA 92807 | |
| **Phone** | (714) 630-7302 | Main |
| | (800) 279-8891 | Toll Free |
| | (714) 630-5451 | Tech Support |
| | (714) 630-6521 | Fax |
| | (714) 630-2570 | BBS |
| **E-Mail** | info@cpx.com | |
| **WWW** | http://www.cpx.com | |

## COMPULAN TECHNOLOGY, INC.

| | | |
|---|---|---|
| **Address** | 1630 Oakland Road<br>Suite A111<br>San Jose, CA 95131 | |
| **Phone** | (408) 432-8899 | Main |
| | (800) 486-8810 | Toll Free |
| | (408) 432-8899 | Tech Support |
| | (408) 432-8699 | Fax |
| | (408) 432-8433 | BBS |
| **WWW** | http://www.cnet.com.tw | |

## COMPUTER MODULES, INC.

| | | |
|---|---|---|
| Address | 2350 Walsh Avenue<br>Santa Clara, CA 95051-1301 | |
| Phone | (408) 496-1881 | Main |
| | (408) 496-1886 | Fax |
| E-Mail | info@compmod.com | |
| WWW | http://www.compmod.com | |

## COMPUTER PERIPHERALS, INC.

| | | |
|---|---|---|
| Address | 7 Whatney<br>Irvine, CA 92618 | |
| Phone | (714) 454-2441 | Main |
| | (800) 854-7600 | Toll Free |
| | (714) 454-2441 | Tech Support |
| | (714) 470-1758 | Fax |
| | (714) 470-1759 | BBS |
| E-Mail | kristajudson@acctechnology.com | |
| WWW | http://www.acctechnology.com | |

## COMPUTONE CORPORATION

| | | |
|---|---|---|
| Address | 1100 Northmeadow Parkway<br>Suite 150<br>Roswell, GA 30076 | |
| Phone | (770) 475-2725 | Main |
| | (800) 241-3946 | Toll Free |
| | (800) 241-3946 X.250 | Tech Support |
| | (770) 664-1510 | Fax |
| | (770) 343-9737 | BBS |
| E-Mail | sales@computone.com | |
| WWW | http://www.computone.com | |

## COMTROL CORPORATION

| | | |
|---|---|---|
| Address | 900 Long Lake Road<br>Suite 210<br>St. Paul, MN 55112 | |
| Phone | (612) 631-7654 | Main |
| | (800) 926-6876 | Toll Free |
| | (800) 926-6876 | Tech Support |
| | (612) 631-8117 | Fax |
| | (612) 631-8310 | BBS |
| WWW | http://www.comtrol.com | |

### CONNECTWARE, INC.
| | | |
|---|---|---|
| **Address** | 1301 E. Arapahoe Road<br>Richardson, TX 75081 | |
| **Phone** | (214) 907-1093 | Main |
| | (800) 357-0852 | Toll Free |
| | (214) 907-1594 | Fax |
| **WWW** | http://www.connectware.com | |

### CONNECTWARE, INC.
| | | |
|---|---|---|
| **Address** | 3300 Gateway Center<br>Morrisville, NC 27560 | |
| **Phone** | (919) 481-5022 | Main |
| | (919) 481-0753 | Fax |

### CONNEXPERTS
| | | |
|---|---|---|
| **Address** | 2626 Lombardy Lane<br>Suite 107<br>Dallas, TX 75220 | |
| **Phone** | (214) 358-4800 | Main |
| | (800) 433-5373 | Toll Free |
| | (214) 351-6741 | Fax |
| | (214) 351-2136 | BBS |

### CREATIVE LABS, INC.
| | | |
|---|---|---|
| **Address** | 1901 McCarthy Blvd.<br>Milpitas, CA 95035 | |
| **Phone** | (408) 428-6600 | Main |
| | (800) 998-1000 | Toll Free |
| | (405) 742-6600 | Tech Support |
| | (408) 428-6611 | Fax |
| | (405) 742-6660 | BBS |
| **WWW** | http://www.creaf.com | |

### CREATIX POLYMEDIA
| | | |
|---|---|---|
| **Address** | 3945 Freedom Circle<br>Suite 670<br>Santa Clara, CA 95054 | |
| **Phone** | (408) 654-9300 | Main |
| | (800) 654-9559 | Tech Support |
| | (408) 654-9423 | Fax |
| | (408) 654-9920 | BBS |

## CSS LABORATORIES, INC.

| | | |
|---|---|---|
| **Address** | 1641 Mcgaw Avenue | |
| | Irvine, CA  92714 | |
| **Phone** | (714) 852-8161 | Main |
| | (800) 966-2771 | Tech Support |
| | (714) 852-0410 | Fax |
| | (714) 852-9231 | BBS |
| **WWW** | http://www.csslabs.com | |

## D-LINK

| | | |
|---|---|---|
| **Address** | 5 Musick | |
| | Irvine, CA  92618 | |
| **Phone** | (714) 455-1688 | Main |
| | (800) 326-1688 | Toll Free |
| | (800) 326-1688 | Tech Support |
| | (714) 455-2521 | Fax |
| | (714) 455-1779 | BBS |
| **E-Mail** | tech@irvine.dlink.com | |
| **WWW** | http://www.dlink.com | |

## DANPEX CORPORATION

| | | |
|---|---|---|
| **Address** | 1342 Ridder Park Drive | |
| | San Jose, CA  95131 | |
| **Phone** | (408) 437-7557 | Main |
| | (800) 452-1551 | Toll Free |
| | (408) 437-7557 | Tech Support |
| | (408) 437-7559 | Fax |
| **WWW** | http://www.danpex.com | |

## DATA RACE

| | | |
|---|---|---|
| **Address** | 12400 Network Blvd. | |
| | San Antonio, TX  78249 | |
| **Phone** | (210) 263-2113 | Main |
| | (210) 263-2075 | Fax |
| **WWW** | http://www.datarace.com | |

## DATAEXPERT CORPORATION

| | | |
|---|---|---|
| **Address** | 1178 Sonora Court | |
| | Sunnyvale, CA  94086 | |
| **Phone** | (408) 737-8880 | Main |
| | (800) 328-2397 | Toll Free |
| | (408) 737-8390 | Fax |
| **WWW** | http://www.dataexpert.com | |

## DATAPOINT CORPORATION
| | | |
|---|---|---|
| **Address** | 8400 Datapoint Drive<br>San Antonio, TX 78229-8500 | |
| **Phone** | (210) 593-7000 | Main |
| | (800) 733-1500 | Toll Free |

## DATUM, INC.
| | | |
|---|---|---|
| **Address** | 6781 Via Del Oro<br>San Jose, CA 95119-1360 | |
| **Phone** | (408) 578-4161 | Main |
| | (800) 348-0648 | Toll Free |
| | (408) 578-4165 | Fax |
| **WWW** | http://www.datum.com | |

## DAVID SYSTEMS, INC.
| | | |
|---|---|---|
| **Address** | 5400 Bayfront Plaza<br>Santa Clara, CA 95052 | |
| **Phone** | (408) 541-6000 | Main |
| | (408) 720-1337 | Fax |
| | (408) 720-0406 | BBS |

## DAYNA COMMUNICATIONS, INC.
| | | |
|---|---|---|
| **Address** | 849 West Levoy Drive<br>Salt Lake City, UT 84123 | |
| **Phone** | (801) 269-7200 | Main |
| | (800) 531-0600 | Toll Free |
| | (801) 269-7200 | Tech Support |
| | (801) 269-7363 | Fax |
| | (801) 269-7398 | BBS |
| **WWW** | http://www.dayna.com | |

## DIGI INTERNATIONAL, INC.
| | | |
|---|---|---|
| **Address** | 11001 Bren Road East<br>Minnetonka, MN 55343 | |
| **Phone** | (612) 912-3444 | Main |
| | (800) 344-4273 | Toll Free |
| | (612) 943-0579 | Tech Support |
| | (612) 912-4991 | Fax |
| | (612) 943-0550 | BBS |
| **E-Mail** | info@dgii.com | |
| **WWW** | http://www.digibd.com | |

## DIGITAL EQUIPMENT CORPORATION

**Address**  111 Powdermill Road
Maynard, MA 01754

**Phone**
| | |
|---|---|
| (508) 493-2211 | Main |
| (800) Digital | Toll Free |
| (800) 354-9000 | Tech Support |
| (508) 493-8780 | Fax |
| (508) 264-7227 | BBS |

**WWW**  http://www.digital.com

## E-TECH RESEARCH, INC.

**Address**  1800 Wyatt Drive
Suite 2
Santa Clara, CA 95054

**Phone**
| | |
|---|---|
| (408) 988-8108 | Main |
| (800) 328-5538 | Toll Free |
| (408) 988-8108 | Tech Support |
| (408) 988-8109 | Fax |
| (408) 988-3663 | BBS |

**WWW**  http://www.e-tech.com

## ECHO COMMUNICATIONS

**Address**  15 East 600 North
Suite 2
Logan, UT 84321

**Phone**
| | |
|---|---|
| (801) 753-9590 | Main |
| (800) 779-8420 | Toll Free |
| (801) 753-9629 | Fax |

**E-Mail**  lturner@onlinex.net

**WWW**  http://www.echousa.com

## EDIMAX COMPUTER COMPANY

**Address**  3350 Scott Blvd., Building. 15
Santa Clara, CA 95054

**Phone**
| | |
|---|---|
| (408) 496-1105 | Main |
| (800) 652-6776 | Toll Free |
| (408) 988-6092 | Tech Support |
| (408) 980-1530 | Fax |
| (408) 988-5904 | BBS |

**WWW**  http://www.edimax.com

## EFA CORPORATION

| | | |
|---|---|---|
| **Address** | 3040 Oakmead Village Drive<br>Santa Clara, CA 95051 | |
| **Phone** | (408) 987-5400 | Main |
| | (800) 800-3321 | Toll Free |
| | (408) 987-5412 | Tech Support |
| | (408) 987-5415 | Fax |
| | (408) 987-5418 | BBS |
| **WWW** | http://www.efacorp.com | |

## EICON TECHNOLOGY CORPORATION

| | | |
|---|---|---|
| **Address** | 14755 Preston Road<br>Suite 620<br>Dallas, TX 75240 | |
| **Phone** | (214) 239-3270 | Main |
| | (800) 803-4266 | Toll Free |
| | (214) 239-3304 | Fax |

## Electronic Industries Association (EIA)

| | | |
|---|---|---|
| **Address** | 2500 Wilson Boulevard #203<br>Arlington, VA 22201 | |
| **Phone** | (703) 907-7545 | Main |
| | (703) 907-7501 | FAX |
| **E-Mail** | prusher@eia.org | |
| **Web** | www.edif.org/edif/eia.html | |

## EMULEX CORPORATION

| | | |
|---|---|---|
| **Address** | 3535 Harbor Blvd.<br>Costa Mesa, CA 92626 | |
| **Phone** | (714) 662-5600 | Main |
| | (800) 854-7112 | Toll Free |
| | (714) 241-0792 | Fax |
| **WWW** | http://www.emulex.com | |

## EVENTIDE, INC.

| | | |
|---|---|---|
| **Address** | One Alsan Way<br>Little Ferry, NJ 07643 | |
| **Phone** | (201) 641-1200 | Main |
| | (800) 446-7878 | Toll Free |
| | (201) 641-1200 | Tech Support |
| | (201) 641-1640 | Fax |
| **WWW** | http://www.eventide.com | |

## EVEREX SYSTEMS, INC.

| | | |
|---|---|---|
| **Address** | 5020 Brandin Court<br>Fremont, CA  94538 | |
| **Phone** | (510) 498-1111 | Main |
| | (800) 821-0806 | Toll Free |
| | (510) 498-4411 | Tech Support |
| | (510) 683-2186 | Fax |
| | (510) 226-9694 | BBS |
| **WWW** | http://www.everex.com | |

## FARALLON COMPUTING, INC.

| | | |
|---|---|---|
| **Address** | 2470 Mariner Square Loop<br>Alameda, CA  94501 | |
| **Phone** | (510) 814-5100 | Main |
| | (510) 814-5000 | Tech Support |
| | (510) 814-5023 | Fax |
| | (510) 865-1321 | BBS |
| **E-Mail** | info@farallon.com | |
| **WWW** | http://www.farallon.com | |

## FORE SYSTEMS, INC.

| | | |
|---|---|---|
| **Address** | 174 Thorn Hill Road<br>Warrendale, PA  15086-7535 | |
| **Phone** | (412) 933-3444 | Main |
| | (412) 933-6200 | Fax |
| **E-Mail** | info@fore.com | |
| **WWW** | http://www.fore.com | |

## FRANKLIN TELECOMMUNICATIONS CORP.

| | | |
|---|---|---|
| **Address** | 733 Lakefield Road<br>Westlake Village, CA  91361 | |
| **Phone** | (805) 373-8688 | Main |
| | (800) 372-6556 | Toll Free |
| | (805) 373-8688 | Tech Support |
| | (805) 373-7373 | Fax |
| | (805) 495-5517 | BBS |
| **WWW** | http://www.ftel.com | |

### GENERAL DATACOMM, INC.
| | | |
|---|---|---|
| **Address** | 1579 Straits Turnpike<br>Po Box 1299<br>Middlebury, CT 06762-1299 | |
| **Phone** | (203) 574-1118 | Main |
| | (203) 758-9129 | Fax |
| **WWW** | http://www.gdc.com | |

### GMX, INC.
| | | |
|---|---|---|
| **Address** | 3223 Arnold Lane<br>Northbrook, IL 60062 | |
| **Phone** | (708) 559-0909 | Main |
| | (708) 559-0942 | Fax |

### GVC TECHNOLOGIES, INC.
| | | |
|---|---|---|
| **Address** | 400 Commons Way<br>Rockway, NJ 07866 | |
| **Phone** | (201) 579-3630 | Main |
| | (201) 579-2702 | Fax |
| **WWW** | http://www.gvc.com | |

### HARMONY MULTIMEDIA
| | | |
|---|---|---|
| **Address** | 1116 South Bixby Drive<br>City of Industry, CA 91745 | |
| **Phone** | (818) 968-8486 | Main |
| | (818) 968-7345 | Fax |

### HAYES MICROCOMPUTER PRODUCTS, INC.
| | | |
|---|---|---|
| **Address** | 5835 Peachtree Corners East<br>Norcross, GA 30092 | |
| **Phone** | (770) 840-9200 | Main |
| | (770) 441-1617 | Tech Support |
| | (770) 441-1238 | Fax |
| | (770) 446-6336 | BBS |
| **WWW** | http://www.hayes.com | |

### HEWLETT PACKARD COMPANY
| | | |
|---|---|---|
| **Address** | 11311 Chindan Blvd.<br>Mail Stop 516<br>Boise, ID 83714 | |
| **Phone** | (208) 396-6192 | Main |
| | (208) 396-6577 | Fax |
| **WWW** | http://www.hp.com | |

## HTI NETWORKS

**Address**  558 Oakmead  
Sunnyvale, CA 94086

**Phone**  (408) 245-3300  Main

**WWW**  http://www.acculan.com

## HYPERMEDIA INTERNATIONAL CORPORATION

**Address**  1523 S. Orchard Hill Lane  
Hacienda Heights, CA 91745

**Phone**  (818) 913-4020  Main  
(818) 913-0920  Fax

**E-Mail**  hypermedia@hypermed.com

**WWW**  http://www.hypermed.com

## I/O MAGIC CORPORATION

**Address**  9272 Jeronimo  
Suite 122  
Irvine, CA 92618

**Phone**  (714) 727-7466  Main  
(800) 607-7466  Toll Free  
(714) 727-3445  Tech Support  
(714) 727-7467  Fax  
(714) 727-3455  BBS

**WWW**  http://www.iomagic.com

## IBM CORPORATION

**Address**  Old Orchard Road  
Armonk, NY 10504

**Phone**  (914) 765-1900  Main  
(800) 426-2255  Toll Free  
(800) 772-2227  Tech Support  
(919) 517-0001  BBS

**WWW**  http://www.ibm.com

## IC INTRACOM USA

**Address**  550 Commerce Blvd.  
Oldsmar, FL 34677

**Phone**  (813) 855-0550  Main  
(800) 881-7325  Toll Free  
(813) 886-2283  Tech Support  
(813) 855-2545  Fax

## IET STARTECH

| | | |
|---|---|---|
| **Address** | 6185-F Jimmy Carter Blvd.<br>Norcross, GA  30071 | |
| **Phone** | (770) 242-5965 | Main |
| | (770) 368-1849 | Tech Support |

## IMC NETWORKS CORPORATION

| | | |
|---|---|---|
| **Address** | 16931 Millikan Avenue<br>Irvine, CA  92606 | |
| **Phone** | (714) 724-1070 | Main |
| | (800) 624-1070 | Toll Free |
| | (714) 724-1070 | Tech Support |
| | (714) 724-1020 | Fax |
| | (714) 724-0930 | BBS |
| **WWW** | http://www.imcnetworks.com | |

## INSTITUTE OF ELECTRICAL AND ELECTRONICS ENGINEERS (IEEE)

| | | |
|---|---|---|
| **Address** | 445 Hoes Lane<br>PO Box 1331<br>Piscataway, NJ 08855-1331 | |
| **Phone** | (800) 678-IEEE | Membership/Sales |
| **WWW** | http://www.ieee.org | |

## INTEL CORPORATION

| | | |
|---|---|---|
| **Address** | P.o. Box 58119<br>Santa Clara, CA  95052-8119 | |
| **Phone** | (408) 765-8080 | Main |
| | (800) 628-8686 | Tech Support |
| **WWW** | http://www.intel.com | |

## INTELLICOM, INC.

| | | |
|---|---|---|
| **Address** | 20415 Nordhoff Street<br>Chatsworth, CA  91311 | |
| **Phone** | (818) 407-3900 | Main |
| | (818) 882-2404 | Fax |
| **WWW** | http://www.intellicom.com | |

## INTERNATIONAL TELECOMMUNICATION UNION (ITU)

| | | |
|---|---|---|
| **Address** | CH-1211 Geneva 20<br>Switzerland | |
| **Phone** | +41 22 730 5852 | Main |
| | +41 22 730 5853 | FAX |
| **E-Mail** | tsbmail@itu.ch | |
| **WWW** | http://www.itu.ch | |

## INTERNATIONAL STANDARDS ORGANIZATION (ISO)

| | | |
|---|---|---|
| **Address** | 1, rue de Varembé<br>Case postale 56<br>CH-1211 Genève 20<br>Switzerland | |
| **Phone** | + 41 22 749 01 11 | Main |
| | + 41 22 733 34 30 | FAX |
| | + 41 22 05 iso ch | Telex |
| **E-Mail** | central@isocs.iso.ch | |
| **WWW** | http://www.iso.ch | |

## INTERPHASE CORPORATION

| | | |
|---|---|---|
| **Address** | 13800 Senlac Drive<br>Dallas, TX 75234 | |
| **Phone** | (214) 654-5000 | Main |
| | (800) 327-8638 | Toll Free |
| | (214) 654-5555 | Tech Support |
| | (214) 654-5500 | Fax |

## INVISIBLE SOFTWARE, INC.

| | | |
|---|---|---|
| **Address** | 1142 Chess Drive<br>Foster City, CA 94404 | |
| **Phone** | (415) 570-5967 | Main |
| | (800) 982-2962 | Toll Free |
| | (415) 570-6017 | Fax |
| | (415) 345-5509 | BBS |
| **WWW** | http://www.invisiblesoft.com | |

## INVISIBLE SOFTWARE, INC.

| | | |
|---|---|---|
| **Address** | 939 Longdale Avenue<br>Longwood, FL 32750 | |
| **Phone** | (407) 260-5200 | Main |
| | (800) 982-2962 | Toll Free |
| | (407) 260-5007 | Tech Support |
| | (407) 260-1841 | Fax |

## JC INFORMATION SYSTEMS CORPORATION

| | | |
|---|---|---|
| **Address** | 4487 Technology Drive<br>Fremont, CA 94538 | |
| **Phone** | (510) 659-8440 | Main |
| | (510) 659-8440 | Tech Support |
| | (510) 659-8449 | Fax |
| **E-Mail** | info@jcis.com | |
| **WWW** | http://www.kcis.com | |

## KATRON TECHNOLOGIES, INC.

| | | |
|---|---|---|
| **Address** | 7400 Harwin Drive<br>Suite 120<br>Houston, TX 77036 | |
| **Phone** | (713) 266-3891 | Main |
| | (800) 275-6387 | Toll Free |
| | (713) 266-3891 | Tech Support |
| | (713) 266-3893 | Fax |
| | (713) 266-3015 | BBS |
| **WWW** | http://www.ktinet.com | |

## KEYSONIC TECHNOLOGY, INC.

| | | |
|---|---|---|
| **Address** | 1040a S. Melrose Street<br>Placentia, CA 92670 | |
| **Phone** | (714) 632-8887 | Main |
| | (800) 779-1704 | Toll Free |
| | (714) 632-8868 | Fax |
| | (714) 632-8853 | BBS |

## KINGSTON TECHNOLOGY CORPORATION

| | | |
|---|---|---|
| **Address** | 17600 Newhope Street<br>Fountain Valley, CA 92708 | |
| **Phone** | (714) 435-2600 | Main |
| | (800) 435-2620 | Toll Free |
| | (714) 438-2796 | Tech Support |
| | (714) 438-2699 | Fax |
| | (714) 435-2636 | BBS |
| **WWW** | http://www.kingston.com | |

## KLEVER COMPUTERS, INC.

| | | |
|---|---|---|
| **Address** | 1841 Zanker Road<br>San Jose, CA 95112 | |
| **Phone** | (408) 467-0888 | Main |
| | (408) 467-0899 | Fax |
| **WWW** | http://www.klever.com | |

## KOUTECH SYSTEMS, INC.
| | | |
|---|---|---|
| **Address** | 9314 Norwalk Blvd.<br>Santa Fe Springs, CA 90670 | |
| **Phone** | (310) 699-5340 | Main |
| | (310) 699-0795 | Fax |
| | (310) 692-6798 | BBS |
| **WWW** | http://www.koutech.com | |

## KOUTECH SYSTEMS, INC.
| | | |
|---|---|---|
| **Address** | 1675 Walsh Avenue<br>Suite A<br>Santa Clara, CA 95050 | |
| **Phone** | (408) 727-8208 | Main |
| | (408) 727-8440 | Fax |

## KYE INTERNATIONAL CORPORATION
| | | |
|---|---|---|
| **Address** | 2605 East Cedar Street<br>Ontario, CA 91761-8511 | |
| **Phone** | (909) 923-3510 | Main |
| | (800) 456-7593 | Toll Free |
| | 909) 923-2417 | Tech Support |
| | (909) 923-5494 | Fax |
| | (909) 923-8454 | BBS |

## LANCAST
| | | |
|---|---|---|
| **Address** | 12 Murphy Drive<br>Nashua, NH 03062 | |
| **Phone** | (603) 880-1833 | Main |
| | (800) 952-6227 | Toll Free |
| | (800) 752-2768 | Tech Support |
| | (603) 881-9888 | Fax |

## LANTECH COMPUTER COMPANY
| | | |
|---|---|---|
| **Address** | 1566 La Pradera Drive<br>Campbell, CA 95008 | |
| **Phone** | (408) 866-8536 | Main |

## LEEMAH DATACOM
| | | |
|---|---|---|
| **Address** | 3948 Trust Way<br>Hayward, CA 94545 | |
| **Phone** | (510) 786-0790 | Main |
| | (510) 786-1123 | Fax |
| **WWW** | http://www.leemah.com | |

## LINKSYS

| | | |
|---|---|---|
| **Address** | 17401 Armstrong Avenue<br>Irvine, CA  92614 | |
| **Phone** | (714) 261-1288 | Main |
| | (800) 546-5797 | Toll Free |
| | (714) 261-1288 | Tech Support |
| | (714) 261-8868 | Fax |
| | (714) 222-5111 | BBS |
| **WWW** | http://www.linksys.com | |

## LITE-ON COMMUNICATIONS, INC.

| | | |
|---|---|---|
| **Address** | 720 S. Hillview Drive<br>Milpitas, CA  95035 | |
| **Phone** | (408) 945-4111 | Main |
| | (800) 785-4831 | Toll Free |
| | (408) 945-4110 | Fax |

## LIUSKI INTERNATIONAL, INC.

| | | |
|---|---|---|
| **Address** | 10 Hub Drive<br>Melville, NJ  11747 | |
| **Phone** | (516) 454-8220 | Main |
| | (800) 347-5454 | Toll Free |
| | (516) 454-8266 | Tech Support |
| | (516) 454-8266 | Fax |
| | (516) 454-8262 | BBS |

## LIUSKI INTERNATIONAL, INC.

| | | |
|---|---|---|
| **Address** | 6585 Crescent Drive<br>Norcross, GA  30071 | |
| **Phone** | (770) 447-9454 | Main |
| | (800) 454-8754 | Toll Free |
| | (800) 347-5454 | Tech Support |
| | (770) 441-1671 | Fax |
| | (770) 447-5454 | BBS |
| **WWW** | http://www.magitronic.com | |

## LOGICODE TECHNOLOGY, INC.

| | | |
|---|---|---|
| **Address** | 1380 Flynn Road<br>Camarillo, CA  93012 | |
| **Phone** | (805) 388-9000 | Main |
| | (800) 735-6442 | Toll Free |
| | (805) 482-0990 | Tech Support |
| | (805) 388-8991 | Fax |
| | (805) 445-9633 | BBS |
| **WWW** | http://www.logicode.com | |

## LONGSHINE MICROSYSTEM, INC.
| | | |
|---|---|---|
| **Address** | 10400-9 Pioneer Blvd. | |
| | Santa Fe Springs, CA  90670 | |
| **Phone** | (310) 903-0899 | Main |
| | (310) 903-0899 | Tech Support |
| | (310) 944-2201 | Fax |
| | (310) 903-4590 | BBS |

## MADGE NETWORKS, LTD.
| | | |
|---|---|---|
| **Address** | 2310 North First Street | |
| | San Jose, CA  95131-1011 | |
| **Phone** | (408) 955-0700 | Main |
| | (800) 876-2343 | Toll Free |
| | (800) 876-2343 | Tech Support |
| | (408) 955-0970 | Fax |
| | (408) 955-0262 | BBS |
| **WWW** | http://www.madge.com | |

## MAGICRAM, INC.
| | | |
|---|---|---|
| **Address** | 1850 Beverly Blvd. | |
| | Los Angeles, CA  90057 | |
| **Phone** | (213) 413-9999 | Main |
| | (800) 272-6242 | Toll Free |
| | (213) 413-9999 | Tech Support |
| | (213) 413-0828 | Fax |
| **WWW** | http://www.magicram.com | |

## MAXTECH CORPORATION
| | | |
|---|---|---|
| **Address** | 13915 Cerritos Corporate Drive | |
| | Cerritos, CA  90703 | |
| **Phone** | (310) 921-1698 | Main |
| **WWW** | http://www.maxcorp.com | |

## MEGAHERTZ CORPORATION
| | | |
|---|---|---|
| **Address** | 605 North 5600 West | |
| | Po Box 16020 | |
| | Salt Lake City, UT  84116 | |
| **Phone** | (801) 320-7000 | Main |
| | (800) LAPTOPS | Toll Free |
| | (801) 320-6020 | Tech Support |
| | (801) 320-6020 | Fax |
| | (801) 320-8840 | BBS |
| **WWW** | http://www.megahertz.com | |

## MICRO DIRECT

| | | |
|---|---|---|
| **Address** | 17895 Sky Park Circle<br>Suite F<br>Irvine, CA  92614 | |
| **Phone** | (714) 251-1818 | Main |
| | (714) 251-1877 | Fax |
| **WWW** | http://www.apache-micro.com | |

## MICRO HOUSE INTERNATIONAL, INC.

| | | |
|---|---|---|
| **Address** | 2477 55$^{th}$ Street<br>Suite 101<br>Boulder, CO  80301 | |
| **Phone** | (303) 443-3388 | Main |
| | (800) 926-8299 | Sales |
| | (303) 443-3389 | Technical Support |
| | (303) 443-3323 | FAX |
| | (617) 443-9957 | BBS |
| **E-Mail** | info@microhouse.com | |
| **WWW** | http://www.microcom.com | |
| **FTP** | ftp.microhouse.com | |

## MICROCOM, INC.

| | | |
|---|---|---|
| **Address** | 500 River Ridge Drive<br>Norwood, MA  02062 | |
| **Phone** | (617) 551-1000 | Main |
| | (800) 822-8224 | Toll Free |
| | (617) 551-1313 | Tech Support |
| | (617) 551-1968 | Fax |
| | (617) 255-1125 | BBS |
| **E-Mail** | sales@microcom.com | |
| **WWW** | http://www.microcom.com | |

## MICRODYNE CORPORATION

| | | |
|---|---|---|
| **Address** | 3601 Eisenhower Avenue, 3rd Floor<br>Alexandria, VA  22304 | |
| **Phone** | (703) 739-0500 | Main |
| | (800) 255-3967 | Toll Free |
| | (800) 255-3967 | Tech Support |
| | (703) 329-3716 | Fax |
| | (703) 960-8509 | BBS |
| **WWW** | http://www.mcdy.com | |

## MICROSOFT CORPORATION

| | | |
|---|---|---|
| **Phone** | (703) 635-2222 | Automated fax-back |
| **WWW** | http://www.microsoft.com | |

## MICROWISE, INC.

| | | |
|---|---|---|
| **Address** | 48057 Fremont Blvd. | |
| | Fremont, CA 94538 | |
| **Phone** | (510) 656-9881 | Main |
| | (510) 656-1996 | Fax |
| **WWW** | http://www.wiseland.com | |

## MITRON COMPUTER, INC.

| | | |
|---|---|---|
| **Address** | 574 Weddell Drive | |
| | Suite 8 | |
| | Sunnyvale, CA 94089 | |
| **Phone** | (408) 752-8989 | Main |
| | (800) 713-6888 | Toll Free |
| | (408) 752-8999 | Fax |
| | (408) 371-9786 | BBS |
| **E-Mail** | mitron@ix.netcom.com | |
| **WWW** | http://www.dus.net/mitron | |

## MNC INTERNATIONAL, INC.

| | | |
|---|---|---|
| **Address** | 2817 Anthony Lane South | |
| | Minneapolis, MN 55418-3254 | |
| **Phone** | (612) 788-1099 | Main |
| | (612) 788-1099 | Tech Support |
| | (612) 788-9365 | Fax |

## MOSES COMPUTERS, INC.

| | | |
|---|---|---|
| **Address** | 15466 Los Gatos Blvd. | |
| | Suite 201 | |
| | Los Gatos, CA 95032 | |
| **Phone** | (408) 358-1550 | Main |
| | (800) 306-6737 | Toll Free |
| | (408) 356-9049 | Fax |
| | (408) 358-3153 | BBS |
| **WWW** | http://www.moses-tm.com | |

## MOTOROLA, INC.
| | | |
|---|---|---|
| **Address** | 20 Cabot Blvd.<br>Mansfield, MA 02048-1193 | |
| **Phone** | (508) 261-4000 | Main |
| | (508) 337-8004 | Fax |
| **WWW** | http://www.mot.com | |

## MOTOROLA, INC.
| | | |
|---|---|---|
| **Address** | 1501 Woodfield Road<br>Suite 120 North<br>Schaumburg, IL 60173-3871 | |
| **Phone** | (847) 576-5000 | Main |

## MULTI-TECH SYSTEMS, INC.
| | | |
|---|---|---|
| **Address** | 2205 Woodale Drive<br>Mounds View, MN 55112 | |
| **Phone** | (612) 785-3500 | Main |
| | (800) 328-9717 | Toll Free |
| | (800) 328-9717 | Tech Support |
| | (612) 785-9874 | Fax |
| | (612) 785-3702 | BBS |
| **WWW** | http://www.multitech.com | |

## MULTIACCESS COMPUTING CORPORATION
| | | |
|---|---|---|
| **Address** | 5350 Hollister Avenue<br>Suite C<br>Santa Barbara, CA 93111 | |
| **Phone** | (805) 964-2332 | Main |
| | (805) 681-7469 | Fax |
| **E-Mail** | multiacc@silcom.com | |

## MYLEX CORPORATION
| | | |
|---|---|---|
| **Address** | 34551 Ardenwood Blvd.<br>Fremont, CA 94555-3607 | |
| **Phone** | (510) 796-6100 | Main |
| | (800) 776-9539 | Toll Free |
| | (800) 776-9539 | Tech Support |
| | (510) 745-7653 | Fax |
| | (510) 793-3491 | BBS |
| **WWW** | http://www.mylex.com | |

## NDC COMMUNICATIONS, INC.

| | | |
|---|---|---|
| **Address** | 265 Santa Ana Court | |
| | Sunnyvale, CA 94086 | |
| **Phone** | (408) 730-0888 | Main |
| | (800) 632-1118 | Toll Free |
| | (800) 323-7325 | Tech Support |
| | (408) 730-0889 | Fax |
| **WWW** | http://www.ndclan.com | |

## NETEDGE SYSTEMS, INC.

| | | |
|---|---|---|
| **Address** | P.o. Box 14993 | |
| | Research Triangle, NC 27709-4993 | |
| **Phone** | (919) 991-9000 | Main |
| | (800) 638-3343 | Toll Free |
| | (919) 991-9060 | Fax |
| **WWW** | http://www.netedge.com | |

## NETFRAME SYSTEMS, INC.

| | | |
|---|---|---|
| **Address** | 1545 Barber Lane | |
| | Milpitas, CA 95035 | |
| **Phone** | (408) 474-1000 | Main |
| | (408) 434-4194 | Tech Support |
| | (408) 434-4190 | Fax |

## NETWORK INTERFACE TECHNOLOGY CORPORATION

| | | |
|---|---|---|
| **Address** | 10200 W 75th Street | |
| | Suite 11a | |
| | Shawnee Mission, Ks 66204 | |
| **Phone** | (913) 789-7120 | Main |
| | (800) 509-3597 | Toll Free |
| | (913) 894-4058 | Fax |
| | (913) 894-8656 | BBS |
| **WWW** | http://www.addtron.com | |

## NETWORK PERIPHERALS

| | | |
|---|---|---|
| **Address** | 1371 Mccarthy Blvd. | |
| | Milpitas, CA 95035 | |
| **Phone** | (408) 321-7300 | Main |
| | (800) 674-8855 | Toll Free |
| | (408) 321-9218 | Fax |
| | (408) 321-9322 | BBS |
| **E-Mail** | boostime@npix.com | |
| **WWW** | http://www.npix.com | |

## NETWORK TECHNOLOGIES, INC.

| | | |
|---|---|---|
| **Address** | 1275 Danner Drive<br>Aurora, OH 44202 | |
| **Phone** | (216) 562-7070 | Main |
| | (216) 562-1999 | Fax |

## NETWORTH, INC.

| | | |
|---|---|---|
| **Address** | 8404 Esters Blvd<br>Irving, TX 75063 | |
| **Phone** | (214) 929-1700 | Main |
| | (800) 544-5255 | Toll Free |
| | (214) 929-6984 | Tech Support |
| | (214) 929-1720 | Fax |
| | (214) 929-4882 | BBS |
| **WWW** | http://www.compaq.com | |

## NETWORTH, INC.

| | | |
|---|---|---|
| **Address** | 61 Daggett Drive<br>San Jose, CA 95134 | |
| **Phone** | (408) 383-9300 | Main |
| | (800) 441-2642 | Tech Support |
| | (408) 303-0136 | Fax |

## NEW MEDIA CORPORATION

| | | |
|---|---|---|
| **Address** | 1 Technology, Building A<br>Irvine, CA 92618 | |
| **Phone** | (714) 453-0100 | Main |
| | (800) 227-3748 | Toll Free |
| | (714) 453-0314 | Tech Support |
| | (714) 453-0114 | Fax |
| | (714) 453-0214 | BBS |
| **WWW** | http://www.newmediacorp.com | |

## NEWBRIDGE NETWORKS, INC.

| | | |
|---|---|---|
| **Address** | 593 Herndon Parkway<br>Herndon, VA 20170 | |
| **Phone** | (703) 834-3600 | Main |
| | (800) 343-3600 | Toll Free |
| | (703) 471-7080 | Fax |
| **WWW** | http://www.vivid.newbridge.com | |

## NORTHERN TELECOM
| | | |
|---|---|---|
| **Address** | 6201 Oakton Street<br>Morton Grove, IL 60053 | |
| **Phone** | (708) 470-2054 | Main |
| | (800) 466-7835 | Toll Free |
| | (708) 470-3400 | Fax |

## NOVELL, INC.
| | | |
|---|---|---|
| **Address** | 1555 North Technology Way<br>Orem, UT 84097 | |
| **Phone** | (708) 470-2054 | Main |
| | (800) 453-1267 | Toll Free |
| **WWW** | http://www.novell.com | |

## OKIDATA
| | | |
|---|---|---|
| **Address** | 532 Fellowship Road<br>Mt. Laurel, NJ 08054 | |
| **Phone** | (609) 235-2600 | Main |
| | (800) OKIDATA | Toll Free |
| | (609) 273-0300 | Tech Support |
| | (609) 778-4184 | Fax |
| **WWW** | http://www.okidata.com | |

## OLIVETTI
| | | |
|---|---|---|
| **Address** | 22425 E. Appleway Avenue<br>Liberty Lake, WA 99019-9534 | |
| **Phone** | (509) 927-5600 | Main |
| | (800) 255-4319 | Toll Free |
| | (509) 927-5774 | Fax |

## OPTICAL DATA SYSTEMS
| | | |
|---|---|---|
| **Address** | 1101 East Arapahoe Road<br>Richardson, TX 75081 | |
| **Phone** | (214) 234-6400 | Main |
| | (214) 301-3873 | Fax |
| **WWW** | http://www.ods.com | |

## OSICOM TECHNOLOGIES, INC.
| | | |
|---|---|---|
| **Address** | 9020 Junction Drive<br>Annapolis Junction, MD 20701 | |
| **Phone** | (800) 367-2729 | Toll Free |
| | (301) 317-7535 | Fax |
| **WWW** | http://www.craycom.com | |

## PC CONNECTLAN

| | | |
|---|---|---|
| **Address** | 12155 Mora Drive<br>Suite 16<br>Santa Fe Springs, CA  90670 | |
| **Phone** | (310) 946-7768 | Main |
| | (310) 946-1929 | Fax |
| **WWW** | http://www.pcconnectlan.com | |

## PENRIL DATABILITY NETWORKS

| | | |
|---|---|---|
| **Address** | 1300 Quince Orchard Blvd.<br>Gaithersburg, MD  20878-4106 | |
| **Phone** | (301) 921-8600 | Main |
| | (800) 473-6745 | Toll Free |
| | (301) 921-8376 | Fax |
| **WWW** | http://www.penril.com | |

## PENRIL DATABILITY NETWORKS

| | | |
|---|---|---|
| **Address** | One Palmer Terrace<br>Carlstadt, NJ  07072 | |
| **Phone** | (201) 438-2400 | Main |
| | (800) 456-7844 | Toll Free |
| | (201) 438-2688 | Fax |

## PINACL COMMUNICATIONS, INC.

| | | |
|---|---|---|
| **Address** | Cross Westchester Executive Park<br>400 Executive Boulevard<br>Elmsford, NY  10523 | |
| **Phone** | (914) 345-8155 | Main |
| | (914) 345-2807 | Fax |

## PLAINTREE SYSTEMS

| | | |
|---|---|---|
| **Address** | Prospect Place<br>9 Hillside Avenue<br>Waltham, MA  02154 | |
| **Phone** | (617) 290-5800 | Main |
| | (800) 370-2724 | Toll Free |
| | (617) 290-0963 | Fax |
| **WWW** | http://www.plaintree.com | |

## POWERCOM AMERICA, INC.

| | | |
|---|---|---|
| Address | 1040a S. Melrose Street<br>Placentia, CA 92670 | |
| Phone | (714) 632-8889 | Main |
| | (800) 666-8931 | Toll Free |
| | (714) 632-8876 | Tech Support |
| | (714) 632-8868 | Fax |
| | (714) 632-8853 | BBS |
| WWW | http://www.powercom-USA.com | |

## PRACTICAL PERIPHERALS, INC.

| | | |
|---|---|---|
| Address | 5854 Peachtree Corner East<br>Norcross, GA 30092-3405 | |
| Phone | (770) 840-9966 | Main |
| | (770) 840-9966 | Tech Support |
| | (805) 734-4601 | Fax |
| | (770) 734-4600 | BBS |
| WWW | http://www.practinet.com | |

## PREMAX ELECTRONICS, INC.

| | | |
|---|---|---|
| Address | 750 North Mary Avenue<br>Sunnyvale, CA 94086 | |
| Phone | (408) 739-7000 | Main |
| | (408) 739-7000 | Tech Support |
| | (408) 739-7001 | Fax |
| | (408) 739-7071 | BBS |
| WWW | http://www.actiontec.com | |

## PROMETHEUS PRODUCTS, INC.

| | | |
|---|---|---|
| Address | 10110 Sw Nimbus Avenue<br>Suite B9<br>Portland, OR 97223 | |
| Phone | (503) 692-9600 | Main |
| | (503) 692-9601 | Tech Support |
| | (503) 691-1101 | Fax |
| | (503) 691-5519 | BBS |

## PROTEON, INC.
| | | |
|---|---|---|
| Address | Nine Technology Drive<br>Westborough, MA 01581-179 | |
| Phone | (508) 898-2800 | Main |
| | (800) 545-7464 | Toll Free |
| | (508) 898-3100 | Tech Support |
| | (508) 366-8901 | Fax |
| | (508) 898-2154 | BBS |
| WWW | http://www.proteon.com | |

## PROXIM, INC.
| | | |
|---|---|---|
| Address | 295 North Bernardo Avenue<br>Mountain View, CA 94043 | |
| Phone | (415) 960-1630 | Main |
| | (800) 229-1630 | Toll Free |
| | (415) 960-1630 | Tech Support |
| | (415) 960-1984 | Fax |
| | (415) 960-2419 | BBS |
| WWW | http://www.proxim.com | |

## PURE DATA, LTD. (SEE WILDCARD TECHNOLOGIES, INC.)

## PURETEK INDUSTRIAL CO., LTD.
| | | |
|---|---|---|
| Address | 44110 Old Warm Springs Blvd<br>Fremont, CA 94538 | |
| Phone | (510) 656-8083 | Main |
| | (510) 656-8085 | Fax |
| WWW | http://www.puretek.com | |

## Q LOGIC CORPORATION
| | | |
|---|---|---|
| Address | 3545 Harbor Blvd.<br>Costa Mesa, CA 92626 | |
| Phone | (800) 662-4471 | Toll Free |
| | (714) 668-5090 | Fax |
| | (714) 708-3170 | BBS |
| WWW | http://www.qlc.com | |

## QUADRANT COMPONENTS, INC.

| | | |
|---|---|---|
| **Address** | 4378 Enterprise Street<br>Fremont, CA 94538 | |
| **Phone** | (510) 656-9988 | Main |
| | (510) 252-6142 | Tech Support |
| | (510) 656-2208 | Fax |
| | (510) 656-6564 | BBS |
| **E-Mail** | quadrant@ix.netcom.com | |

## QUICKPATH SYSTEMS, INC.

| | | |
|---|---|---|
| **Address** | 46723 Fremont Blvd.<br>Fremont, CA 94538 | |
| **Phone** | (510) 440-7288 | Main |
| | (800) 995-8828 | Toll Free |
| | (510) 440-7285 | Tech Support |
| | (510) 440-7289 | Fax |
| | (510) 440-7284 | BBS |
| **E-Mail** | qpinfo@quickpath.com | |
| **WWW** | http://www.quickpath.com | |

## RACAL-DATACOM

| | | |
|---|---|---|
| **Address** | 1601 N. Harrison Parkway<br>P.o. Box 407044<br>Fort Lauderdale, FL 33340-7044 | |
| **Phone** | (305) 846-1601 | Main |
| | (800) Racal-55 | Toll Free |
| | (305) 846-5510 | Fax |
| **WWW** | http://www.racal.com | |

## RACAL INTERLAN, INC.

| | | |
|---|---|---|
| **Address** | 1025 Timber Creek Drive<br>Annapolis, MD 21403 | |
| **Phone** | (800) Lantalk | Main |
| | (410) 280-0471 | Fax |
| **WWW** | http://www.interlan.com | |

### RACORE COMPUTER PRODUCTS, INC.

| | | |
|---|---|---|
| **Address** | 2355 South 1070 West<br>Salt Lake City, UT 84119 | |
| **Phone** | (801) 973-9779 | Main |
| | (800) 635-1274 | Toll Free |
| | (801) 973-9779 | Tech Support |
| | (801) 973-2005 | Fax |
| | (801) 973-4228 | BBS |
| **WWW** | http://www.xmission.com/~racore | |

### RAD DATA COMMUNICATIONS

| | | |
|---|---|---|
| **Address** | 900 Corporate Drive<br>Mahwah, NJ 07430 | |
| **Phone** | (201) 529-1100 | Main |
| | (201) 529-5777 | Fax |
| **E-Mail** | market@radusa.com | |

### RAGULA SYSTEMS

| | | |
|---|---|---|
| **Address** | 404 East, 4500 South<br>Suite A22<br>Salt Lake City, UT 84107 | |
| **Phone** | (801) 281-3434 | Main |
| | (800) Ragula1 | Toll Free |
| | (801) 281-0317 | Fax |
| **WWW** | http://www.ragula.com | |

### RAYLAN CORPORATION

| | | |
|---|---|---|
| **Address** | 989 East Hillsdale Blvd.<br>Suite 290<br>Foster City, CA 94404 | |
| **Phone** | (415) 341-1376 | Main |
| | (415) 494-7844 | Fax |
| **WWW** | http://www.connectware2.com | |

### RHETOREX, INC.

| | | |
|---|---|---|
| **Address** | 151 Albright Way<br>Los Gatos, CA 95030 | |
| **Phone** | (408) 370-0881 | Main |
| | (408) 370-1171 | Fax |
| **WWW** | http://www.rhetorex.com | |

## SBE, INC.

| | | |
|---|---|---|
| **Address** | 4550 Norris Canyon Road | |
| | San Ramon, CA 94583-1369 | |
| **Phone** | (510) 355-2000 | Main |
| | (800) 925-2666 | Toll Free |
| | (800) 444-0990 | Tech Support |
| | (510) 355-2020 | Fax |
| | (510) 355-2048 | BBS |
| **E-Mail** | Netxpand@sbei.com | |
| **WWW** | http://www.sbei.com | |

## SIEMENS NIXDORF INFORMATIONS SYSTEMS, INC.

| | | |
|---|---|---|
| **Address** | 200 Wheeler Road | |
| | Burlington, MA 01803 | |
| **Phone** | (617) 273-0480 | Main |
| | (617) 221-0231 | Fax |

## SIIG, INC.

| | | |
|---|---|---|
| **Address** | 6078 Stewart Avenue | |
| | Fremont, CA 94538 | |
| **Phone** | (510) 657-8688 | Main |
| | (510) 353-7542 | Tech Support |
| | (510) 657-5962 | Fax |
| | (510) 353-7532 | BBS |

## SILICOM CONNECTIVITY SOLUTIONS, INC.

| | | |
|---|---|---|
| **Address** | 15311 Ne 90th Street | |
| | Redmond, WA 98052 | |
| **Phone** | (206) 882-7995 | Main |
| | (800) 474-5426 | Toll Free |
| | (206) 882-7995 | Tech Support |
| | (206) 882-4775 | Fax |
| | (206) 883-3274 | BBS |
| **E-Mail** | stephanie@silicom.wa.com | |
| **WWW** | http://www.silicom.wa.com | |

## SMART MODULAR TECHNOLOGIES

| | | |
|---|---|---|
| **Address** | 4305 Cushing Parkway | |
| | Fremont, CA 94538 | |
| **Phone** | (510) 623-1231 | Main |
| | (800) 956-Smart | Toll Free |
| | (510) 623-1434 | Fax |
| **WWW** | http://www.smartm.com | |

## SOLECTEK COMPUTER SUPPLY, INC.

| | | |
|---|---|---|
| **Address** | 6370 Nancy Ridge Drive<br>Suite 109<br>San Diego, CA 92121-3212 | |
| **Phone** | (619) 450-1220 | Main |
| | (800) 437-1518 | Toll Free |
| | (619) 457-2681 | Fax |
| **WWW** | http://www.solectek.com | |

## SONIC SYSTEMS, INC.

| | | |
|---|---|---|
| **Address** | 575 N. Pastoria Avenue<br>Sunnyvale, CA 94086 | |
| **Phone** | (408) 736-1900 | Main |
| | (800) 535-0725 | Toll Free |
| | (408) 736-1900 X105 | Tech Support |
| | (408) 736-7228 | Fax |
| **E-Mail** | Info@sonicsys.com | |
| **WWW** | http://www.sonicsys.com | |

## STANDARD MICROSYSTEMS CORPORATION

| | | |
|---|---|---|
| **Address** | 80 Arkay Drive<br>Hauppauge, NY 11788 | |
| **Phone** | (516) 435-6000 | Main |
| | (800) Smc-4you | Toll Free |
| | (800) 992-4762 | Tech Support |
| | (516) 273-1803 | Fax |
| | (516) 434-3162 | BBS |
| **E-Mail** | techsupt@ccmail.west.smc.com | |
| **WWW** | http://www.smc.com | |

## STAR LOGIC, INC.

| | | |
|---|---|---|
| **Address** | 830 Stewart<br>Sunnyvale, CA 94086 | |
| **Phone** | (408) 730-6938 | Main |
| | (800) 339-0920 | Toll Free |
| | (408) 730-8142 | Fax |
| **E-Mail** | sales@relia1.relia.com.tw | |
| **WWW** | http://relia1.relia.com.tw | |

## SUPRA DIAMOND

| | | |
|---|---|---|
| **Address** | 312 SE Stone Mill Drive, Suite 150<br>Vancouver, WA 98684 | |
| **Phone** | (360) 604-1400 | Main |
| | (800) 774-4965 | Toll Free |
| | (360) 604-1499 | Tech Support |
| | (360) 604-1401 | Fax |
| | (503) 967-2444 | BBS |
| **E-Mail** | supratech@supra.com | |
| **WWW** | http://www.supra.com | |

## SUPRA DIAMOND

| | | |
|---|---|---|
| **Address** | 7101 Supra Drive SW<br>Albany, OR 97321 | |
| **Phone** | (541) 967-2400 | Main |

## SVEC COMPUTER CORPORATION

| | | |
|---|---|---|
| **Address** | 2691 Richter Avenue, Suite 130<br>Irvine, CA 92714 | |
| **Phone** | (714) 756-2233 | Main |
| | (800) 756-Svec | Toll Free |
| | (714) 756-1340 | Fax |
| | (714) 724-9229 | BBS |
| **E-Mail** | Infor@svec.com | |
| **WWW** | http://www.svec.com | |

## SYSKONNECT, INC.

| | | |
|---|---|---|
| **Address** | 1922 Zanker Road<br>San Jose, CA 95112 | |
| **Phone** | (408) 437-3800 | Main |
| | (800) 752-3334 | Toll Free |
| | (408) 437-3857 | Tech Support |
| | (408) 437-3866 | Fax |
| | (408) 437-3869 | BBS |
| **E-Mail** | Info@syskonnect.com | |

## TARGET TECHNOLOGIES, INC.

| | | |
|---|---|---|
| **Address** | 6714 Netherlands Drive<br>Wilmington, NC 28405 | |
| **Phone** | (910) 395-6100 | Main |
| | (910) 395-6108 | Fax |
| **WWW** | http://www.cphone.com | |

## TCL, INC.
| | | |
|---|---|---|
| **Address** | 41829 Albrae Street<br>Fremont, CA  94538 | |
| **Phone** | (510) 657-3800 | Main |
| | (510) 490-5814 | Fax |
| **E-Mail** | 75070.660@compuserve.com | |

## TDK ELECTRONICS CORPORATION
| | | |
|---|---|---|
| **Address** | 1600 Feehanville Drive<br>Mount Prospect, IL  60056 | |
| **Phone** | (708) 803-6100 | Main |
| | (708) 803-6100 | Fax |

## TECHNOLOGY WORKS
| | | |
|---|---|---|
| **Address** | 4030 W. Braker Lane<br>Suite 350<br>Austin, TX  78759 | |
| **Phone** | (512) 794-8533 | Main |
| | (800) 688-7466 | Toll Free |
| | (512) 794-8520 | Fax |
| | (512) 794-9329 | BBS |
| **WWW** | http://www.techworks.com | |

## TEKNIQUE, INC.
| | | |
|---|---|---|
| **Address** | 911 North Plum Grove Road<br>Schaumburg, IL  60173 | |
| **Phone** | (847) 706-9700 | Main |
| | (847) 706-9735 | Fax |
| **WWW** | http://www.mmss.com | |

## TELEBIT CORPORATION
| | | |
|---|---|---|
| **Address** | One Executive Drive<br>Chelmsford, MA  01824 | |
| **Phone** | (508) 441-2181 | Main |
| | (800) 835-3248 | Toll Free |
| | (800) 835-3248 | Tech Support |
| | (508) 441-9060 | Fax |
| | (408) 745-3861 | BBS |
| **WWW** | http://www.telebit.com | |

## TELECOMMUNICATIONS INDUSTRY ASSOCIATION (TIA)

| | | |
|---|---|---|
| **Address** | 2500 Wilson Blvd | |
| | Arlington, VA 22201 USA | |
| **Phone** | (703) 907-7700 | Main |
| | (703) 907-7727 | FAX |

## THE NETWORKING COMPANY, PTE., LTD.

| | | |
|---|---|---|
| **Address** | 1840 County Line Road | |
| | Building #300 | |
| | Huntingdon Valley, PA 19006 | |
| **Phone** | (215) 355-4600 | Main |
| | (215) 953-5269 | Fax |
| **WWW** | http://www.mpcdirect.com | |

## THOMAS-CONRAD CORPORATION

| | | |
|---|---|---|
| **Address** | 12301 Technology Blvd. | |
| | Austin, TX 78727 | |
| **Phone** | (512) 433-6000 | Main |
| | (512) 433-6822 | Tech Support |
| | (512) 433-6153 | Fax |
| | (512) 433-6156 | BBS |
| **WWW** | http//www.tci.com | |

## TIARA COMPUTER SYSTEMS, INC.

| | | |
|---|---|---|
| **Address** | 2302 Walsh Avenue | |
| | Santa Clara, CA 95051 | |
| **Phone** | (408) 327-2389 | Main |
| **WWW** | http://www.internex.net | |

## TOP MICROSYSTEMS, INC.

| | | |
|---|---|---|
| **Address** | 3320 Victor Court | |
| | Santa Clara, CA 95054 | |
| **Phone** | (408) 980-9813 | Main |
| | (800) 827-8721 | Toll Free |
| | (408) 980-1183 | Tech Support |
| | (408) 980-8626 | Fax |

## TRANCELL SYSTEMS, INC.

| | | |
|---|---|---|
| **Address** | 3180 De La Cruz Blvd.<br>Suite 200<br>Santa Clara, CA 95054 | |
| **Phone** | (408) 988-5353 | Main |
| | (408) 988-5353 | Tech Support |
| | (408) 988-6363 | Fax |
| **E-Mail** | Info@trancell.com | |
| **WWW** | http://www.trancell.com | |

## TRANSITION ENGINEERING, INC.

| | | |
|---|---|---|
| **Address** | 6475 City West Parkway<br>Minneapolis, MN 55344 | |
| **Phone** | (612) 941-7600 | Main |
| | (800) 325-2725 | Toll Free |
| | (800) 260-1312 | Tech Support |
| | (612) 941-2322 | Fax |
| | (612) 941-9304 | BBS |
| **WWW** | http://www.transition.com | |

## TRENDWARE INTERNATIONAL, INC.

| | | |
|---|---|---|
| **Address** | 2421 W. 205th Street<br>Suite D102<br>Torrance, CA 90501 | |
| **Phone** | (310) 328-7795 | Main |
| | (310) 328-7798 | Fax |
| | (310) 328-8191 | BBS |
| **WWW** | http://www.trendware.com | |

## TTC COMPUTER PRODUCTS

| | | |
|---|---|---|
| **Address** | 3244 North Skyway Circle<br>Suite 102<br>Irving, TX 75038 | |
| **Phone** | (214) 594-8103 | Main |
| | (214) 594-8103 | Tech Support |
| | (214) 255-3174 | Fax |
| | (214) 594-6837 | BBS |
| **WWW** | http://www.infooncall.com | |

## TULIP COMPUTERS

| | |
|---|---|
| **Address** | Industriezone Mechelen Noord<br>Zandvoorstraat 4<br>2800 Mechelen, Belgium |

## TUT SYSTEMS

| | | |
|---|---|---|
| **Address** | 2495 Estand Way<br>Pleasant Hill, CA 94523 | |
| **Phone** | (510) 682-6510 | Main |
| | (800) 998-4888 | Toll Free |
| | (510) 682-4125 | Fax |
| **WWW** | http://www.tutsys.com | |

## U.S. ROBOTICS, INC.

| | | |
|---|---|---|
| **Address** | 7770 N. Frontage Road<br>Skokie, IL 60077-2690 | |
| **Phone** | (847) 676-7010 | Main |
| | (800) 342-5877 | Toll Free |
| | (847) 982-5151 | Tech Support |
| | (847) 676-7323 | Fax |
| | (847) 982-5092 | BBS |

## U.S. ROBOTICS, INC.

| | | |
|---|---|---|
| **Address** | 2070 Chain Bridge Road<br>Suite 100<br>Vienna, VA 22182 | |
| **Phone** | (703) 848-7700 | Main |
| | (703) 848-7888 | Fax |

## UNGERMANN-BASS NETWORKS, INC.

| | | |
|---|---|---|
| **Address** | 3900 Freedom Circle<br>Santa Clara, CA 95054 | |
| **Phone** | (408) 496-0111 | Main |
| | (800) 777-4lan | Toll Free |
| | (408) 970-7386 | Fax |
| **WWW** | http://www.ub.com | |

## UNICOM ELECTRIC, INC.

| | | |
|---|---|---|
| **Address** | 11980 Telegraph Road<br>Suite 103<br>Santa Fe Springs, CA 90670 | |
| **Phone** | (310) 946-9650 | Main |
| | (800) 346-6668 | Toll Free |
| | (310) 946-9167 | Fax |
| **WWW** | http://www.unicomlink.com | |

## UPDATE TECHNOLOGY, INC.

| | | |
|---|---|---|
| **Address** | 6800 Orangethorte Avenue<br>Suite A<br>Buena Park, CA 90620 | |
| **Phone** | (714) 522-2210 | Main |
| | (714) 522-0990 | Fax |

## VERILINK

| | | |
|---|---|---|
| **Address** | 145 Baytech Drive<br>San Jose, CA 95134 | |
| **Phone** | (408) 945-1199 | Main |
| | (408) 262-6260 | Fax |
| **WWW** | http://www.verilink.com | |

## VISIONTEK

| | | |
|---|---|---|
| **Address** | 1175 Lakeside Drive<br>Gurnee, IL 60031 | |
| **Phone** | (847) 360-7500 | Main |
| | (800) 726-9695 | Toll Free |
| | (847) 360-7144 | Fax |
| | (847) 360-7432 | BBS |
| **WWW** | http://www.visiontek.com | |

## WESTERN DATACOM CO., INC.

| | | |
|---|---|---|
| **Address** | 959 Bassett Road<br>Westlake, OH 44145-0113 | |
| **Phone** | (216) 835-1510 | Main |
| | (800) 262-3311 | Toll Free |
| | (216) 835-9146 | Fax |
| **WWW** | http://www.western-data.com | |

## WILDCARD TECHNOLOGIES, INC.

| | | |
|---|---|---|
| **Address** | 180 West Beaver Creek Road<br>Richmond Hill on Canada L4B 1B4 | |
| **Phone** | (905) 731-6444 | Main |
| | (800) 661-8210 | Toll Free |
| | (905) 731-4533 | Tech Support |
| | (906) 731-7017 | Fax |
| | (905) 731-4679 | BBS |
| **E-mail** | support@wildcardtech.com | |
| **WWW** | http://www.wildcardtech.com | |

## XINETRON, INC.

| | | |
|---|---|---|
| Address | 3022 Scott Blvd.<br>Santa Clara, CA 95054 | |
| Phone | (408) 727-5509 | Main |
| | (800) 345-4415 | Toll Free |
| | (408) 727-6499 | Fax |

## XIRCOM, INC.

| | | |
|---|---|---|
| Address | 2300 Corporate Center Drive<br>Thousand Oaks, CA 91320-1420 | |
| Phone | (805) 376-9300 | Main |
| | (800) 438-4526 | Toll Free |
| | (805) 376-9200 | Tech Support |
| | (805) 376-9311 | Fax |
| | (805) 376-9130 | BBS |
| E-Mail | cs@xircom.com | |
| WWW | http://www.xircom.com | |

## XYLOGICS, INC.

| | | |
|---|---|---|
| Address | 53 Third Avenue<br>Burlington, MA 01803 | |
| Phone | (617) 272-8140 | Main |
| | (800) 225-3317 | Toll Free |
| | (617) 273-5392 | Fax |
| E-Mail | sales@xylogics.com | |
| WWW | http://www.bayweb.com | |

## ZEITNET, INC.

| | | |
|---|---|---|
| Address | 5150 Great America Parkway<br>Santa Clara, CA 95054 | |
| Phone | (408) 986-9100 | Main |
| | (408) 562-1889 | Fax |
| | (408) 986-9833 | BBS |
| E-Mail | info@zeitnet.com | |
| WWW | http://www.zeitnet.com | |

## ZENDEX CORPORATION

| | | |
|---|---|---|
| Address | 6780 Sierra Court<br>Suite A<br>Dublin, CA 94568 | |
| Phone | (510) 828-3000 | Main |
| | (510) 828-1574 | Fax |
| WWW | http://www.zendex.com | |

## ZENITH DATA SYSTEMS

**Address**   Linke Wienzeile 192
              1150 Wien, Austria

## ZERO ONE NETWORKING

**Address**   4920 East La Palma Avenue
              Anaheim, CA 92807

**Phone**     (714) 693-0804        Main
              (800) 255-4101        Toll Free
              (714) 693-0804        Tech Support
              (714) 693-8811        Fax
              (714) 693-0762        BBS

**WWW**       http://www.zyxel.com

## ZNYX CORPORATION

**Address**   48501 Warm Springs Blvd.
              Suite 107
              Fremont, CA 94539

**Phone**     (510) 249-0800        Main
              (800) 724-0911        Tech Support
              (510) 656-2460        Fax
              (510) 656-7969        BBS

**WWW**       http://www.znyx.com

## ZOLTRIX, INC.

**Address**   47273 Fremont Blvd.
              Fremont, CA 94538

**Phone**     (510) 657-1188        Main
              (510) 657-1280        Fax
              (510) 657-7413        BBS

**WWW**       http://www.zoltrix.com

## ZOOM TELEPHONICS, INC.

**Address**   207 South Street
              Boston, MA 02111

**Phone**     (617) 423-1072        Main
              (800) 666-6191        Toll Free
              (617) 423-1076        Tech Support
              (617) 423-5536        Fax
              (617) 423-3733        BBS

**APPENDIX D**

# Bonus
# CD-ROM Programs

## Appendix D Contents:

Part 1: Micro House Technical Library ........................ 458

Part 2: FCC ID Finder ................................................... 459

Part 3: Micro House Multimedia Product Demo ........... 460

This appendix covers general installation and set up instructions for the programs contained in the bundled CD-ROM.

There are three programs included on the CD-ROM:

- The Micro House Technical Library of Network Cards.
- FCC ID Finder
- Micro House Multimedia Product Demonstration

These programs require Windows 3.1x or above. Please read the following sections for details on installation and use of these programs.

# The Micro House Technical Library of Network Interface Cards

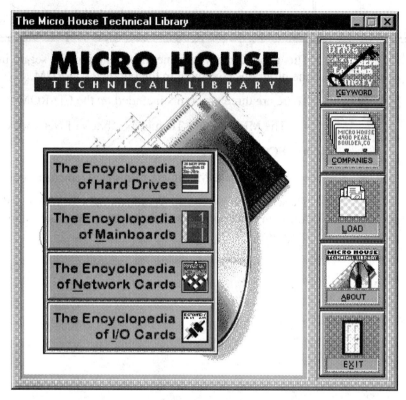

The Technical Library of NICs is a fully operational, NIC only version of the acclaimed MTL hardware support tool.

This program requires Windows 3.1x or above. To install, access the CD-ROM, enter the root directory, and run INSTALL. For example, if your CD is drive D:, then call up the Windows Run dialog and type D:\INSTALL in the entry field. Click on the OK button and the setup process will begin.

For more information on this product, please consult the user's manual, which is contained in the MANUAL directory in Adobe Acrobat™ format (.PDF), along with the Acrobat Reader. Once installed, extensive online help is also available.

# FCC ID Finder

The FCC ID Finder provides easy reference to electronic hardware manufacturers. Simply type in the FCC identification number for an item (printed directly on the PCB) – the FCC ID Finder will return information on the manufacturer of the item.

The MTL installation process (see the "Micro House Technical Library of Network Interface Cards" section) will present you with the option to create an FCC ID Finder icon in the Micro House Technical Library program group. Check the box to enable this option. If you do not wish to install the MTL altogether, the FCC ID Finder may still be accessed under the \FCCID directory of the CD-ROM.

This program requires Windows 3.1x or above. To use the FCC ID Finder, access the CD-ROM, enter the \FCCID directory, and run FCCID16.EXE (16-bit version) or FCCID32.EXE (32-bit version). For example, if your CD is drive D:, then call up the Windows Run dialog and type D:\FCCID\FCCID16 or D:\FCCID\FCCID32. Click on the OK button to execute the FCC ID program.

# Micro House Multimedia Product Demonstration

The Micro House Product demonstration is a multimedia introduction to Micro House and its award-winning products. To access the demos, access the root directory of the CD-ROM and run `DEMO.EXE`. For example, if your CD is drive `D:`, then call up the Windows Run dialog and type `D:\DEMO` in the entry field. Click on the OK button to begin the demonstration program.

# Glossary of Terms

# 1

**1Base5**

An IEEE 802.3 1Mbps CSMA/CD LAN standard using twisted pair wiring and linear bus or star topology. 1Base5 was developed by AT&T under the name Starlan.

*Also see CSMA/CD, IEEE 802.x*

**10Base2**

An IEEE 802.3 10Mbps CSMA/CD LAN standard using thin coaxial cable and linear bus or star topology. 10Base2 is also known as Cheapernet or ThinNet.

*Also see CSMA/CD, IEEE 802.x*

**10Base5**

An IEEE 802.3 10Mbps CSMA/CD LAN standard using 50 Ohm coaxial cabling and linear bus or star topology. 10Base5 is also known as Thicknet.

*Also see CSMA/CD, IEEE 802.x*

**10BaseF**

An IEEE 802.3 10Mbps CSMA/CD LAN standard using fiber-optic cabling and star topology. 10BaseF further divided into three subsections: 10BaseFB covers backbone hubs, 10BaseFL covers active hubs segments and transceivers, and 10BaseFP covers passive hub segments and transceivers.

*Also see CSMA/CD, Fiber-Optic, IEEE 802.x*

**10BaseT**

An IEEE 802.3 10Mbps CSMA/CD LAN standard using twisted pair wiring and star topology.

*Also see CSMA/CD, IEEE 802.x*

**100BaseT**

An IEEE 802.3 100Mbps CSMA/CD LAN standard using twisted pair wiring and star topology. 100BaseT is also known as Fast Ethernet.

*Also see CSMA/CD, IEEE 802.x*

**100VG-AnyLAN**

An IEEE 802.12 100Mbps LAN standard developed by Hewlett Packard designed to be compatible with Token Ring and Ethernet networks. 100VG-AnyLAN is also known as 100BaseVG.

*Also see CSMA/CD, IEEE 802.x*

# A

**ACCESS METHOD**

A protocol that establishes how a network device interfaces with the physical media in order to transmit information. For example, CSMA/CD is the access method used with IEEE 802.3 Ethernet networks.

**ADF**

Adapter Definition File

An adapter definition file having the file name extension .ADF. It contains base address, interrupts, and other pertinent information about a Micro Channel adapter that is required to configure a device on the Micro Channel Bus.

**ADDRESS**

A location in memory, or a node location on the network.

**ACTIVE HUB**

A device used in network topologies to amplify transmission signals. It is used when expanding the distance of a network or bridging to another network.

**ACTIVE MONITOR**

A node in a Token Ring network that originates the token.

**ANSI**

American National Standards Institute.

An organization that establishes standards over a wide range within the computer industry, from programming languages to network communications. It is the United States representative to ISO.

*See ISO.*

**APPLICATION LAYER**

*See OSI Model.*

## APPLICATION SERVER

A computer on the network that shares (provides access to) software programs residing on its mass storage media. Database applications are often run on these servers.

## APPLETALK

A comprehensive networking system developed by Apple Computer, Inc. for use in their computers. AppleTalk works with the proprietary LocalTalk access method, as well as some of the more popular standards. Appletalk communication protocols are built into the Macintosh operating system.

*Also see LocalTalk*

## ARCNET

A popular token-passing LAN architecture developed by the Datapoint Corporation. Standard data transfer rate is 2.5 Mbps.

## ARPA

Advanced Research Projects Agency

The U.S. government agency credited with laying the foundation for the modern Internet by establishing ARPANET and the TCP/IP data communication standards. ARPA was reformed into the Defense Advanced Research Projects Agency (DARPA)

*Also see TCP/IP.*

## ASCII, Extended ASCII

American Standard Code for Information Interchange

An eight bit binary code set commonly used to represent alphanumeric and special characters on personal computers. Each binary code corresponds to one of 256 possible characters.

*Also see EBCDIC.*

## ASYNCHRONOUS

Data that is delimited with start and stop bits instead of a timing signal.

*Also see synchronous.*

### ATM

Asynchronous Transfer Mode

A high-speed cell-switched communications protocol. ATM is part of the Broadband ISDN (PRI) specification.

*Also see cell switching, ISDN, PRI.*

### ATM FORUM

An ad hoc group of private corporations that focuses on developing and promoting ATM networking standards.

*Also see ATM.*

### ATTENUATION

Signal weakening over the length of the transmission medium.

### AUI

Attachment Unit Interface

The DB-15 port on an Ethernet card. The AUI is used as a universal connector to any Ethernet cabling media through the use of an appropriate external transceiver. The AUI ports most common use is to connect to 10Base5 (thick-net) and FOIRL (fiber optic), although it can be used to connect to 10BaseT (UTP) and 10Base2 (thin-net). The maximum distance between the NIC AUI port and the external transceiver is 164 feet (50m).

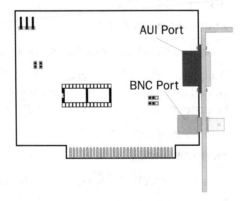

*NIC with AUI Port*

### AWG

American Wire Gauge

A specification for the diameter of wire. Wider diameters correspond to lower AWG numbers.

# B

### B-CHANNEL

A channel used by ISDN for data.

*Also see BRI, ISDN, D-Channel.*

### BABY BELLS

The regional telephone companies that were formed from the government-ordered breakup of AT&T.

### BACKBONE

The core transmission elements of a network, usually referring to the cabling.

### BALUN

Stands for "Balanced/Unbalanced." This is a connector that matches the impedance between balanced media, like twisted pair, and unbalanced media, like coax. Balanced means that electrical elements in the cable have equal characteristics.

### BANDWIDTH

The capacity of a network to carry information. The usual unit for measuring bandwidth is the number of bits per second (bps) that the network can transmit. The higher the bandwidth, the greater the capacity of the network, and the faster data can be transmitted from one node to another.

### BASE ADDRESS

The location in the host systems memory area used to establish communications with the NIC.

### BASEBAND

A method of data transmission in which information is encoded and impressed on the transmission medium without altering the frequency of the information. Baseband transmissions occupy the entire bandwidth of the medium, using TDM when line sharing is required.

*Also see broadband, TDM.*

### BASEBAND ISDN

*See ISDN.*

**BIT**

BInary digiT

The smallest unit of digital information. A bit can only be a 1 or 0.

**BNC**

British Naval Connector

A connector used with most networks that utilize coaxial cable.

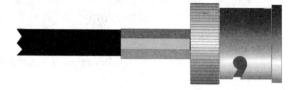

*BNC*

**BNC-T CONNECTOR**

A connector used with coaxial cable. In most network installations BNC-Ts are used to connect to the individual NICs.

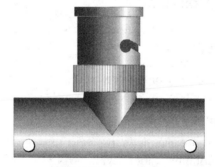

*BNC-T Connector*

**BOOT ROM**

An optional non-volatile memory chip that is used on diskless workstations to load the operating system from a remote server.

**BPS**

Bits Per Second. A measurement of data transfer speed or throughput.

### BRI

Basic Rate Interface

Part of the Baseband ISDN specification. The name for the version of ISDN that uses two "B" channels for data and one "D" channel for control signals.

*Also see ISDN, PRI.*

### BRIDGE

Software and hardware used to connect two or more separate networks.

*Also see internal and external bridge.*

### BROADBAND

In contrast to baseband, media can carry multiple carrier signals simultaneously. Each signal is transmitted over the medium with a different frequency range. This is the technique used by cable television companies. Each channel is broadcast over the coax at a different frequency.

*Also see baseband, FDM.*

### BROADBAND ISDN

*See ISDN.*

### BUFFER

A temporary storage area in random access memory (RAM) where the NIC or computer stores information (usually while transmitting or receiving network traffic).

### BUS

A path or channel with one or more conductors through which all devices communicate.

### BUSY TOKEN

A token (frame) that is passed around the network.

*Also see token passing.*

### BYTE

A logical grouping of eight bits. In most systems a byte is roughly equivalent to one character.

# C

**CCITT**

Consultive Committee in International Telegraphy and Telephony

*See ITU.*

**CELL**

A fixed-length data packet.

**CELL SWITCHING**

A method of transmitting data using fixed-length data packets.

**CENTRAL OFFICE**

A station used by the telephone company for switching local voice and data communications. Sometimes called an *exchange*.

**CENTRALIZED PROCESSING**

A networking system in which the workstations contain no native logic (dumb terminal), instead depending on a central computer, typically a mainframe, to perform all processing tasks. Contrast with *Distributed Processing*.

**CHANNEL**

The path used by a signal for transmission between two points

**CIRCUIT**

A closed-loop electrical connection.

**CIRCUIT SWITCHING**

A method of signal transmission in which a dedicated point-to-point circuit is established for the duration of an information exchange session. Telephone connections are circuit switched.

*Also see switching, packet switching.*

**CLIENT**

A device on a network that requests some service to be performed by another device on a network.

**CLADDING**

The layer encasing the inner glass core of the fiber optic cable. This is typically glass with a low refractive index that makes it an excellent reflective medium.

## COAXIAL CABLE

An electrical cable consisting of a wire surrounded by a cylindrical conductor, both of which have the same center or axis, hence the term "coaxial." Separated from the central wire by insulation, the cylindrical conductor forms a shield for the electronic impulses traveling along the central wire. The cylindrical conductor is also surrounded by insulation.

## COLLISION

A network condition that occurs when two devices attempt to transmit simultaneously on a single communications channel.

## CONCENTRATOR

A type of hub that is a central wiring center for 10BaseT (twisted pair Ethernet) networks. All workstations connect to this wiring center to form a star topology.

## CONTENTION PROTOCOLS

Protocols that control access to the network cable when all workstations are competing to use the same line.

*See CSMA/CD.*

## CPS

Characters Per Second. A measurement of speed or throughput.

## CROSSTALK

Interference caused by the leakage or overlap of electronic signals from one channel to another, usually measured in decibels. Crosstalk often occurs when cables are crossed over each other.

## CRC

Cyclic Redundancy Checking

A data-link level error checking method in which error detection information is calculated by dividing the data stream by an agreed upon value. The receiving node compares its calculation to that transmitted by the sending node, with a match indicating error-free transmission.

*Also see Logical Redundancy Checking (LRC), Vertical Redundancy Checking (VRC) and Data Link Layer.*

## CSMA/CA

Carrier Sense Multiple Access with Collision Avoidance

A network data control method that allows multiple stations (multiple access) to gain access to a network by listening until no signals are detected (carrier sense), and then signaling their intent to transmit before actually transmitting. If two or more devices try to transmit at the same time, then they cease transmitting and re transmission occurs after a randomly selected delay (collision avoidance).

## CSMA/CD

Carrier Sense Multiple Access with Collision Detection

A network data control method that allows multiple stations to gain access to a network (multiple access) and each device monitors its own transmission (carrier sense). If it does not receive what it sent then it decides that a collision must have occurred and then ceases its transmissions (collision detection). It will then retry after a randomly selected delay. The IEEE 802.3 Ethernet protocol uses the CSMA/CD technique.

# D

## D-CHANNEL

A channel used by ISDN for control signals.

*Also see BRI, ISDN, B-Channel.*

## DAISY CHAIN TOPOLOGY

A variation of the linear bus topology in which workstations tap into the linear cable using twisted pair wiring with modular jacks rather than coaxial cable with T-connectors.

## DATA LINK LAYER

*See OSI Model.*

## DATA TRANSFER RATE

The rate, expressed in bits per second (BPS), that information is transmitted across a network. Also called throughput and data rate.

## DB-9/DB-15/DB-25

Connectors that are formed in a "D" shape. The number after "DB" indicates the number of pins used in the connector.

## DARPA

Defense Advanced Research Projects Agency

*See ARPA.*

## DEFACTO STANDARD

A standard established by dominant use within the marketplace rather than by standards organizations. TCP/IP is the defacto standard for the Internet.

## DEVICE DRIVER

A file that loads in the CONFIG.SYS file that controls the interface between DOS and external devices such as NICs. Only some NICs require a device driver to be loaded.

## DIP SWITCH

Dual In-line Poles

## DISKLESS WORKSTATION

A low-end workstation that has minimal native capabilities, relying almost entirely on network resources.

## DISTRIBUITED PROCESSING

A networking system in which the workstations are able to perform local processing chores. Contrast with *Central Processing*.

## DIX (CONNECTOR)

Named after the three companies that cooperated to develop the original Ethernet standard (DEC, Intel, and Xerox). This is the same connector that is referred to as the AUI port, a DB-15 connector.

*Also see AUI.*

## DMA

Direct Memory Access

Used by NICs and other peripherals to directly access system memory without system processor intervention.

## DOWN

Not functioning.

## DROP CABLE

Cable that connects the main trunk cabling to hubs and workstations. The main trunk cabling is often concealed in the ceilings, thus the cables tapping into it are "dropped" from the ceiling.

# E

## E-MAIL

Electronic Mail

A method of transmitting messages over the network. One popular e-mail protocol is SMTP, an Internet standard.

*See SMTP.*

## EBCDIC

Extended Binary Coded Decimal Interchange Code

Created by IBM in 1960, this is an eight-bit code set used to represent alphanumeric and special characters. Similar to ASCII but generally found on mainframes.

*See ASCII.*

## EEPROM

Electronically Erasable Programmable Read Only Memory

A type of memory that can be reprogrammed through software. When power is removed from the EEPROM it will still retain its memory.

## EIA/TIA

Electronics Industries Association / Telecommunications Industries Association

An organization of American manufacturers that establishes electrical standards and wiring specifications.

## EISA

Extended Industry Standard Architecture. A type of bus used in PCs. It is 32-bits wide, and is also compatible with ISA cards. An alternative to IBM's MCA.

## ETHERNET

A media access protocol for LANs based on a collision detection method. Ethernet also defines the physical transmission medium and the method of packet signaling. It is most often used with the linear bus network topology. Ethernet is standardized under the IEEE 802.3 specification.

*Also see collision, CSMA/CD.*

## ETHERNET CABLE

A term that refers to thick or thin coaxial cable used in an Ethernet network.

## ETHERTALK

Apple Computer's data link product that allows an AppleTalk network to use Ethernet cable.

## EXTERNAL BRIDGE

A bridge created in a workstation by joining like or unlike network topologies. This is usually accomplished by placing two NICs into one workstation.

*Also see bridge and internal bridge.*

# F

## FCC

Federal Communications Commission

A government organization that regulates network pricing, profits and company mergers both in and across states to protect consumers and encourage open system competition.

## FDDI

Fiber Distributed Data Interface

A high-speed LAN access standard (OSI layers 1 and 2) that uses fiber optic cable and a token-passing protocol. **FDDI-2** adds sound and video channels in addition to data.

*Also see OSI Model.*

## FDDI-2

*See FDDI*

**FDM**

Frequency Division Multiplexing

A method of dividing available bandwidth in a transmission medium by modulating different signals into separate frequency ranges. Broadband uses this method for creating separate channels in the same physical media.

*Also see baseband, broadband, TDM.*

**FFDT**

FDDI Full-Duplex Technology

A full-duplex, 200Mbps version of FDDI.

*Also see FDDI.*

**FIBER CHANNEL**

A proposed standard for a common internal and external computer data bus. Fiber Channel can be used to connect peripherals, as well as perform network backbone functions.

**FIBER-OPTIC**

A data transmission medium in which information is transmitted through glass fiber cables as pulsed light signals.

Fiber-optic cable transmits at over 100 Mbps through single fiber (**monomode**) or several fibers (**multimode**). Standard multimode fiber (**step index fiber**) is the most widely used, however a variation (**graded-index fiber**) transmits at high rates over greater distances.

**FILE SERVER**

A computer that stores common files and applications used over a network. It is equipped with relatively large amounts of mass storage and provides file-handling services that allow it to regulate access to these files.

**FIRMWARE**

A program that is in permanent Read Only Memory on the NIC.

**FLOW CONTROL**

Data-link level protocols that coordinate data packet transmission to avoid overloading the receiving node. A dedicated communication link between sending and receiving nodes is required so data transmissions can be acknowledged or retransmission requested.

## FOIRL

Fiber Optic Inter-Repeater Link

An IEEE Ethernet standard that governs how AUI signals are converted into light pulses. The standard is for NICs that have fiber optic signaling capacities, but no AUI port.

## FRAME

A basic unit of data transmission in some network protocols. For example, in a Token ring network a frame is created when data and control information are appended to a token.

*Also see cell, packet.*

## FTP

One of many task-oriented TCP/IP protocols that operates at the Application Level. It is the standard for file transfer across the Internet.

*See TCP/IP.*

# G

## GATEWAY

A network station that interconnects two incompatible networks or devices.

## GRADED-INDEX FIBER

*See Fiber Optic.*

## GSTN

General Switched Telephone Network

*See PSTN*

# H

## HDLC

High Level Data Link Control

The serial data link protocol used in X.25 networks. Produced by International Standards Organization (ISO), it provides error correction at the data link layer.

## HOST

The central computer in a network system, usually more powerful than all the other on the network, that provides services to the other network devices.

## HUB

A multiport repeater for twisted pair networks.

*See also repeater and MAU.*

## Hz

Hertz

A measurement of frequency in cycles per second. A KHz is one thousand cycles per second, a MHz is one million cycles per second, and a GHz is one billion cycles per second.

# I

## IEEE

Institute of Electrical and Electronics Engineers

An organization that maintains standards for many of the protocols used for network communications.

## IEEE 802.X

The standards created by IEEE to establish interface and protocol specifications for various LAN topologies. 802.2 deals with logical link control. IEEE 802.3 (Ethernet) and 802.5 (Token Ring) are the most popular parts; both correspond to the physical layer and part of the data link layer (layer 2) of the OSI model.

*Also see CSMA/CD, Ethernet, MAC, OSI Model, Token Ring*

## IMPEDANCE

The electrical property of network media (cable) that combines capacitance, inductance, and resistance. Measured in ohms.

## INTERNET

A wide area network (WAN) originally developed by the Department of Defense (DOD) to link various research sites over phone lines. Today, it links computers around the globe for both commercial and personal use.

## INTRANET

A private office data communications system based on TCP/IP (Internet) applications. Intranets are sometimes called office or private internets.

## INTERNAL BRIDGE

A bridge created in a file server joining like or unlike network topologies. This is usually accomplished by installing two or more NICs into the file server.

*Also see bridge and external bridge.*

## INTER-REPEATER LINK

A cable segment with no workstations attached.

## INTERRUPT

A signal that causes the hardware to transfer program control to some aspect of the CPU. This signal can be generated by hardware or software and indicates that a task needs to be performed by the CPU. This is how the computer can be told to temporarily stop what it's doing and handle another task such as servicing a NIC.

## I/O PORT

The port that is assigned a NIC (and all other interface cards on a system bus) for all communications between the computer system and NIC.

## IRQ

Interrupt Request

A method used by NICs and other peripheral cards to inform the CPU that it is in need of servicing. When an IRQ is triggered on the NIC the CPU stops what its doing to service it. Each peripheral card, including each NIC, installed into a system must have a unique IRQ setting.

### IP

Internet Protocol

*See TCP/IP.*

### ISDN

Integrated Services Digital Network

The planned all-digital replacement for analog telephone lines. ISDN can use existing telephone wiring (baseband). Broadband ISDN uses fiber optic cabling and ATM switching technology for extremely high throughput.

*Also see B-Channel, D-Channel, ATM, BRI, PRI.*

### ISO

International Standards Organization

Based in Geneva, the ISO develops international data communications standards. Its members include representatives from many countries. Among their achievements is the Open System Interconnection Model, a standard against which the network industry compares product functions and protocols.

*See OSI.*

### ITU

International Telecommunications Union

The ITU, formerly the CCITT (Consultative Committee for International Telephony and Telegraphy), is an international organization that sets communications standards. Examples of their protocols include V.27ter, V.32bis and V.34.

# J

### JABBER

Continuous defective transmissions caused by faulty wiring or node connections.

# L

### LAN

Local Area Network

A group of individual computers that share resources across a limited geographic area, communicating through a dedicated hardware and software connection.

### LED

Light-Emitting Diode

A type of diode that gives off light when power is applied. Typically used for displays and indicators, as well as signaling devices in fiber-optic networks.

### LEGACY

A computer industry term used to describe a predecessor to more current technology. Usually used in the context of upgrades in which the new technology features backwards-compatibility with the predecessor.

### LINEAR BUS TOPOLOGY

The arrangement of nodes on a network along one continuous, unbroken length of cable. Workstations tap into the cable, which is typically coaxial, using T-connectors.

### LOBE (CABLE)

In Token Ring topologies, a cable that connects a node to a multistation access unit (MAU).

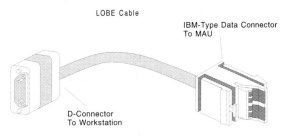

*Lobe Cable*

### LAN

Local Area Network

A system where numerous computers and various devices communicate in a relatively small geographical area (no more than three miles).

### LOCALTALK

Apple Computer's proprietary LAN access method using a daisy-chain topology and twisted pair cable.

*Also see AppleTalk.*

### LLC

Logical Link Control

A layer within the IEEE 802 standard that provides the interface between protocols at the Media Access Control (MAC) layer and network operating systems.

*Also see IEEE 802.x, MAC.*

### LRC

Logical Redundancy Checking

A data link level error checking method in which parity information is added to each block of characters, rather than to each single character as done in Vertical Redundancy Checking (VRC).

*Also see Cyclic Redundancy Checking (CRC), Vertical Redundancy Checking (VRC), Parity Checking and Data Link Layer.*

# M

### MAC

Media Access Control

The rules followed by network nodes in order to access the transmission medium. There are two basic types of MACs:
- Contention-type MACs are passive, depending on statistical probability to avoid or resolve conflict between two network nodes transmitting simultaneously (data collision).
- Non-contention-type MACs use an active protocol to determine the priority of nodes, precluding the possibility of data collision.

*Also see collision.*

### MAN

Metropolitan Area Network

A WAN that is confined to a limited geographic area, such as a city.

*Also see LAN, WAN.*

### MANCHESTER ENCODING

A method used to encode data bits of "0" and "1" as electrical signals on the network cable. A current transition from high to low within a fixed time interval represents one data bit while a transition from low to high represents another.

### MAU

Multistation Access Unit

A Token Ring hub. A MAU usually has eight or more ports for workstations and two ports for connecting to other MAUs. A MAU can automatically isolate a problem connection (workstation or another MAU) and prevent the entire network from failing.

### MAU TRANSCEIVER

Media Attachment Unit Transceiver

An external transceiver used to connect AUI drop-cable to thick Ethernet coaxial cable. MAUs are also available for connection to thin Ethernet, twisted-pair, and fiber optic cable.

### MEDIA

The network cable such as coaxial, twisted pair, or fiber optic. Plural of medium.

### MEDIA ACCESS CONTROL (MAC)

The IEEE 802 protocol level that determines how a physical medium will be shared. MAC rules and functions are similar to those defined within the lowest two layers in the OSI model.

*Also see Data-Link Layer and Physical Layer.*

### MII

Media Independent Interface

The MII is used as a universal connector to any Fast Ethernet (100BaseT) cabling media through the use of an appropriate external transceiver.

**MONOMODE**

See Fiber Optic.

**MULTIMODE**

See Fiber Optic.

**MULTIPLEX**

To share a communications channel by dividing individual lines of data and interleaving the resulting segments in the channel.

*Also see baseband, broadband.*

# N

**NETWORK**

A collection of connected computers which have the ability to communicate and share peripherals.

**NETWORK CONTROLLER**

A general term describing a network node that enables communications among all other member nodes.

**NETWORK LAYER**

See OSI Model.

**NOS**

Network Operating System

The "brains" of a network, controlling and coordinating workstation access to shared resources. Typical network operating systems include Novell Netware, Windows NT and AppleTalk.

**NIC**

Network Interface Card

A peripheral card that occupies an expansion slot on the workstation or file server, providing the physical link between the computer and the LAN cable. It allows the computer to communicate with devices on the network.

**NLIM**

Network adapter Link Interface Module

A Gateway Communications term for their proprietary NIC.

### NODE

Any terminal, workstation, or other communications terminal on the network. Each station on the network must have a unique node identification number (node ID). This is usually accomplished by setting jumpers on the NIC or is pre configured by the manufacturer.

### NODE ADDRESS

The unique address of the NIC in the workstation on a network. Each workstation must have a unique Node Address (also referred to as node ID).

### NODE ID

*See node address.*

### NON-CONTENTION PROTOCOLS

Protocols that control access to a network cable when all workstations on the cable participate to avoid collision problems.

*See CSMA/CA.*

### NRC

NonReturn to Zero

A method used to encode data bits of "0" and "1" as electrical signals on the network cable. A low-level current over a fixed time interval represents one data bit while a high-current over the same interval represents another.

### N-SERIES CABLE CONNECTOR

A type of connector used with thick-Ethernet networks.

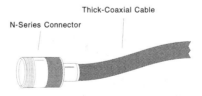

*N-Series Connector*

## N-SERIES CABLE TERMINATOR

A device used to terminate thick coaxial cable.

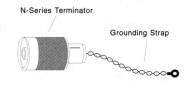

*N-Series Terminator*

# O

## OSI-MODEL

*Open System Interconnection Model*

A standard developed by the International Standards Organization (ISO) for the design and implementation of a network. It defines a set of layered guidelines which allows equipment from different vendors to communicate with each other. Each of this hierarchical model's seven layers provide a specific service and define a particular interface to the layers above and below within the same device, such that there can be peer-to-peer communications between the layers in different devices without regard to how each layer is implemented on the other device.

The layers are specified as:

(7) Application layer

This layer consists of drivers that interact with (and may be embedded in) the user's application programs.

(6) Presentation layer

This layer performs code conversions, encryption, data compression and other translations.

(5) Session layer

This layer establishes and manages a session. It verifies access permissions, interprets logical and physical addresses, and performs application-oriented data integrity functions.

(4) Transport layer

This layer ensures the quality of data transmission.

(3) Network layer

This layer forms packets, administers flow control, and, in more extensive networks, dynamically determines the route data packets will take in light of current traffic on the network.

(2) Data Link layer

This layer forms the data frames for transmission by the physical layer, and processes the acknowledgment frames sent back by the receiver.

(1) Physical layer

This layer establishes the physical connection between a computer and the network, that is, it transmits the raw bits over the network media (cable), and provides hardware controls over the transmission of information.

# P

### PACKET

A bundle of data organized in a specific way for transmission over a network. For example, an Ethernet packet consists of a header (which contains the destination node address, source node address, and the Ethernet protocol type), data, and padding as may be needed to fill up the minimum packet size of 60 bytes. Packet is sometimes used interchangeably with frame.

*Also see cell, frame.*

### PACKET SWITCHING

A method of transmitting data (normally over wide area connections) in which data is broken up into packets that are all tagged with their destination address.

*Also see packet.*

### PAD

Packet Assembler/Disassembler

A device that encapsulates data in a packet for transmission over a packet-switched network. A PAD also strips the data out of the packet at the receiving node.

*Also see X.25, packet, packet switching*

## PAL

Programmed Array Logic

A small computer logic chip that can only be programmed once. The Node ID on NICs is usually stored in a PAL by the manufacturer.

## PARITY CHECKING

A error checking method using an extra bit to make the total number of binary 1's (or 0's) in a byte always odd or always even; thus, in a parity scheme, every byte has eight bits of data and one parity bit. If using odd parity and the number of 1 bits comprising the byte of data is not odd, the $9^{th}$ or parity bit is set to 1 to create odd parity. In this way, a byte of data can be checked for accurate transmission by simply counting the bits for an odd parity indication. If the count is ever even, an error is indicated.

*See Vertical Redundancy Checking (VRC) and Longitudinal Redundancy Checking (LRC).*

## PARALLEL

Data that is transmitted across multiple lines simultaneously.

*Also see serial.*

## PASSIVE HUB

A junction box used by some network topologies to extend the number of workstations on a network. This type of hub does not boost the network signals. *See also Active Hub.*

## PATCH CABLE

The cable which connects the Ring Out port on one Token Ring MAU to the Ring In port on another MAU.

## PBX

Private Branch Exchange

A private telephone switching system, usually confined within an office or building.

## PDN

Public Data Network

An WAN using the X.25 packet-switching protocol.

*Also see X.25, packet-switching.*

## PEER-TO-PEER COMMUNICATIONS

Within the OSI model, 'peer-to-peer' describes communications between a specific layer at one node with the same layer at other nodes within the network.

## PEER-TO-PEER NETWORKING

Peer-to-peer networking describes communications among two or more nodes in a network which may be initiated by any of the participants, and any of the nodes may take on the role of client or server.

## PERIPHERAL

A device attached to a computer to perform an input/output function such as NICs, modems, and printers.

## PERMISSION TOKEN

The empty token (frame) that is passed around the network.

*See token passing.*

## PHYSICAL LAYER

*See OSI Model.*

## POLLING METHOD

A method used to control data transmission in client/server networks, wherein the controlling station polls each node to see whether it needs to send or receive data.

## PORT

A connector that allows two devices to attach to one another, usually through a cable.

## POTS

Plain Old Telephone System

*See PSTN.*

## PRESENTATION LAYER

*See OSI Model.*

**PRI**

Primary Rate Interface

Part of the Baseband ISDN specification. The name for the version of ISDN that uses twenty-three "B" channels for data and one "D" channel for control signals.

*Also see BRI, ISDN.*

**PRINT SERVER**

A computer and one or more attached printers that provide printing services to a network.

**PROPAGATION DELAY**

The delay between when a signal is transmitted and when it is received. Normally not of great importance in LANs, except when using unusually long cable runs between nodes, or satellite communications.

**PROTOCOL**

A set of rules or conventions that define how different types of computer applications or equipment communicate with one another.

**PROTOCOL STACK**

Layers of protocols through which data must pass on its way to/from the network medium.

**PROTOCOL SUITE**

A collection of protocols that operate at various functional levels. Some protocols in the collection may operate at the same level.

**PSTN**

Public (General) Switched Telephone Network

Also called the GSTN or POTS. The PSTN is the conventional analog switched telephone system.

# Q

**QUEUE**

A line. A temporary holding place for user data or jobs, such as print jobs.

# R

**RAM**

Random Access Memory

This type of memory is temporary and is lost when power fails or the system is rebooted.

**REMOTE SYSTEM RESET**

A feature that allows a workstation to boot from files on the server without the use of a local boot device such as a floppy or hard disk. Most NICs must have an optional Boot ROM installed to enable this feature.

**REPEATER**

Used to connect cable segments together. The repeater re-times and transmits data signals as they pass from one segment to another. This helps to boost the network signal strength and increases the distance that the network data can travel. Single and multiport repeaters are available for coaxial, fiber optic, and twisted pair wiring.

**RING TOPOLOGY**

In networking, a topology in which signals transmitted from one node are relayed through each subsequent node in the network. The cable is one continuous length with both ends connected to form a loop.

**RING IN (RI) AND RING OUT (RO) PORT**

A cable in Token ring networks that connects other multistation access units (MAUs) to the ring.

**RJ-11 / RJ-45**

Modular connectors used with unshielded twisted pair wiring. RJ-11 connectors have 4-pins and RJ-45 connectors have 8-pins. RJ-11 connectors are also used on modern telephones.

**ROM**

Read Only Memory

A type of non-volatile memory that is permanent and is not lost when power is removed.

### ROUTER

An interface between two LANs of different types.

### RPL

Remote Program Load

The process of loading the operating system from a server to a diskless workstation via an optional Boot ROM.

*Also see Boot ROM*

### RX-NET

Novell's ARCNET-compatible network interface cards.

# S

### SDLC

Synchronous Data-Link Control

The data-link protocol for IBM's SNA network.

### SDH

Synchronous Digital Hierarchy

The European standard for telecommunications digital networks using fiber-optic media. SDH is the counterpart to the U.S. SONET.

### SERVER

A computer on the network that provides shared services, such as access to files, peripherals, and networked applications.

### SESSION

A physical and logical connection between two devices on a network.

### SESSION LAYER

*See OSI Model.*

### SERIAL

Data that is transmitted sequentially, one bit at a time.

*Also see parallel.*

## SIGNALING PROTOCOL

A set of rules that dictate how data is to be transmitted across a physical medium.

## SHARED BANDWIDTH

The traditional method of allocating media bandwidth in LANs in which bandwidth is distributed among nodes simultaneously.

*Also see circuit switching, FDM, packet switching, switching, TDM*

## SLIDING WINDOW CONTROL

A flow control method that supports simultaneous two-way data transmission. Frames include sequence numbers, enabling receiving nodes to acknowledge receipt of specific packets while sending nodes transmit.

## SMDS

Switched Multimegabit Data Service

A high-speed packet-switched wide area data connection.

*Also see packet switching.*

## SMTP

Simple Mail Transfer Protocol

Part of the TCP/IP family of protocols used for transmitting e-mail.

*Also see TCP/IP.*

## SNA

Systems Network Architecture

A proprietary network architecture introduced by IBM in 1973. Its seven functional layers influenced development of the OSI Model.

*See OSI Model.*

## SOCKET

Not only can Apple's AppleTalk address nodes, but also specific parts of a node. This is accomplished by the use of sockets, which are logical, addressable entity within a node in an AppleTalk network, owned and used by a software process in the node known as the socket client.

### SOCKET NUMBER

The numbers used to address a socket on a node in an AppleTalk network. Static socket numbers 1-127 are internal to AppleTalk protocols and other reserved uses. Socket numbers 128-254 are assigned dynamically as needed.

*See socket.*

### SONET

Synchronous Optical NETwork

The U.S. standard for telecommunications digital networks using fiber-optic media.

### STAR TOPOLOGY

A network topology in which nodes are physically connected to a central wiring concentrator. Each workstation has its own cable segment connecting it to the central concentrator.

### STAR WIRING (TOKEN RING)

A topology of a Token Ring LAN where all devices are connected to a central hardware device and the logical topology is a continuous loop. Information also travels in a circle around the logical loop.

### STATION

An individual computer that is connected to the network.

*See node.*

### STEP-INDEX FIBER

*See Fiber Optic.*

### STM

Synchronous Transfer Mode

A circuit-switched method of transmitting data in packet form. Long-distance phone companies use STM for voice transmission.

*Also see circuit-switching.*

### STOP AND WAIT CONTROL

A flow control method that supports transmission in only one direction at a time. A receiving node must acknowledge successful transmission of one frame before another is transmitted.

**STP**

Shielded Twisted Pair

Twisted-pair wiring that comprises twisted-pair conductors within a foil or braided shield.

*Also see Twisted Pair (TP), Unshielded Twisted Pair UTP.*

**Switched 56**

A switched data communications service, similar to the analog PSTN, with the exception that it is used for digital traffic. The Switched 56 is so-named because it is capable of conveying data at 56 Kbps.

**SWITCHING**

A method of allocating media bandwidth that allows each node the full available bandwidth of a segment for a short period of time.

*Also see circuit switching, packet switching, shared bandwidth*

**SYNCHRONOUS**

Data that is sequenced to a clock.

*Also see asynchronous.*

# T

**T-CONNECTOR**

A coaxial cable connector used to connect drop cables to a trunk.

**TAP**

A connection to the network media without blocking or interfering with the passage of signals.

**TCP**

Transmission Control Protocol

TCP/IP protocol that controls communications between receiving and sending nodes during transmissions.

*See TCP/IP, UDP.*

## TCP/IP

Transmission Control Protocol/Internet Protocol.

Defines specific networking rules originally developed for the U.S. Department of Defense. TCP and IP are specific protocols within the larger set of Defense Data Network protocols. Its purpose is to connect widely different computers while providing for data errors, security, and line failure.

## TDM

Time Division Multiplexing

A method of sharing a transmission medium among multiple channels by interleaving segments of different signals.

*Also see baseband, broadband FDM.*

## TELNET

Telnet is a part of the TCP/IP family of protocols designed to allow a remote terminal to connect to computer and use its resources.

*Also see TCP/IP*

## TERMINAL

An individual computer that is connected to the network.

*See station or node.*

## TERMINATOR

A device that is used to mark the end of (or "terminate") a linear bus network. These are usually a single resistor. Each end of the linear bus cable must have a terminator installed.

## THICK ETHERNET

Cable whose electrical characteristics are specified in IEEE 802.3 Type 10Base5. In practice it is 10mm thick and is also known as RG-11 coaxial cable.

## THIN ETHERNET

Cable whose electrical characteristics are specified in IEEE 802.3 Type 10Base2. In practice it is 5mm thick and is also known as RG-58A/U coaxial cable. It is used with BNC connectors.

Note: Coaxial cable designated as RG58U is similar in appearance to RG58A/U cable, but does not comply with the IEEE thin Ethernet standard.

## TOKEN

A sequence of bits that is passed from one node to another node in a network.

## TOKEN PASSING

A type of LAN in which a token (frame) is passed around the network, permitting a station on the network to transmit data only when that station has possession of an empty token. If a passing token contains data and the station is the intended recipient, it removes the data and marks the token empty before passing it on.

*See Token Ring.*

## TOKEN RING

A LAN, developed by IBM, that uses a token passing protocol on a ring topology. The official IEEE version is number 802.5

## TOKENTALK

Apple Computer's data link product that allows AppleTalk to function as a token-passing network.

## TOPOLOGY

The arrangement of nodes on a network. Some of the topologies are linear bus, star, and ring.

## TP

Twisted Pair

*See Twisted Pair Cable, Shielded Twisted Pair (STP), and Unshielded Twisted Pair (UTP).*

## TRANSCEIVER

A device that can both transmit and receive data. The word is a combination of transmitter and receiver.

*Also see MAU Transceiver.*

## TRANSPORT LAYER

*See OSI Model.*

## TRUNK CABLE

The medium that connects to switching nodes, such as a T1 line connecting an office LAN to a central telephone office.

Trunk is sometimes used to mean the same thing as *backbone*.

### TWISTED PAIR CABLE

A type of cable that consists of two paired wires twisted together two or more times per inch to help cancel out "noise." The idea is that if the two wires are twisted they will pickup the same amount of noise thereby canceling it out. Twisted pair cable usually consists of two or more pairs are bundled together within a shield (Shielded Twisted Pair or STP) or without a shield (Unshielded Twisted Pair or UTP).

### TWISTED PAIR ETHERNET

Cable whose electrical characteristics are specified in IEEE 802.3 Type 10BaseT. In practice it is an 8-conductor telephone wire and is also known as TP (Twisted Pair) or UTP (Unshielded Twisted Pair). It is used by RJ-45 connectors.

# U

### UART

Universal Asynchronous Receiver/Transmitter.

A device that converts parallel data into serial form for transmission along a serial interface and vice versa.

### UDP

User Data-gram Protocol

TCP/IP protocol that transmits without communications between receiving and sending nodes. It functions at the same level as TCP but provides no reliability.

*See TCP/IP, UDP.*

### UTP

Unshielded Twisted Pair

Twisted-pair wiring in which each pair of conductors is not within a common shield.

*Also see Twisted Pair (TP) and Shielded Twisted Pair (STP).*

# V

### VAX

A line of minicomputers from Digital Equipment Corp. (DEC).

### VIRTUAL CIRCUIT

A type of packet-switched communications scheme in which the packets follow the same physical path through a network, but no dedicated bandwidth is allocated to the connected nodes. The circuit only exists logically.

*Also see circuit switching and packet switching*

### VLAN

Virtual LAN

A logical segmenting (grouping) of elements in a physical LAN or WAN.

### VRC

Vertical redundancy checking

A data-link level error checking method in which a parity bit is added to each byte (bit string representing each character) that is sent over the network.

*Also see Logical Redundancy Checking (LRC), Cyclic Redundancy Checking (CRC), Parity Checking, and Data Link Layer.*

# W

### WAN

Wide Area Network

A network of computers that share resources outside the scope of a physical LAN connection. Several LANs connected together would be considered a WAN.

*Also see LAN.*

### WIRELESS LAN

A LAN network with connections that utilize infrared or microwave technology.

**WORKSTATION**

An individual computer that is connected to a network.

*Also see file server.*

# X

**X.25**

An ITU-T standard for packet-switched networks.

# Index of Terms and Topics

## 1

10Base2, *see Ethernet*
10Base5, *see Ethernet*
10BaseF, *see Ethernet*
10BaseFL, *see Ethernet*
10BaseFP, *see Ethernet*
10BaseT, *see Ethernet*
100BaseT, *see Ethernet*
100BaseVG, *see 100VG-AnyLAN*
100VG-AnyLAN, 71, 464

## A

Access Control Method (see also Media Access), 32, 44, 46, 65, 66, 72, 464
Adapter (see also Network Interface Adapter), 7, 16, 28, 36, 40, 72, 92, 101, 116, 135, 139, 141, 373, 458, 459
Adapter Definition File (ADF), 464
Addressing, 7, 15, 24, 25, 29, 30, 35, 61-66, 73, 82, 83, 96, 97, 99, 130, 137-139, 464
ADF, *see Adapter Definition File*
Advanced Research Projects Agency (ARPA), 96, 465, 473
Advanced Research Projects Agency Network (ARPANET), 96, 465
American National Standards Institute (ANSI), 17, 50, 61, 65, 87, 392, 408, 464
American Standard Code for Information Interchange (ASCII), 11, 19, 465
American Telegraph & Telephone (AT&T), 46, 78, 463, 467
American Wire Gauge (AWG), 51, 106, 370, 371, 373, 376, 382, 383, 385, 390, 391, 393, 397, 466
Amplification, 108
Analog,
 Lines, 82, 91, 92, 101
 Signals, 91, 100, 109

ANSI, *see American National Standards Institute*
Apple, 121, 359
Appletalk, 465
Application Layer, 18, 19, 22, 464
ARCNET, 30, 51, 61-65, 98, 141, 359-365, 369, 393-400, 410, 465, 492
ARPA, s*ee Advanced Research Projects Agency*
ARPANET, *see Advanced Research Projects Agency Network*
ASCII, *see American Standard Code for Information Interchange*
Asynchronous Transmissions, 30, 89, 465
Asynchronous Transfer Mode (ATM), 45, 70, 71, 85, 87-90, 92, 94, 95, 99, 466
AT&T, *see American Telegraph & Telephone*
ATM, *see Asynchronous Transfer Mode*
ATM Forum, 70, 90, 466
Attachment Unit Interface (AUI), 48-50, 139, 371, 373, 379, 466, 473, 483
Attenuation, 109, 376, 382, 385, 390, 391, 466
AUI, *see Attachment Unit Interface (AUI)*
AWG, *see American Wire Gauge (AWG)*

## B

B-channel (see *ISDN*)
Backbone (see *Campus Backbone*)
BALanced/Unbalanced (BALUN), 467
Bandwidth, 7, 11, 31, 32, 44, 45, 70, 71, 73, 77, 78, 82-86, 88-91, 93, 94, 96-98, 101, 110, 467, 476, 493
Banyan VINEs, 125
Base address, 467
Baseband, 7, 8, 34, 44, 45, 92, 107, 108, 467, 476, 484, 496
Basic Rate Interface (BRI), 92, 467, 469
BISDN, *see Broadband ISDN*
Bit, 33, 34, 107, 127, 359, 459, 468
Bits Per Second (BPS), 468

502  Index

BNC connector, 374, 468
BPS, *see Bits Per Second*
BRI, *see Basic Rate Interface*
Bridges, 10, 97-99, 469, 475, 479
Broadband, 7, 34, 44, 45, 92, 107, 108, 114, 469, 476, 484, 496
Broadband ISDN (BISDN), 70, 466, 480
Broadcast, 29, 31, 32, 62, 65, 71, 108, 359
Bus, 39, 40, 43, 61, 372, 374, 469, 472, 475, 476, 479, 496, 497
Busy Token, 52, 54, 469
Byte, 33, 95, 469

# C

Cable Length, 9, 51, 105, 116, 375, 379, 384, 392
Cabling,
   Coaxial, 8, 48, 65, 107, 108, 116, 117, 370, 373, 374, 379, 380, 393, 394, 463, 468, 471, 475, 481, 483, 486
   Fiber Optic, 48, 49, 390, 466, 470, 476, 483, 491, 494
   Twisted-Pair, 8, 40, 46, 48, 51, 59, 65, 141, 371, 376, 379, 382, 383, 385, 390, 391, 397, 400, 463, 471, 472, 478, 483, 495, 498
   Wireless, 79, 111
Campus Backbone, 78, 467
Capacitance, 370, 371, 373, 382, 383, 385, 390, 391, 393
Carrier Sense Multiple Access with Collision Detection (CSMA/CD), 46, 48, 72, 73, 369, 463, 464, 472
CCITT (see *ITU*)
Cell Switching, 84, 85, 88, 95, 102, 466
Cells, 30, 71, 84, 88, 89, 95
Centralized Processing, 5, 12, 470
Channel, 81, 82, 84, 85, 87-90, 92, 102, 107, 108, 116, 469, 470
Channel Service Unit (CSU), 90, 101
Characters Per Second (CPS), 471
Circuit, 31, 45, 56, 67, 80-84, 88, 90-92, 102, 470
Circuit Switching, 31, 81, 82, 90, 470
Cladding, 109, 110, 470
Client/Server, 5, 27, 122, 125, 132

Coaxial cable, 8, 48, 65, 107, 108, 114, 116, 117, 374, 379, 394, 463, 468, 471, 475, 481, 483, 486
Collision, 32, 44, 47, 48, 55, 61, 141, 471, 485
Collision Avoidance, 32, 55, 72, 472
Collision Detection, 46, 47, 472
Collision Signal, 47
Conductors, 107, 371, 469, 471, 495, 498
Configuration, 8, 141, 372, 374, 380
Conflicts, 135
Connectors, 21, 39, 40, 43, 136, 139, 371, 374, 376, 380, 387-389, 400, 466-468, 473, 474, 481, 485, 488, 489, 491, 495, 498
Contention, 32, 44, 46, 55, 72, 471
Control Network Access, 122
Cost, 6, 44, 48, 65, 83, 86, 92, 94, 95, 112, 113, 116, 117
CRC, *see Cyclic Redundancy Check*
Crosstalk, 116, 378, 387, 391, 471
CSMA/CD, *see Collision Detection*
Cyclical Redundancy Check (CRC), 33, 47, 53, 68, 471

# D

D-channel (see *ISDN*)
DAS, *see Dual Attached Station*
Data compression, 123, 486
Data Link Layer, 21-23, 28, 34, 35, 472, 478, 482
Datagram, 27, 82, 84
DIX, 50, 371, 473
Dedicated Bandwidth, 45, 61, 71, 83
Demand Priority, 72
Department of Defense, 27, 479, 496
Diagnostics, 101, 139
Digital, 81, 82, 85, 91-93, 100, 101, 107, 109
Digital Service Unit (DSU), 90, 101
Digital Signal (DS), 85, 86
Digital transmission, 81
Diode, 109, 481
DIP Switch, 62, 473
Direct Memory Access (DMA), 473

Disk Operating System (DOS), 7, 19, 128, 138, 473
Distributed Processing, 5
DMA, *see Direct Memory Access*
DOS, *see Disk Operating System*
Drivers, 28, 141, 473, 486
Drop Cable, 371, 374, 494
DS Hierarchy:
   DS0, 85,
   DS1, 85,
   DS4, 85
DS, *see Digital Signal*
DSU, *see Digital Service Unit*
Dual Attached Station (DAS), 69, 392
Duplexing, 33, 476

# E

EBCDIC, 19, 474
EIA/TIA (see Electronics Industry Association/Telecommunications Industry Association), 106, 474
Electromagnetic, 111
Electronic Industries Association, 423
Electronic Mail (E-mail), 474
Electronically Erasable Programmable Read-Only-Memory (EEPROM), 474
Electronics Industry Association/Telecommunications Industry (EIA/TIA), 106, 474
Encoding, 21, 34,
   Manchester, 34
   NRZ, 34
Encryption, 123, 486
Error Checking, 20, 30, 33, 53, 68, 471, 482, 488, 499
Error Correction, 21
Error Detection, 21, 29, 33, 123, 471
Ethernet, 30, 32, 44, 46-48, 50, 51, 55, 61, 62, 70-73, 98, 107, 113, 139, 141, 369, 370, 372, 373, 376, 378, 379, 464, 466, 471-473, 475, 477, 478, 483, 487;
   100BaseT, 72, 369, 379, 463
   10Base2, 46, 48, 49, 369, 373-375, 463, 466
   10Base5, 46, 48, 49, 369-372, 463, 466, 496
   10BaseF, 46, 369, 463
   10BaseT, 46, 48-50, 139, 369, 376, 378, 463, 466, 471, 498

# F

Failure, 69, 121, 483, 491, 496
Fast Ethernet, 72
FDDI, *see Fiber Data Distributed Interface*
FDM, *see Frequency Division Muliplexing*
Federal Communications Commission (FCC), 18, 457, 459, 475
Fiber Data Distributed Interface (FDDI), 17, 30, 65-71, 78, 369, 392, 475, 476
Fiber Optic, 49, 109, 390, 466, 470, 477, 480, 483, 491
Fiber Optic Repeater Link (FOIRL), 48, 49, 466, 476, 477
File Server, 7, 8, 39, 41, 42, 372, 476, 479, 484
File Transfer Protocol (FTP), 27, 433, 477
Firmware, 124, 476
Flag, 54
Flexibility, 48, 89, 112, 113, 126, 129, 130, 376, 397
Flow control, 29, 32, 33, 93, 94, 476, 487, 493, 494
FOIRL, *see Fiber Optic Repeater Link*
Fractional T1, 90, 95
Frame Relay, 92, 94, 95
Frame Relay Forum, 94
Frame Stamp, 54
Frames, 21, 22, 24, 29, 30, 33, 52-56, 66-68, 71, 72, 87-89, 94, 95, 469, 477, 487, 489, 493, 497
Frequency, 7, 8, 18, 44, 91, 108, 115, 467, 469, 476, 478
Frequency Division Muliplexing (FDM), 45, 469, 476, 493, 496
FTP, *see File Transfer Protocol*

# G

Gateway, 99, 419, 477
Graded Index, 390, 477

## H

Host, 7, 100, 137, 478
Host Operating System, 7
HTML, *see HyperText Markup Language*
HTTP, *see Hypertext Transport Protocol*
Hub, 9, 43, 57, 61, 65, 72, 77, 378, 379,
   394-396, 399, 400, 431, 471, 474, 478,
   483, 488, 494
   Active Hub, 9, 65, 394-396, 399, 464
   Passive Hub, 9, 65, 394-396, 488
HyperText Markup Language (HTML), 130,
   131
Hypertext Transport Protocol (HTTP), 27

## I

I/O Address, 137, 138
IBM, *see International Business Machines*
IEEE, *see Institute of Electrical and
   Electronic Engineers*
IEEE 802 Series
   802.12, 72, 464
   802.3, 32, 46, 48, 72, 73, 369, 463, 464,
      472, 475, 478, 498
   802.4, 32
   802.5, 32, 50, 73, 380
Incompatibility, 128, 477
Institute of Electrical and Electronic
   Engineers (IEEE), 17, 29, 32, 34-36, 46,
   48, 50, 61, 72, 73, 95, 369, 378-380, 427,
   463, 464, 472, 475, 477, 478, 482, 483,
   497, 498
Integrated Services Digital Network (ISDN),
   70, 82, 85, 91, 92, 95, 101, 466, 467, 469,
   472, 480, 490
Interference, 101, 106, 107, 112, 114-116,
   376, 397, 471
International Business Machines (IBM), 6,
   8, 26, 29, 46, 50, 51, 121, 128, 380, 382,
   383, 385, 388-391, 474, 492, 493, 497
International Standards Organization (ISO),
   6, 16, 17, 65, 428, 480, 486
International Telecommunications Union
   (ITU), 17, 70, 86, 87, 90, 91, 93, 428, 480
Internet, 4, 11, 27, 91, 96, 97, 125, 128, 130,
   465, 473, 474, 477, 479, 480, 496

Internet Protocol (IP), 11, 26, 27, 33, 96, 97,
   125, 130, 465, 473, 477, 479, 480, 493,
   496, 498
Internet Protocol Exchange (IPX), 20, 125
Interrupt Request (IRQ), 137, 138, 479
Intranet, 11, 130, 131, 479
IP, *see Internet Protocol*
IPX, *see Internet Protocol Exchange*
IRQ, see Interrupt Request
ISDN, *see Integrated Services Digital
   Network*
ISO, *see International Standards
   Organization*
ITU, *see International Telecommunications
   Union*
ITU Standards: V.34, 91

## J

Jabber, 140, 480
Jumper, 135-137
Jumper Settings, 141-356

## L

LAN, *see Local Area Network*
Leased Lines, 79, 81, 90
Light Emitting Diode (LED), 109, 139, 141,
   481
Limits, 77, 91, 112
Linear Bus Topology, 39-41, 43, 48, 61,
   372, 374, 393, 463, 472, 475, 481, 496,
   497
Linear Topology, 40, 481
Linkage, 4, 41, 59, 77-79, 86, 96, 102, 111,
   121, 122, 124, 126, 372, 373, 375, 379,
   479
LLC, *see Logical Link Control*
Lobe Ports, 57, 59, 481
Local Area Network (LAN), 3-12, 25, 29-
   31, 34-36, 39, 43-46, 48, 61, 62, 70, 71,
   73, 75, 77, 78, 90, 94, 95, 97-100, 102,
   106-108, 110, 111, 121, 127, 129, 369,
   463-465, 475, 478, 48, 482, 490, 492-494,
   497, 499
Local Operating System, 7, 125, 129
Logical LANs, 78

Logical Link Control (LLC), 35, 53, 68, 482
Logical Ring, 57, 61
Longitudinal Redundancy Check (LRC), 33, 482
Loopback Tests, 101
LRC, *see Longitudinal Redundancy Check*

# M

MAC, *see Media Access Control*
MAN, *see Metropolitan Area Network*
Manchester Encoding, 34, 483
Manufacturers: (directory 403-453), (hardware settings 141-356)
   3COM Corporation, 141, 403
   Accton Technology Corporation, 148, 404
   Acculogic, Inc., 404
   Aceex Corporation, 404
   Actiontec Electronics, Inc., 404
   Adak Communications Corporation, 405
   Adaptec, Inc., 405
   ADC Kentrox, 405
   Addtron Technology Co., Ltd., 153, 405
   Advanced Integration Research, Inc., 406
   Advanced Interlink Corporation, 161, 406
   Advanced Logic Research, Inc., 406
   Advanced Telecommunications Modules Ltd., 406
   Aironet Wireless Communications, Inc., 407
   Alfa Netcom, Inc., 407
   Allied Telesyn, Inc., 407
   Alta Research Corporation, 163, 407
   American Research Corporation, 165, 408
   Amkly Systems, Inc., 408
   Amquest Corporation, 408
   Ancor Communications, Inc., 409
   Andrew Corporation, 409
   Angia Communications, Inc., 409
   Ansel Communications, Inc., 167, 409
   Archtek Corporation, 410
   Ark Pc Technology, 410
   Artisoft, Inc., 410
   Asante Technologies, Inc., 410
   Askey Computer Corporation, 411
   Aspen Technologies, Inc., 411
   Asus Computer International, 411
   AT-LAN-Tec, 169
   AT&T Paradyne, 412
   Aurora Technologies, Inc., 412
   Aztech Labs, Inc., 412
   Bay Networks, Inc., 413
   Best Data Products, Inc., 413
   Black Box Corporation, 413
   Boca Research, Inc., 414
   Broadtech International Company, 414
   Brooktrout Technology, Inc., 414
   Buslogic, Inc., 414
   Cabletron Systems, Inc., 171, 414
   Cardinal Technologies, Inc., 415
   Cardware Lab, Inc., 415
   Castelle, Inc., 415
   Celan Technology, Inc., 415
   Cisco Systems, Inc., 416
   Claflin & Clayton, Inc., 416
   CNET Technology, Inc., 175
   Codenoll Technology Corporation, 416
   Cogent Data Technologies, Inc., 181, 417
   Compex, Inc., 184, 417
   Compulan Technology, Inc., 417
   Computer Modules, Inc., 418
   Computer Peripherals, Inc., 418
   Computone Corporation, 418
   Comtrol Corporation, 418
   Connectware, Inc., 419
   Connexperts, 419
   Creative Labs, Inc., 419
   Creatix Polymedia, 419
   CSS Laboratories, Inc., 420
   D-Link, 185
   Danpex Corporation, 189, 420
   Data Race, 420
   Dataexpert Corporation, 420
   Datapoint Corporation, 61, 196, 421, 465
   Datum, Inc., 421
   David Systems, Inc., 421
   Dayna Communications, Inc., 421
   Digi International, Inc., 421
   Digital Communications Association, 198
   Digital Equipment Corporation, 422
   E-Tech Research, Inc., 422
   Eagle Technology, 200
   Echo Communications, 422
   Edimax Computer Company, 204, 422
   EFA Corporation, 423

Eicon Technology Corporation, 423
Electronic Industries Association (EIA), 106, 423, 474
Emulex Corporation, 423
Eventide, Inc., 423
Everex Systems, Inc., 424
Farallon Computing, Inc., 424
Fore Systems, Inc., 424
Franklin Telecommunications Corp., 424
General Datacomm, Inc., 425
GVC Technologies, Inc., 206, 425
Harmony Multimedia, 425
Hayes Microcomputer Products, Inc., 425
Hewlett-Packard Company, 204
HTI Networks, 426
Hypermedia International Corporation, 426
I/O Magic Corporation, 426
IBM Corporation, 426
IC Intracom USA, 426
ICL, 210
IET Startech, 427
IMC Network Corporation, 212
Intel Corporation, 214, 427
Intellicom, Inc., 427
Interphase Corporation, 428
Invisible Software, Inc., 428
JC Information Systems Corp., 429
Katron Technologies, Inc., 216, 429
Keysonic Technology, Inc., 429
Kingston Technology Corporation, 218, 429
Klever Computers, Inc., 220, 429
Koutech Systems, Inc., 430
Kye International Corporation, 430
Lancast, 430
Lantech Computer Company, 430
Leemah Datacom, 430
Linksys, 431
Lite-On Communications, Inc., 431
Liuski International, Inc., 431
Logicode Technology, Inc., 431
Longshine Microsystem, Inc., 223, 432
Madge Networks, Ltd., 225, 432
Magicram, Inc., 432
Maxtech Corporation, 234, 432
Megahertz Corporation, 432
Micro Direct, 433

Micro House International, 433
Microcom, Inc., 433
Microdyne Corporation, 237, 433
Microsoft Corporation, 434
Microwise, Inc., 434
Mitron Computer, Inc., 434
MNC International, Inc., 434
Moses Computers, Inc., 434
Motorola, Inc., 435
Multi-Tech Systems, Inc., 239
Multiaccess Computing Corporation, 435
Mylex Corporation, 435
NDC Communications, Inc., 241
Net Edge Systems, Inc., 436
Net Frame Systems, Inc., 436
Network Interface Technology Corporation, 436
Network Peripherals, 436
Network Technologies, Inc., 437
Networth, Inc., 242, 437
New Media Corporation, 437
Newbridge Networks, Inc., 437
Northern Telecom, 438
Novell, Inc., 438
Okidata, 438
Olicom International, 245
Olivetti, 438
Optical Data Systems, 438
OSICOM Technologies, Inc., 438
PC Connectlan, 439
Penril Datability Networks, 439
Pinacl Communications, Inc., 439
Plaintree Systems, 439
Powercom America, Inc., 440
Practical Peripherals, Inc., 440
Premax Electronics, Inc., 440
Prometheus Products, Inc., 440
Proteon, Inc., 250, 441
Proxim, Inc., 441
Pure Data, Ltd., 253
Puretek Industrial Co., Ltd., 441
Q Logic Corporation, 441
Quadrant Components, Inc., 442
Quickpath Systems, Inc., 442
Racal Interlan, Inc., 264, 442
Racore Computer Products, Inc., 277, 443
Ragula Systems, 443
Raylan Corporation, 443

Rhetorex, Inc., 443
SBE, Inc., 444
Siemens Nixdorf, 283
Siig, Inc., 444
Silcom Manufacturing Technology, 288
Silicom Connectivity Solutions, Inc., 444
Smart Modular Technologies, 444
Solectek Computer Supply, Inc., 445
Sonic Systems, Inc., 445
Standard Microsystems Corporation, 291, 445
Star Logic, Inc., 445
Supra Diamond, 446
SVEC Computer Corporation, 297, 446
Syskonnect, Inc., 299, 446
Target Technologies, Inc., 446
TCL, Inc., 447
TDK Electronics Corp., 447
Technology Works, 447
Teknique, Inc., 447
Telebit Corporation, 447
Telecommunications Industry Association (TIA), 106, 474
The Networking Company, Pte., Ltd., 448
Thomas-Conrad Corporation, 305
Tiara Computer Systems, Inc., 318, 448
Top Microsystems, Inc., 327, 448
Trancell Systems, Inc., 449
Transition Engineering, Inc., 329, 449
Trendware International, Inc., 449
TTC Computer Products, 333
Tulip Computers, 335
Tut Systems, 450
U.S. Robotics, Inc., 450
Ungermann-Bass Networks, Inc., 338
Unicom Electric, Inc., 340, 450
Update Technology, Inc., 451
Verilink, 451
Visiontek, 451
Western Datacom Co., Inc., 451
Wildcard Technologies, Inc., 451
Xinetron, Inc., 344, 452
Xircom, Inc., 452
Xylogics, Inc., 452
Zeitnet, Inc., 452
Zendex Corporation, 452
Zenith Data Systems, 346
Zero One Networking, 355, 453

Znyx Corporation, 453
Zoltrix, Inc., 453
Zoom Telephonics, Inc., 453
MAU, *see Multistation Access Unit*
Media Access, 28, 31, 36, 44, 46, 475
Media Access Control (MAC), 7, 8, 15, 28, 31, 34-36, 39, 44, 46, 53, 55, 61, 68, 369, 380, 392, 393, 482, 483
Media Independent Interface (MII), 379, 483
Metropolitan Area Network (MAN), 78, 94, 95, 483
Micro House Technical Library (MTL), 135, 457, 458, 459
Microsoft, 8, 121, 126, 127, 128, 130, 131, 132
MII, *see Media Independent Interface*
Modem, 11, 91, 92, 100
Monomode, 109, 110, 476
MTL, *see Micro House Technical Library*
Multiplexing, 7, 8, 44, 84, 476, 484, 496
Multistation Access Unit (MAU), 9, 57, 58, 59, 380, 384, 386, 389, 478, 483, 488, 497

# N

NetBIOS, 20, 125
Netware, 484
Network Controller, 62, 63, 64, 484
Network Interface Card (NIC) 7, 25, 47, 62, 63, 65, 138, 139, 359, 378, 388, 398, 458, 466, 467, 469, 476, 479, 484, 485
Network Layer, 20, 21, 23, 96, 486
Network Operating System (NOS), 8, 15, 27, 35, 121-132, 482, 484
   Netware, 484
   OS/2, 7, 8, 19, 128, 129
   Vines, 125
   Windows for Workgroups, 27, 127
   Windows NT, 8, 126-129, 132, 484
Node Address, 61-65, 485, 487
Nodes, 7-9, 15, 16, 18-24, 27-33, 35, 36, 39, 42-48, 51, 52, 54-56, 58-65, 68, 69, 71-73, 77, 78, 80-82, 84, 92, 96-98, 107, 108, 116, 141, 357, 359-365, 370, 372, 373, 376, 393, 397, 464, 467, 476, 481, 484, 485, 487-491, 493-499

Non-Return to Zero (NRZ), 34, 485
NOS, see Network Operating System
Novell, 8, 27, 121, 125-127, 130-132, 484, 492
NRZ, *see Non-Return to Zero*

# O

Open Systems Interconnect (OSI) Model, 6, 10, 15-22, 24-26, 28, 35, 36, 46, 65, 93, 96, 99, 124, 125, 480, 484, 486, 489, 497
   Application Layer, 18, 19, 22
   Data Link Layer, 21-23, 28, 34, 35, 482, 499
   Network Layer, 20, 21, 23, 96
   Physical Layer, 18, 21, 22, 25, 28, 34-36
   Presentation Layer, 19
   Session Layer, 20
   Transport Layer, 20, 21, 24, 96
Optical Repeater, 109
OS/2, 7, 8, 19, 128, 129
OSI, *see Open Systems Interconnect*
OSI Guidelines, 18, 36

# P

Packet Switching, 82-84, 94, 102, 470, 487, 493, 495, 499
Packets, 11, 15, 20-22, 24, 29, 30, 32, 47, 53, 68, 71, 80, 82-84, 88-90, 93-96, 102, 125, 470, 475-477, 487, 493-495, 499
PAL, 488
Passive, 9, 65, 394-396
PDN, *see Public Data Network*
Peer-to-Peer Model Network, 26, 29, 132, 489
Performance, 30, 45, 65, 112, 114, 117, 127, 128
Permission Token, 32, 63, 64, 489
Physical Layer, 18, 21, 22, 25, 28, 34-36, 486
Pin Assignments, 369, 371, 377, 386-389, 397
Polling Method, 31, 489
Ports, 49, 50, 57, 59, 111, 139, 373, 395, 396, 399, 466, 473, 477, 479, 483, 489

PPSDN, *see Public Packet-Switched Data Network*
Primary Rate Interface (PRI), 85, 92, 466, 469, 480, 490
Private Virtual Circuit (PVC), 376, 382, 383, 385, 390, 391
Private Virtual Circuits (PVC), 82, 376, 382, 383, 385, 390, 391
Programmable, 129, 488
Protocol Stack, 25-28, 124
Protocol Suite, 27
PSTN, *see Public Switched Telephone Network*
Public Data Network (PDN), 94
Public Packet-Switched Data Network (PPSDN), 83
Public Switched Telephone Network (PSTN), 82, 91, 92, 100, 101, 490, 495
PVC, *see Private Virtual Circuit*

# R

Random Access Memory (RAM), 8, 469, 491
Rays, 110
Read Only Memory (ROM), 135, 138, 139, 455, 457-460, 468, 491, 492
Redundant Ring, 69, 392
Refraction, 109, 110
Remote Access, 11, 100
Remote Program Load (RPL), 138, 491, 492
Repeaters, 49, 109, 370, 372-376, 379, 390, 399, 400, 492, 477, 478, 491
Resistors, 40, 496
Ring Topology, 42, 43, 61, 497
ROM, *see Read Only Memory*
Routers, 99
Routing, 53

# S

Safety, 112, 114
SAS, *see Single Attached Station (SAS)*
Satellites, 79, 86, 490
SCLC, *see Synchronous Data Link Control*
SDH, *see Synchronous Digital Hierarchy*

Security, 6, 11, 20, 111, 112, 115, 116, 132, 496
Serial, 101, 107, 126,
Servers, 5-8, 10, 26-28, 39, 41, 42, 122, 125, 126, 131, 132, 138, 372, 465, 492
Session Layer (see *OSI Model*)
Shared Bandwidth, 44, 45, 77, 493
Sheath, 110
Shielded Twisted Pair (STP), 51, 106, 114, 382, 383, 388, 390, 495, 498
Shielding, 51, 106, 112, 114-116
Shorting, 389
Signaling, 7-9, 15-17, 21, 22, 24, 28, 30, 34, 35, 46, 47, 49, 59, 63, 89, 91, 92, 100, 101, 107-111, 116, 395, 396, 493
Simple Mail Transfer Protocol (SMTP), 27, 493
Single Attached Station (SAS), 69, 392
Sliding Window Control, 33, 493
SMDS, *see Switched Multi-Megabit Data Service*
SMTP (see *Simple Mail Transfer Protocol*)
SNA, *see Systems Network Architecture*
Sockets, 138, 493, 494
Software, 3, 5, 6, 11, 15, 25, 28, 124, 125, 129-131, 135, 465, 474, 479, 481
SONET, *see Synchronous Optical Network*
Splice, 109, 110
Standards Organizations, 15, 16
Star Topology, 41, 398, 494
Starlan, 46, 48, 463
Station, 9, 31, 32, 52, 53, 56, 66, 68, 92, 97, 380, 384, 386, 494
Step Index Fiber, 110, 476
STM, *see Synchronous Transfer Mode*
STP, *see Shielded Twisted Pair*
Stranded, 51, 390
Strap, 372, 374
Switched Multi-Megabit Data Service (SMDS), 82, 95, 493
Switches, 10, 62, 71, 89, 93, 97-99, 102, 135, 136, 359, 398
Switching, 31, 45, 56, 70, 71, 72, 78-85, 87, 88, 90-95, 99, 102, 470, 487, 488, 495
Synchronous, 30, 81, 88, 89
Synchronous Data Link Control (SDLC), 28, 29, 33, 492

Synchronous Digital Hierarchy (SDH), 86, 492
Synchronous Optical Network (SONET), 86, 492, 494
Synchronous Transfer Mode (STM), 81, 82, 85-89, 494
Systems Network Architecture (SNA), 26, 28, 33, 46, 492, 493

# T

Tapping, 39, 108, 115, 472, 474, 481
TCP, *see Transmission Control Protocol*
TCP/IP, *see Transmission Control Protocol/Internet Protocol*
TDM, *see Time Division Multiplexing*
Telephones, 51, 79, 82, 91, 113, 116, 376, 378, 383, 397, 467, 470, 480, 488, 490, 498
Television, 107, 108, 469
Terminals, 6, 12, 92-94, 101, 126, 470, 485, 496
Termination, 31, 93, 101, 398-400, 486, 496
Terminators, 40, 92, 372, 374, 398, 496
Throughput, 7, 32, 48, 61, 65, 81, 95, 384, 386, 392, 465
Time Division Multiplexing (TDM), 8, 45, 87, 476, 493, 496
Token Passing, 44, 61, 63, 64, 67, 73, 380, 392, 393, 469, 489, 497
Token Ring, 9, 29, 30, 32, 50-57, 59, 60, 62, 66-68, 71, 72, 141, 359, 369, 380, 381, 386, 387, 389, 464, 477, 478, 481, 483, 488, 491, 494, 497
Topology, 8, 29, 39, 40-43, 45, 57, 61, 73, 126, 370, 373, 376, 378, 380, 392, 393, 397, 398, 400, 463, 464, 471, 472, 475, 478, 479, 481, 482, 488, 491, 494, 497
TP, *see Twisted Pair*
Transceivers, 49, 50, 139, 372, 373, 379, 466, 483, 497
Transferring, 7, 27, 32, 61, 71, 84, 106, 107, 110, 121, 122, 465, 477, 479
Transmission Control Protocol (TCP), 11, 20, 26, 27, 33, 96, 97, 130, 465, 473, 477, 479, 480, 493, 495, 496, 498

Transmission Control Protocol/Internet Protocol (TCP/IP), 11, 26, 27, 33, 96, 97, 130, 465, 473, 477, 479, 480, 493, 496, 498
Transmitters, 111, 497, 498
Transport Layer (see *OSI Model*)
Trunk, 90, 92, 372, 474
Trunk Cable, 370, 372, 497
Twisted Pair, 8, 40, 46, 48, 51, 59, 65, 141, 371, 376, 379, 382, 383, 385, 386, 390, 391, 393, 397-400, 463, 467, 471, 472, 478, 482, 483, 491, 495, 497, 498

# U

UART, *see Universal Asynchronous Receiver Transmitter*
UDP, *see User-Datagram Protocol*
Universal Asynchronous Receiver Transmitter (UART), 498
UNIX, 19, 128, 129, 132
Unshielded Twisted Pair (UTP), 48, 51, 65, 72, 106, 113, 376, 380, 385, 393, 466, 491, 495, 497, 498
User-Datagram Protocol (UDP), 27, 498
Utilities, 27
UTP, *see Unshielded Twisted Pair*

# V

VCI, *see Virtual Circuits Identifier*
Vertical Redundancy Check (VRC), 33, 499
Virtual Circuits, 82
Virtual Circuits Identifier (VCI), 89, 499
VRC, *see Vertical Redundancy Check*

# W

Wide Area Network (WAN), 3-5, 11, 12, 29, 31, 71, 77-79, 85, 90, 93-96, 99, 101, 102, 129, 479, 483, 499
Windows:
    Windows 3.11, 129
    Windows 95, 27, 127-129
    Windows for Workgroups, 27, 127

Windows NT, 8, 126, 127, 128, 129, 132, 484
Wireless, 79, 111, 499
Workstation, 5-8, 10, 11, 39-42, 47, 58, 65, 107, 108, 123, 125, 128, 132, 138, 372, 375, 378, 379, 389, 392, 394-396, 399, 468, 470-475, 479, 483-485, 488, 491, 492, 494, 500
World Wide Web (WWW), 27, 403-453

# X

X.25, 82, 92-95, 418, 500
Xerox, 46, 50, 121, 473